Time-Domain Synthesis of Linear Networks

PRENTICE-HALL NETWORKS SERIES

Robert W. Newcomb, Editor

ANDERSON & MOORE *Linear Optimal Control*
ANNER *Elementary Nonlinear Electronic Circuits*
CARLIN & GIORDANO *Network Theory: An Introduction to Reciprocal and Nonreciprocal Circuits*
CHIRLIAN *Integrated and Active Network Analysis and Synthesis*
DEUTSCH *Systems Analysis Techniques*
HERRERO & WILLONER *Synthesis of Filters*
HOLTZMAN *Nonlinear System Theory: A Functional Analysis Approach*
HUMPHERYS *The Analysis, Design, and Synthesis of Electrical Filters*
KIRK *Modern Control Theory*
MANASSE, ECKERT, & GRAY *Modern Transistor Electronics Analysis and Design*
NEWCOMB *Active Integrated Circuit Synthesis*
SAGE *Optimum Systems Control*
SU *Time-Domain Synthesis of Linear Networks*
VAN VALKENBURG *Network Analysis, 2d ed.*

Time-Domain Synthesis of Linear Networks

KENDALL L. SU

School of Electrical Engineering
Georgia Institute of Technology

PRENTICE-HALL, INC., Englewood Cliffs, New Jersey

Printed in the United States of America

13-921668-5

Library of Congress Catalog Card No.: 71-138819

Current printing (last digit):
10 9 8 7 6 5 4 3 2 1

PRENTICE-HALL INTERNATIONAL, INC., London
PRENTICE-HALL OF AUSTRALIA, PTY. LTD., Sydney
PRENTICE-HALL OF CANADA, LTD., Toronto
PRENTICE-HALL PRIVATE LIMITED, New Delhi
PRENTICE-HALL OF JAPAN, INC., Tokyo

Preface

This volume presents a treatise of some theories of and techniques for specifying a network function such that a desired time-domain response will be produced by the network with such a function when subjected to a certain time-domain excitation. It is, therefore, dealing with a very unique engineering approximation problem. This area has traditionally been known as the *time-domain synthesis*.

In the past two decades, numerous authors have contributed to the area of time-domain synthesis of linear lumped networks. Two interesting phenomena in the development of this area might be worth noting. One is the fact that the development has been rather gradual. As we look over the literature in this area, each author seems to have bitten off a small chunk of the problem at a time. The other is that very few authors contributed more than once to this area. However, the accumulated work done by these authors collectively is quite substantial and significant.

Most of the contributors in this area either are from academic institutions or did their work in the area while in an academic institution. Hence much of their work has been regarded as rather of pure academic interest. However, practicing engineers are becoming more and more in need of knowledge of the pulse-shaping capabilities of linear lumped networks and techniques of designing these networks, particularly in the fields of data transmission, digital communication systems, and digital-system design. This volume is intended to provide the basic mathematical tools for, to give a unified survey of some important techniques in, and to present an introduction of some computer-aided applications to the design of a lumped linear network in the time domain.

This book grew out of a set of notes prepared for groups of students who have specialized interest in network theory at Georgia Tech. Typically these students have already had one to two years of courses in network theory, although a few first-year graduate students had enrolled and found the material digestable. Hence it is fair to say that the material in this book can be taught at any level in a graduate program. Actually, most seniors will find it readable and interesting if not for the fact that the subject matter might be too specialized at that level. Most practicing engineers should find the level of presentation quite agreeable.

The presentation assumes only the usual mathematical background of an undergraduate electrical-engineering curriculum. To pursue the material, the reader should have some proficiency in calculus, matrix algebra, Laplace transform, and differential equations. Some knowledge in the theory of functions of a complex variable and some familiarity with difference equations will be helpful. Special mathematical tools beyond these topics will be developed as the need arises. A background in network synthesis is not absolutely necessary to understanding the developments of the various methods presented herein. However, a background in network synthesis in general will enable a student to appreciate more the context in which the material in this volume might fit. In other words, such a background will help him to see not only *how*, but also *why*, these methods are used.

Chapter 1 gives a brief introductory explanation of the subject area. Chapter 2 includes several mathematical topics that are pertinent to time-domain problems. Chapters 3 through 6 constitute the main body of the subject area to which this volume is devoted—that of how to specify a network function to produce a certain desired response in the time domain. Chapter 7, the longest and possibly the most useful and interesting chapter, deals with a very specialized topic within time-domain synthesis—that of linear lumped constant-delay networks. Chapter 8 gives a few examples of how computer-oriented optimization techniques might be applied to some time-domain problems.

By the author's own estimation, the entire volume can be covered in detail in a three-hour course in one semester. The book can be used in a three-hour course in one quarter by judiciously leaving out a few minor topics. Also, it is believed that the arrangement of the material is sufficiently flexible to allow for the book to be used in a part of a network-theory course by choosing only a few selected topics to be covered as time permits.

The author is indebted to the administration of the Georgia Institute of Technology for the numerous considerations given him during the preparation of this book. He is also grateful to the many former students who have contributed to the manuscript in various ways. Among these he wishes especially to acknowledge the assistance given him by Douglas R. Cobb, N. Walter Cox, Jr., Chung Duk Kim, Te-son Kuo, and Maw Huei Lee. During his senior year at Georgia Tech, Stephen W. Kelley contributed a great deal to the computer programming efforts that are so important in many parts of this book. The author also wishes to thank Professor Robert W. Newcomb, the Prentice-Hall Network Series Editor, for his careful reading of the manuscript and numerous valuable comments.

Kendall L. Su

Contents

Time-Domain Synthesis of Linear Networks

1

Introduction

In the general area of electric network synthesis and design, there exists a very special class of problems commonly known as *time-domain synthesis problems*. This branch of network theory deals with the finding of a network to give a prescribed response for a given excitation, both of which are specified as functions of time. This class of problems arises in pulse-forming circuits, pulse transmission systems, delay networks, transmission-line equalizations, video coupling systems, and the like.

A time-domain synthesis problem is usually characterized by an excitation and a response, both of which are specified in terms of time. However, this does not mean that the entire synthesis procedure must be carried out in terms of time; in fact, except for very simple cases and computer applications, the necessary synthesis procedures are seldom performed in the time domain.

Modern network synthesis procedures are generally divided into two separate steps—the approximation and the realization. The former deals with finding a mathematical network function that (under certain criteria) best meets the requirement for which the synthesis is being performed. The latter deals with finding a network or networks that realize the network function. Frequently, the realization step is loosely described as "synthesis." The meaning of the two uses of the term "synthesis" can be distinguished by the context.

The term "time-domain synthesis" is something of a misnomer. The synthesis of a network to perform the desired time-domain functions actually deals with only the approximation part of the problem. That is, the term "time-domain synthesis" is commonly used to mean that part of the problem which enables us to determine a network function such that the time-domain performance of the network with this function meets the requirement. Once this function is found, the realization part is frequently done in the frequency domain. Hence time-domain synthesis is really a special form of mathematical approximation technique which enables us to determine the approximating network function in the time-domain while ensuring that such a function is physically realizable by a network in the frequency domain.

The realizability of a network function by a physical network depends on the types of networks permitted as well as other constraints, such as whether the network must have a common ground or not and whether the network is terminated or open circuited. There is a wealth of literature that deals with the realizability of various types of networks and the various techniques of realization [BA2, CA2, CH1, GU1, HA2, KU3, KU6, MI1, NE3, NE4, ST3, SU2, TU2, VA5, WE2, YE2], so for our purposes a time-domain synthesis problem will be considered solved once the approximation problem has been satisfactorily solved.

Since the eventual realization of a network is assumed to be done in the frequency domain, the type of networks that can be used in connection with the particular network function in question is only limited by the frequency-domain methods. Since most of the developments in the frequency-domain methods have been done for the linear, lumped, and time-invariant networks, we shall limit our development to this class of networks. Although some of the principles presented in this volume are also applicable to linear, distributed networks (especially in Chapters 2, 3, and 8), we shall not deal with them specifically. Aside from being linear, lumped, and time-invariant, our networks can be either passive or active, either reciprocal or nonreciprocal.

The time-domain synthesis methods described here are certainly also applicable to physical systems other than electric networks as long as these systems are describable by the same mathematical formalism as the various classes of electric networks. Many linear physical systems such as mechanical, acoustical, hydraulic, pneumatic, fluidic systems, or various hybrid versions of these systems can concievably make use of the approximation techniques treated in this volume.

1.1 COMPARISON OF A NETWORK PROBLEM TREATED IN THE TIME DOMAIN AND IN THE FREQUENCY DOMAIN

The reader is undoubtedly familiar with the network analysis and synthesis problems in the frequency domain. In a similar manner, a network analysis or synthesis problem can also be formulated in the time-domain. The time-domain analysis is the problem of finding a set of responses when the network and a set of excitations are given. Both the responses and the excitations are time functions of either voltages or currents, or combinations of some currents and some voltages.

A synthesis procedure is the systematic method by which a network can be found when both the set of responses and the set of excitations are given. In the time-domain, substantial results have been obtained only when there are only a single excitation and a single response. In most situations, the response and the excitation take place at different locations in the network. These situations correspond to the transfer function synthesis in the frequency domain. In other situations, the excitation and the response may occur at the same locality of a network. This corresponds to the driving-point immittance synthesis in the frequency domain.

In this volume the subscript 1 shall be used to signify that the quantity is an excitation and 2 to signify a response. Generally, a capital letter shall be used to denote a quantity expressed in the frequency domain and a lowercase letter to denote one in the time-domain. The letter F or f shall be used to represent either a voltage (E or e) or a current (I or i). For convenience, all time-domain quantities in this volume shall be assumed to be zero for $t < 0$.†

The analysis of a linear network in the frequency domain is a relatively easy task in circuit theory. Once a network is given and the electric quantity (voltage or current) and its location are specified, it is a routine matter to write out the necessary equations and then to reduce these equations to obtain the desired *network function* in terms of the frequency variable s. The network function $H(s)$ is the ratio of the response $F_2(s)$ to the excitation $F_1(s)$, or

$$H(s) = \frac{F_2(s)}{F_1(s)} \tag{1.1}$$

The characterizing function useful in the time-domain is the response due to a unit-impulse‡ excitation. Just as the network function can be viewed as the frequency response when the frequency-domain excitation is unity, or

$$H(s) = F_2(s)|_{F_1(s)=1} \tag{1.2}$$

the unit-impulse response is the response when the excitation is a unit impulse, or

$$h(t) = f_2(t)|_{f_1(t)=\delta(t)} \tag{1.3}$$

It is well known that for an arbitrary $f_1(t)$, the response $f_2(t)$ can be evaluated exactly by the superposition integral,

†This amounts to assuming that $f_1(t) = 0$, for $t < 0$, and that the network is causal and initially idle [NE3, p. 31].

‡A unit impulse occurring at $t = 0$ is denoted by $\delta(t)$ and is defined by the equations

$$\int_{0^-}^{0^+} \delta(t)\,dt = 1$$

$$\delta(t) = 0, \qquad \text{for } t \neq 0$$

$$
\begin{aligned}
f_2(t) &= \int_0^t f_1(\tau)h(t-\tau)d\tau \\
&= \int_0^t h(\tau)f_1(t-\tau)d\tau \\
&= f_1(t) \circledast h(t)
\end{aligned}
\tag{1.4}
$$

The last expression states that $f_2(t)$ is the convolution of $f_1(t)$ and $h(t)$.

The analysis of a network in the time domain is generally a much more difficult problem than one in the frequency domain. The equations that govern the time-domain behavior of a network are differential equations. The solution of these equations can be very difficult when the excitation to a network is allowed to assume any shape of time variation. In this respect we can consider the frequency-domain analysis as a special case of time-domain analysis in which the excitation is restricted to sinusoidal (damped or undamped) variations with time.

In a typical frequency-domain analysis problem we have the advantage that once a network function is obtained, the analysis is done for all frequencies. This simplicity clearly does not exist in a time-domain problem. Furthermore, in the frequency domain the response of a network at a particular frequency is dependent solely on the excitation of that frequency. This is not true with time-domain problems. Clearly, an excitation applied at any time will have effects on the response at all times after the application of the excitation. Conversely, the response at any particular instant depends on all the excitations that have been applied to the network prior to that instant.

For these reasons, it is usually more convenient to transform a time-domain problem to the frequency domain and solve it in the frequency domain. The Fourier transform method relates the time-domain behavior of the network on the real-frequency axis. The Laplace transform method relates the time-domain behavior of a network to its behavior on a contour in the complex-frequency plane.

An analysis problem in the time domain can first be transformed into the frequency domain by either of the two transforms mentioned above and the answers found by the simpler frequency-domain analysis methods. The corresponding answers in the time domain are found by the corresponding inverse transform.

For exactly the same reasons, parts of the synthesis problems in the time domain are more conveniently solved in the frequency domain. This not only enables us to circumvent a very difficult problem in the time domain but lets us take advantage of the great wealth of knowledge that has been developed throughout the years both in the determination of various realizability conditions and in the methods of realization of various classes of linear networks.

Although a problem in the time domain can, in principle, be transformed into one in the frequency domain, the mechanics of this transform and its inverse are not always straightforward. In fact, the bulk of this volume deals exactly with the various details of implementing the mechanics for accomplishing these transformations.

1.2 NATURE OF A TIME-DOMAIN SYNTHESIS PROBLEM

A time-domain synthesis problem can typically be represented by the diagram of Fig. 1.1. A desired time-domain response $f_2(t)$ is specified for a corresponding given excitation $f_1(t)$. Although representation of $f_1(t)$ and $f_2(t)$ in Fig. 1.1

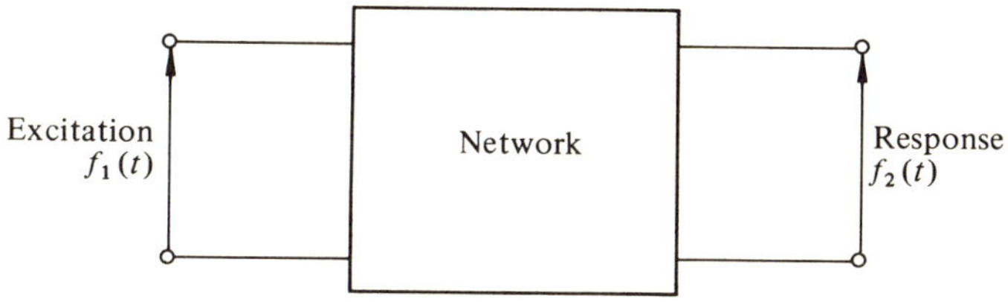

Figure 1.1 A typical arrangement in a time-domain problem.

strongly suggests that they are voltages, they are not limited to voltages; each could be a current as well. Also, $f_1(t)$ and $f_2(t)$ do not have to occur at two different parts of a network, as explained earlier.

The specification of $f_1(t)$ or $f_2(t)$ can generally take one of three forms:

1. Analytical expressions
2. Graphical representations
3. Tabulated data

If the analytical expression of a time function is given, its graphical representation and its point-by-point values can certainly be generated without any difficulty. The converse of this statement is not always true, however. In fact, to find an analytical expression to coincide with the given graphical or tabulated data is sometimes a major part of the approximation problem.

Even if a specified function is known in analytical form, this analytical expression may or may not be suitable for the ensuing synthesis steps. If the analytical expresssion is not suitable for this purpose, one that is suitable must first be found. This is another possible major step in the approximation procedure. As explained earlier, all functions specified in the time domain are assumed to begin at $t = 0$, unless otherwise stated. Hence, when we have given $f_1(t) = \sin t$, it is implied that

$$\begin{aligned} f_1(\mathrm{t}) &= 0, & t &< 0 \\ &= \sin t, & t &\geq 0 \end{aligned} \tag{1.5}$$

Equation (1.5) is sometimes written as $f_1(t) = (\sin t)u(t)$, where $u(t)$ denotes a unit step function† at $t = 0$, for emphasis.

When the time-domain expressions of the response and the excitation are known and have the correct form, the synthesis problem can sometimes be transformed into the frequency domain and becomes a rather simple problem. For example, if the Laplace transforms of $f_1(t)$ and $f_2(t)$ exist and can be obtained, we call them $F_1(s)$ and $F_2(s)$, respectively. The network that gives this pair of excitation and response will have a network function

$$H(s) = \frac{F_2(s)}{F_1(s)} \tag{1.6}$$

in the frequency domain, and the problem is solved.

For example, suppose it is desired to find a network such that the current is

$$f_2(t) = i(t) = (1 - \epsilon^{-t})u(t) \tag{1.7}$$

for a unit step applied voltage, or

$$f_1(t) = e(t) = u(t) \tag{1.8}$$

We find

$$I(s) = \frac{1}{s(s+1)}$$

$$E(s) = \frac{1}{s}$$

Hence

$$Y(s) = \frac{I(s)}{E(s)} = \frac{1}{s+1} \tag{1.9}$$

which can be readily identified as the series combination of a 1-ohm resistance and a 1-henry inductance.

Some time-domain synthesis problems can be solved this way. However, problems of this type are extremely rare and are not our major concern here. When these problems do arise, they usually are sufficiently obvious and no complicated method of solution will be sought. We are more interested in problems that do not have simple outright solutions.

†A unit step function occurring at $t = 0$ is defined by

$$u(t) = 0, \quad \text{for } t < 0$$
$$= 1, \quad \text{for } t \geq 0$$

A typical time-domain synthesis procedure may take one of the several possible paths shown in Fig. 1.2 in going from a given pair of $f_1(t)$ and $f_2(t)$ to a realizable frequency-domain network function $H^*(s)$. [Of course, it is possible that $H^*(s) = H(s)$ in some special cases.] In Fig. 1.2, as in the re-

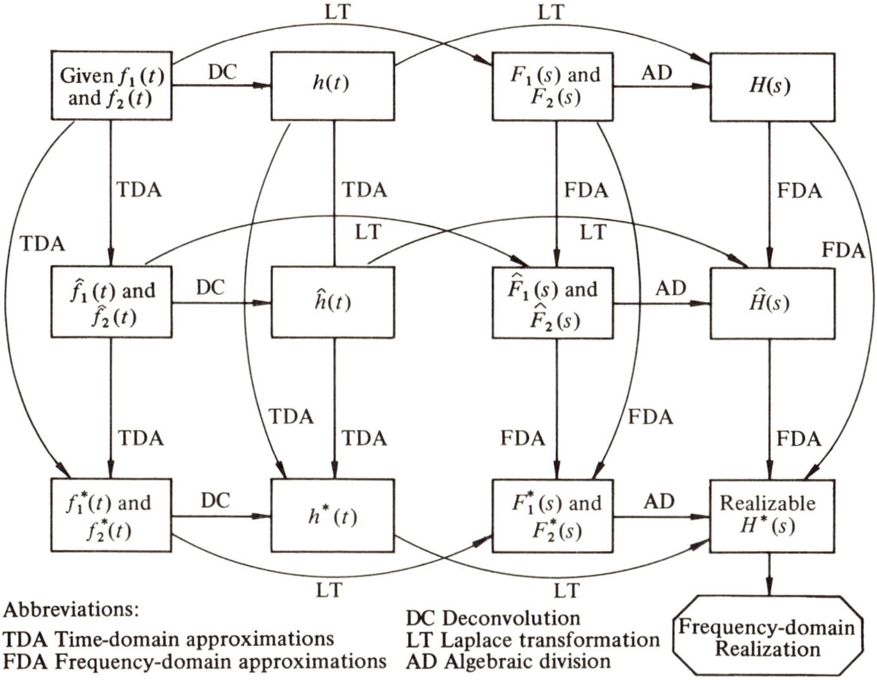

Figure 1.2 Some possible intermediate steps for transforming a time-domain synthesis problem into the frequency domain.

mainder of this volume, a circumflexed (hatted) quantity [for example, $\hat{h}(t)$] denotes an approximate representation of the corresponding quantity denoted by the same letter [$h(t)$] without any consideration of its network realizability. An asterisked quantity [$h^*(t)$] denotes an approximating function of the corresponding quantity denoted by the same letter and is realizable by a linear lumped network.

Depending either on the given $f_1(t)$ and $f_2(t)$ or on the desired form of solution, we may be forced to take certain limited numbers of possible paths. For instance, when $f_1(t)$ and $f_2(t)$ are given in the form of tabulated data, their Laplace transforms are not readily obtainable. In that case, we are forced to take one of the paths on the extreme left. In other situations, $F_1(s)$ and $F_2(s)$ may be easily obtainable, but $H(s)$ is not realizable. In that case, some form of frequency-domain approximation scheme must be employed. For example, suppose we have

$$f_1(t) = u(t) \tag{1.10}$$

with the response a delayed replica

$$f_2(t) = u(t - T_d), \qquad T_d > 0 \tag{1.11}$$

which is another unit step occurring at $t = T_d$. We can easily obtain

$$F_1(s) = \frac{1}{s} \tag{1.12}$$

$$F_2(s) = \frac{\epsilon^{-T_d s}}{s} \tag{1.13}$$

These require

$$H(s) = \epsilon^{-T_d s} \tag{1.14}$$

which is not realizable by a lumped network. A method for finding a realizable $H^*(s)$ that approximates the $H(s)$ of (1.14) is discussed in Chapter 7.

It should be noted that Fig. 1.2 is included here merely as an aide to delineate how a time-domain problem can be related to a frequency-domain problem through various combinations of smaller steps. The figure certainly does not include all possible paths and detours we may take. Many methods described in the subsequent chapters either bypass some of these steps or take completely different intermediate steps than those indicated in Fig. 2.1. In some instances (for example, Chapter 8), even the frequency domain is completely bypassed.

2

Some Basic Mathematical Principles

In this chapter we shall present several mathematical techniques and principles that are the basis of some of the developments in subsequent chapters. The presentation also serves to introduce some basic definitions and notation to be used in this volume.

To begin with, a brief discussion on the philosophical ideas behind the approximation problem is given. The techniques of data smoothing are then presented. These techniques serve as a link between a function that is only known at discrete points and an analytical expression. Finally, two most powerful approximation methods—the orthogonal functions and the Padé approximants—are discussed.

2.1 PHILOSOPHY OF MATHEMATICAL APPROXIMATION

As explained earlier, a time-domain synthesis problem may be considered solved once a realizable network function $H^*(s)$ has been found such that its corresponding impulse response $h^*(t)$ is an *acceptable* approximation of a given $h(t)$. Here the word *acceptable* refers to the accuracy by which the artifically constructed $h^*(t)$ approximates the given $h(t)$. For instance, the specification on the accuracy may be stated in terms of the amount of deviation to be no greater than a certain amount, absolute or relative. In other words, the requirement may read

$$|h(t) - h^*(t)| \leq M \tag{2.1}$$

or

$$\frac{|h(t) - h^*(t)|}{|h(t)|} \leq N \tag{2.2}$$

for all $t > 0$, where M and N are some prescribed real numbers. On other occasions, criteria that are mathematically more convenient are made.

Several mathematical techniques are frequently used to achieve the best approximation possible. These techniques are usually based on certain assumed criteria of measuring the error and then the error is either minimized or successively reduced. The resulting approximating function is then checked against the specified acceptability specification. If the function is not acceptable, a better approximation must be sought.

One of the most frequently used criteria is the least-mean-square-error approximation. In this scheme the error integral

$$E_2 = \int_0^\infty [h(t) - h^*(t)]^2 r(t)\, dt \tag{2.3}$$

where $r(t)$ is a weighting function, is minimized.

In other instances, the error integral may be replaced by either

$$|E|_m = \int_0^\infty |h(t) - h^*(t)|^m r(t)\, dt \tag{2.4}$$

or

$$E_m = \int_0^\infty [h(t) - h^*(t)]^m r(t)\, dt \tag{2.5}$$

Another widely used criterion is the minimum–maximum deviation approximation, in which the quantity

$$\epsilon_\infty = \max |h(t) - h^*(t)| \tag{2.6}$$

is minimized. This criterion frequently leads to a resulting $h^*(t)$ such that ϵ_∞ occurs at several values of the variable t. When this feature results, the approximation is said to be of the equiripple-error (or Chebyshev) type.

Another important scheme of approximation is the Taylor's approximation, in which the functions are matched term for term in their Taylor's expansions about a common point. This method of approximation is quite useful in the frequency domain. It can also be useful in a time-domain problem. One such scheme is given in Section 2.5.

In addition to these more popular methods of minimizing the error, many other specialized methods exist. In fact, part of the purpose of this volume is to explore the applicability of some of the more popular methods mentioned above and to present other more specialized methods developed for time-domain approximations.

In many respects, a time-domain approximation problem is exactly the same as one in the general theory of mathematical approximation. However, one most important aspect that is peculiar to the time-domain synthesis problem is that the approximating $h^*(t)$ must lead to a physically realizable network which is dynamically stable. The topics of realizability and stability are discussed in detail in Chapter 3.

2.2 SMOOTHING OF DATA

Frequently, a given time-domain function is specified in graphical form or tabulated data. In either case, some preliminary consideration of the accuracy and suitability of such information must first be given before a sound mathematical problem can be meaningfully formulated. With the wide availability of electronic computers, there is usually little difficulty in the carrying out of a numerical procedure itself with sufficient accuracy, except in occasions such as when a large matrix or a high-order rational function is involved. In fact, when a given function is not given in analytic form, the given data will usually be far less accurate than the ensuing numerical procedures that are used to carry out the required mathematical operations computationally.

The reasons for this type of inaccuracy could be many. Principally, when the data are obtained experimentally, either measured or simulated, the accuracy of the readings is necessarily limited. On other occasions, some of the functional values must be estimated by indirect means. Even when the functional values are known exactly, roundoff errors can still be present when the values are entered into a computational procedure. This is the result of the impossibility of representing an arbitrary mathematical value by a number with only a finite number of digits.

As an example, suppose we are given the response $i(t)$ in the circuit of Fig. 2.1 with switch S closed at $t = 0$. We have

$$i(t) = (1 - \epsilon^{-t})u(t) \tag{2.7}$$

The values of $i(t)$ for several values of t are listed in Table 2.1 to two de-

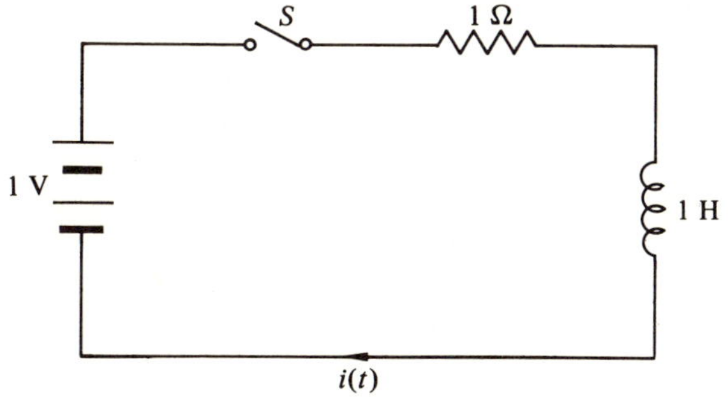

Figure 2.1 Circuit whose current is given by (2.7).

grees of accuracy. The four-place values have random deviations from the more accurate nine-place values. These deviations are shown in the error column. If the four-place values of $i(t)$ from Table 2.1 are used in a computational procedure that uses nine-place accuracy, five zeros are added to each value and this value is treated as if it were the true numerical value.

In a typical engineering situation in which the data are read from a recorded curve or a meter, the accuracy of these readings is likely to be much worse than the rounded-off data of Table 2.1. When such a set of data is given, we should

TABLE 2.1

Roundoff Errors of $i(t)$ of Equation (2.7)

t	$i(t)$ (9 places)	$i(t)$ (4 places)	Roundoff error
0.2	0.181269247	0.1813	0.30753×10^{-4}
0.4	0.329679954	0.3297	0.20046
0.6	0.451188364	0.4512	0.11636
0.8	0.550671036	0.5507	0.28964
1.0	0.632120669	0.6321	−0.20559
1.2	0.698805788	0.6988	−0.05788
1.4	0.573403036	0.7534	−0.03036
1.6	0.798103482	0.7981	−0.03482
1.8	0.834701112	0.8347	−0.01112
2.0	0.864664717	0.8647	0.35283
2.2	0.889196842	0.8892	0.03158
2.4	0.909282048	0.9093	0.17952
2.6	0.925726422	0.9257	−0.26422
2.8	0.939189937	0.9392	0.10063
3.0	0.950212932	0.9502	−0.12932

not treat these readings as absolutely exact values. Rather, we should recognize that they are only the best estimate obtainable with the available instrumentation.

Figure 2.2 shows a set of readings (in dots) whose deviations from the true values (solid curve) are greatly exaggerated. In most situations we are not as interested in finding a curve (dashed) that passes through all the points corresponding to the readings (as if they were exact) as we are in assuming a smooth curve that passes by the vicinity of all these readings. Chances are that the smooth curve is not only easier to handle mathematically, but also represents to a greater degree the most likely values of the true functional values of $f(t)$.

The procedure of finding an analytic function to approximate a given set of data is known in classical numerical analysis as the "analytic substitution." Once the analytic substitution is performed, the analytic function will be operated on, approximated by, and so on, as if it were the given function. For many problems we are going to consider here, it is far better to assume the analytically substituted function to be the given function than the raw data. In some situations it might be more advantageous to generate a new set of data based on this

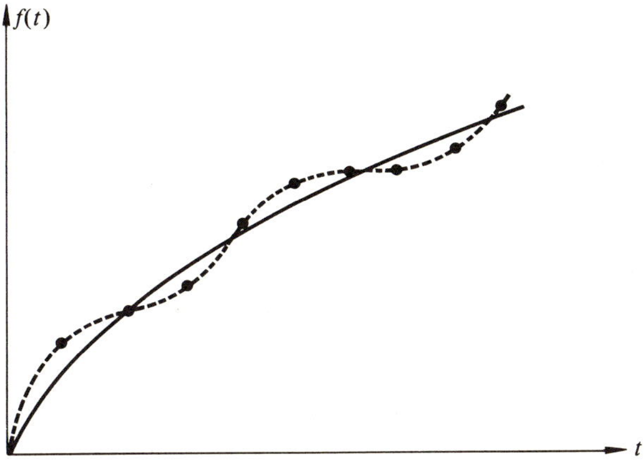

Figure 2.2 Example of random error and data smoothing.

analytic function to take the place of the original set. Since the new set is generated analytically, it can be made to have as many significant figures as can be meaningfully assumed in conjunction with the computational methods used.

In most of the classically developed methods of analytic substitution, the functions used are polynomials or rational functions. In each method, some form of artificial criterion is chosen. Then schemes are found to minimize the error between the given data and the substituting function. Finally, the "goodness of fit" is checked to see if the accuracy of this substitution is acceptable.

In essense, the whole problem of time-domain approximation is the problem of analytically substituting a given $h(t)$, in any of the three forms, by a function with the form

$$h^*(t) = \sum_i (A_0^{(i)} + A_1^{(i)}t + \dots + A_m^{(i)}t^m)A_i\epsilon^{s_it} \tag{2.8}$$

to within a prescribed tolerance. It is the type of analytic functions we use, not the procedure, that distinguishes the time-domain synthesis problems from other problems of analytic substitution.

The procedure of analytic substitution sometimes assumes a different form. That is, a set of points is given and regarded as accurate, and an analytic function is to be found such that it passes through these points exactly. This problem is called the interpolation problem. The functions used are again most commonly the polynomials or the rational functions.

The interpolation methods are also important for our purposes. They may be used to generate a set of equally spaced data points from a set of unequally spaced points. Also, they are useful when a polynomial is to be found to represent a given graphically described curve.

In persuing the following chapters, the reader will do well to keep these aspects of data-smoothing procedures in mind. Although no loss of appreciation of the time-domain synthesis problem will result without any familiarity with the various numerical techniques, many substeps of the approximation procedure and the implementation of these substeps can be much better understood in the light of the philosophy of analytic substitution.

In the chapters to follow, when a time-domain function is specified, it is understood that one of the following situations has been assumed:

1. The problem is specified analytically.
2. The procedure of analytic substitution has judiciously been applied, so the given function or data are a sufficiently accurate representation of the desired functions.
3. The given data are a true desired set of points to be approximated. For instance, if the points corresponding to the dots of Fig. 2.2 are to be approximated, then the dashed curve is the one to be approximated, not the solid one.

In other words, when a problem is specified in the following chapters, the data are not to be regarded as raw data. Some form of engineering or numerical modification has already been applied to the raw data and the problem has been appropriately reformulated.

For detailed discussions on analytic substitution and interpolation, the reader may wish to refer to many textbooks on numerical analysis [for example, HA1].

2.3 TECHNIQUE FOR DATA SMOOTHING

One simple and effective method of smoothing a given set of equally spaced data, in which random inaccuracy and local fluctuation are known to be present, is described here. In this method, a point on the smoothed curve is located on a parabola that fits the given data at and in the vicinity of that point in the least-mean-square-error sense. For many practical purposes either a five-point or a seven-point fit is sufficient, although the method can be extended to any number of points.

1. Five-Point Smoothing

Suppose $f(t)$ is given at a series of equally spaced points along the t axis Δt apart. For convenience and without loss of generality, we shall assume that for each group of five consecutive points $[t = (k-2)\,\Delta t, (k-1)\,\Delta t, k\Delta t, (k+1)\Delta t, (k+2)\,\Delta t]$, the translation of origin and scale normalization,

$$x = \frac{t - k\,\Delta t}{\Delta t} \tag{2.9}$$

has been applied and

$$y_x = y(x) = f[(k - x)\,\Delta t], \qquad x = -2, -1, 0, 1, 2 \tag{2.10}$$

and the parabola

$$\hat{y}(x) = a + bx + cx^2 \tag{2.11}$$

is to be found such that

$$E_2 = \sum_{k=-2}^{2} (\hat{y}_k - y_k)^2 \tag{2.12}$$

where $\hat{y}_k = \hat{y}(k)$, is a minimum. This procedure can obviously be repeated for any series of five consecutive points. Usually this procedure is applied to all possible groups of points and only the center points in each group retained, except the two end groups, in which all three end points are retained.

Substituting (2.10) and (2.11) into (2.12) gives

$$\begin{aligned} E_2 = {} & (a - 2b + 4c - y_{-2})^2 + (a - b + c - y_{-1})^2 + (a - y_0)^2 \\ & + (a + b + c - y_1)^2 + (a + 2b + 4c - y_2)^2 \end{aligned} \tag{2.13}$$

To minimize E_2, $\partial E_2/\partial a = 0$ gives

$$5a + 10c = \sum_{k=-2}^{2} y_k \tag{2.14}$$

$\partial E_2/\partial b = 0$ gives

$$10b = \sum_{k=-2}^{2} k y_k \tag{2.15}$$

and $\partial E_2/\partial c = 0$ gives

$$10a + 34c = \sum_{k=-2}^{2} k^2 y_k \tag{2.16}$$

Solution of (2.14) and (2.16) gives

$$\hat{y}_0 = a = \frac{-3y_{-2} + 12y_{-1} + 17y_0 + 12y_1 - 3y_2}{35} \tag{2.17}$$

Thus the value of $\hat{y}_0$ is found. Equation (2.17) is independent of the spacing of the data points as long as they are equally spaced and consecutive.

For the two end groups, either $\hat{y}_{-2}$ and $\hat{y}_{-1}$ or $\hat{y}_1$ and $\hat{y}_2$ must be calculated. From (2.14), (2.15), and (2.16) we have

$$b = \frac{-2y_{-2} - y_{-1} + y_1 + 2y_2}{10} \tag{2.18}$$

$$c = \frac{2y_{-2} - y_{-1} - 2y_0 - y_1 + 2y_2}{14} \tag{2.19}$$

and

$$\hat{y}_{-2} = a - 2b + 4c = \frac{31y_{-2} + 9y_{-1} - 3y_0 - 5y_1 + 3y_2}{35} \tag{2.20}$$

$$\hat{y}_{-1} = a - b + c = \frac{9y_{-2} + 13y_{-1} + 12y_0 + 6y_1 - 5y_2}{35} \tag{2.21}$$

Since these formulas are symmetric, expressions for $\hat{y}_2$ and $\hat{y}_1$ are readily obtained by reversing the signs of all subscripts in (2.20) and (2.21).

Equation (2.17) is sometimes expressed in terms of finite differences. The notation used here is best explained by referring to the arrangement in Table 2.2, in which each additional δ operator represents the evaluation of the increment of the two quantities to its immediate left. For example,

$$\delta y_{k-(3/2)} = y_{k-1} - y_{k-2} \tag{2.22}$$

TABLE 2.2

Notation for the Finite Differences

$y(x)$	First difference	Second difference	Third difference	Fourth difference
y_{k-2}				
	$\delta y_{k-(3/2)}$			
y_{k-1}		$\delta^2 y_{k-1}$		
	$\delta y_{k-(1/2)}$		$\delta^3 y_{k-(1/2)}$	
y_k		$\delta^2 y_k$		$\delta^4 y_k$
	$\delta y_{k+(1/2)}$		$\delta^3 y_{k+(1/2)}$	
y_{k+1}		$\delta^2 y_{k+1}$		
	$\delta y_{k+(3/2)}$			
y_{k+2}				

In general,

$$\delta^n y_i = \delta^{n-1} y_{i+(1/2)} - \delta^{n-1} y_{i-(1/2)} \tag{2.23}$$

It is easy to show that

$$\delta^4 y_k = y_{k+2} - 4y_{k+1} + 6y_k - 4y_{k-1} + y_{k-2} \tag{2.24}$$

which is the fourth difference. Equation (2.17) thus becomes

$$\hat{y}_0 = y_0 - \tfrac{3}{35}\delta^4 y_0 \tag{2.25}$$

and the five-point smoothing method is sometimes known as *smoothing by fourth difference.*

Computation of the finite differences is systematic and easy to follow. When a large number of points are to be smoothed, the arrangement of Table 2.2 is easier to use than the repeated application of (2.17).

2. Seven-Point Smoothing

The foregoing method can readily be extended to include seven consecutive points at a time. Using the same notation, we are attempting to find the $y(x)$ of (2.11) for a set of given y_k, $k = -3, -2, -1, 0, 1, 2, 3$, such that the error

$$E_2 = \sum_{k=-3}^{3} (\hat{y}_k - y_k)^2$$

is minimum. Following the same procedure for the five-point smoothing we find

$$\hat{y}_{-3} = \frac{32y_{-3} + 15y_{-2} + 3y_{-1} - 4y_0 - 6y_1 - 3y_2 + 5y_3}{42} \tag{2.26}$$

$$\hat{y}_{-2} = \frac{5y_{-3} + 4y_{-2} + 3y_{-1} + 2y_0 + y_1 - y_3}{14} \tag{2.27}$$

$$\hat{y}_{-1} = \frac{y_{-3} + 3y_{-2} + 4y_{-1} + 4y_0 + 3y_1 + y_2 - 2y_3}{14} \tag{2.28}$$

$$\hat{y}_0 = \frac{-2y_{-3} + 3y_{-2} + 6y_{-1} + 7y_0 + 6y_1 + 3y_2 - 2y_3}{21} \tag{2.29}$$

Formulas for $\hat{y}_3$, $\hat{y}_2$, and $\hat{y}_1$ can be obtained by reversing the signs of all subscripts in (2.26), (2.27), and (2.28), repectively. In terms of the finite differences, the seven-point smoothing equation may be written as

$$\hat{y}_0 = y_0 - \tfrac{9}{21}\delta^4 y_0 - \tfrac{2}{21}\delta^6 y_0 \tag{2.30}$$

Example. As a numerical example of these smoothing formulas, a set of artifical data is given in the second column of Table 2.3. Columns 3 and 4 give the smoothed values using the five-point and the seven-point formulas. The fluctuations of values of y_k have been exaggerated so that the effects of these formulas can be discerned more easily. These data are also plotted in Figure 2.3.

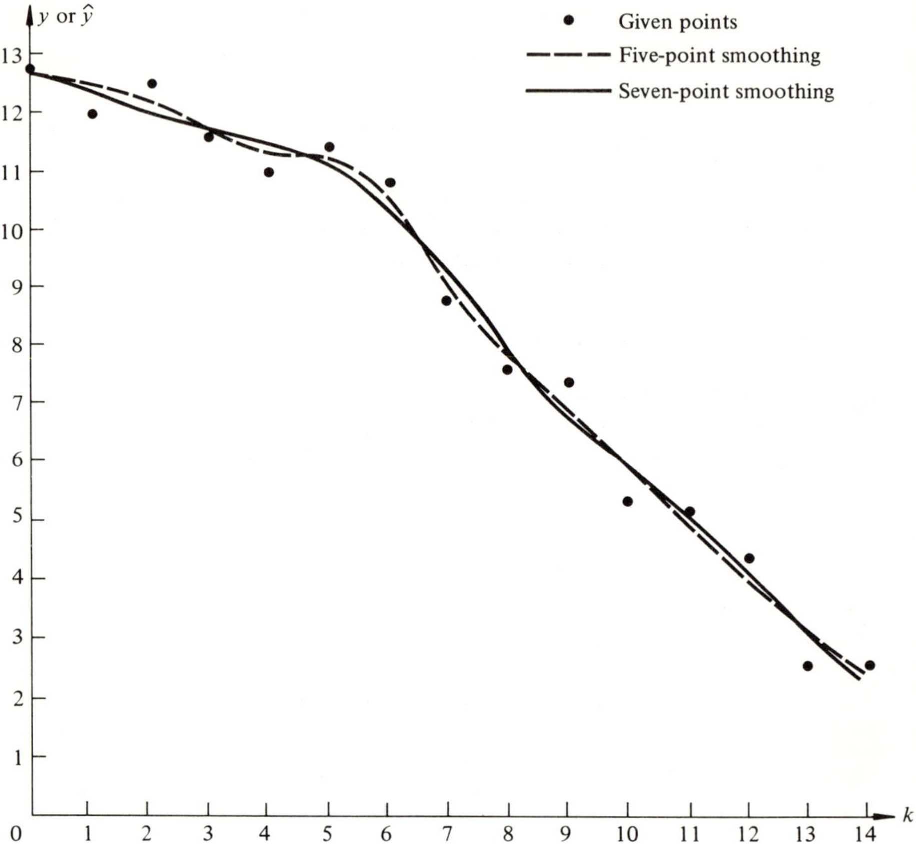

Figure 2.3 An example of five-point and seven-point smoothing.

It is seen that the five-point formulas tend to smooth out fluctuations that alternate between two sides of a normally smooth curve. The seven-point formulas are able to smooth out readings that vary in a more erratic manner.

If desired, these smoothing procedures may be reapplied to the smoothed data. However, the smoothing effect of these formulas is generally reduced in the subsequent applications of these formulas.

These smoothing techniques serve a very similar function to that of an engineer taking a French curve and drawing a compromise curve among a group of noisy readings. However, the newly computed points obtained from the formulas are usually more reliable than a curve drawn by judgment, because a curve is still subject to reading error even after it is judiciously drawn. Generally, the computed numbers furnish several more places that are more dependable than the raw data. In many (but not all) engineering problems, it is more meaningful to consider the smoothed curve to be the exact one than to consider the individual readings to be exact.

TABLE 2.3

Numerical Example of Data Smoothing

k	y_k	$\hat{y}_k$ (five-point smoothing)	$\hat{y}_k$ (seven-point smoothing)
0	12.8	12.637	12.726
1	12.0	12.451	12.350
2	12.5	12.123	12.000
3	11.6	11.686	11.676
4	11.0	11.231	11.586
5	11.4	11.263	11.029
6	10.8	10.577	10.219
7	8.8	8.971	9.314
8	7.6	7.857	7.895
9	7.4	6.851	6.733
10	5.4	5.914	6.029
11	5.2	5.029	4.990
12	4.4	4.126	4.085
13	2.6	3.263	3.200
14	2.6	2.314	2.333

3. Extension of the Smoothing Method

The five point and seven-point smoothing methods are generally sufficient in dealing with most engineering problems. There are two ways these methods can be extended.

(a) Replace the parabolic equation (2.11) by a higher-degree polynomial.
(b) Increase the number of points included in each smoothing range.

In each of these extensions, the principle and procedure needed are identical to the simpler cases described above.

The present method of using a parabolic equation to obtain local smoothing for individual points is a special form of analytic substitution by a polynomial. In fact, when this method is extended to including all the points given and using only one high-degree polynomial, the procedure is precisely a form of the analytic substitution by polynomials.

2.4 ORTHOGONAL-FUNCTION APPROXIMATION

A set of complex-valued functions of the real variable t,

$$\phi_n(t), \qquad n = 0, 1, 2, \ldots$$

is said to be *orthogonal* over the range $t = a$ and $t = b$ with respect to the real weighting function $r(t)$ if

$$\int_a^b \phi_m(t)\bar{\phi}_n(t)r(t)\,dt = 0, \qquad m \neq n$$
$$\neq 0, \qquad m = n \tag{2.31}$$

where $\bar{\phi}_n$ designates the complex conjugate of ϕ_n. Further, if

$$\int_a^b \phi_m(t)\bar{\phi}_m(t)r(t)\,dt = 1, \qquad \text{for all } m \tag{2.32}$$

then the set is said to be *orthonormal* with respect to $r(t)$.

The usefulness of the orthogonal functions to approximation problems lies in the fact that the error minimization with the use of a set of orthogonal functions is mathematically easiest to trace. This fact is due to the simplicity in the calculus and algebra in the error-minimization procedure when the least-mean-square-error criterion is used.

Suppose it is desired to approximate a function $f(t)$ by $\hat{f}_N(t)$ in the least-mean-square-error sense, where

$$\hat{f}_N(t) = \sum_{i=0}^{N} C_i\phi_i(t) \tag{2.33}$$

in which C_i are generally complex numbers. The squared error is

$$\begin{aligned} E_2 &= \int_a^b |f(t) - \hat{f}_N(t)|^2 r(t)\,dt \\ &= \int_a^b |f(t)|^2 r(t)\,dt - \sum_{i=0}^{N} C_i \int_a^b \phi_i(t)\bar{f}(t)r(t)\,dt \\ &\quad - \sum_{i=0}^{N} \bar{C}_i \int_a^b \bar{\phi}_i(t)f(t)r(t)\,dt + \sum_{i=0}^{N} |C_i|^2 \int_a^b |\phi_i(t)|^2 r(t)\,dt \end{aligned} \tag{2.34}$$

Let $C_i = a_i + jb_i$. Then set $\partial E_2/\partial a_i = 0$ and $\partial E_2/\partial b_i = 0$ to find a set of C_i for minimum E_2. We have

$$\begin{aligned} \frac{\partial E_2}{\partial a_i} &= -\int_a^b \phi_i(t)\bar{f}(t)r(t)\,dt - \int_a^b \bar{\phi}_i(t)f(t)r(t)\,dt \\ &\quad + 2a_i \int_a^b |\phi_i(t)|^2 r(t)\,dt = 0 \end{aligned}$$

Thus

$$a_i = \frac{\int_a^b \operatorname{Re}\,[\bar{\phi}_i(t)f(t)]r(t)\,dt}{\int_a^b |\phi_i(t)|^2 r(t)\,dt} \tag{2.35}$$

Also, we have

$$\frac{\partial E_2}{\partial b_i} = j\int_a^b \bar{\phi}_i(t)f(t)r(t)\,dt - j\int_a^b \phi_i(t)\bar{f}_i(t)r(t)\,dt$$
$$+ 2b_i\int_a^b |\phi_i(t)|^2 r(t)\,dt = 0$$

Thus

$$b_i = \frac{\int_a^b \operatorname{Im}[\bar{\phi}_i(t)f(t)]r(t)\,dt}{\int_a^b |\phi_i(t)|^2 r(t)\,dt} \tag{2.36}$$

Combining (2.35) and (2.36) yields

$$C_i = \frac{\int_a^b f(t)\bar{\phi}_i(t)r(t)\,dt}{\int_a^b |\phi_i(t)|^2 r(t)\,dt} \tag{2.37}$$

Thus C_i can be evaluated directly once $\phi_i(t)$, $r(t)$, and $f(t)$ are given. The denominator of (2.37) is independent of $f(t)$. If ϕ_i has already been normalized to be an orthonormal set, then the denominator is unity.

Equation (2.37) also indicates that each C_i is independent of other functions in the set. Thus once the coefficient of a particular ϕ_i has been evaluated, it is unique for this ϕ_i. No further change in this particular C_i can improve the least-squares approximation. The only way the approximation can be improved with this orthogonal set is to include more ϕ_i terms than the ones that have been evaluated.

The weighting function $r(t)$ generally determines the corresponding ϕ_i and thus controls the outcome of the approximation. This function determines which portion of the axis and the extent over which the approximation is emphasized. The choice of an $r(t)$ for a particular approximation problem is a matter of judgment. Usually, the result of a particular choice is not known until the approximation procedure has been completed. If the first approximation is unsatisfactory, one can either include more terms of the same set or modify the weighting function and switch to another set of orthogonal functions.

An important property of an orthogonal function set is its *completeness*. A set of orthogonal functions $\phi_i(t)$ is said to be complete in a function space if there exists no function $g(t)$ in that space, except the identically zero function and linear combinations of members of the set, that satisfies

$$\int_a^b g(t)\bar{\phi}_i(t)r(t)\,dt = 0 \tag{2.38}$$

for all values of i. The orthogonal-function series expansion of a given function $f(t)$ is expressed in (2.33) with C_i evaluated according to (2.37) may not converge to $f(t)$ if the set of $\phi_i(t)$ is not complete.

An example of an incomplete set of orthogonal functions in the real function space in the range $t = 0$ and $t = 2\pi$ is

$$\phi_i = \sin it$$

An example of a nonzero function orthogonal to this set is $g(t) = \cos kt$. When a set of incomplete orthogonal functions is used in an approximation problem, it is very likely that by increasing the number of terms included, the sequence may not improve in approximation, or it may improve too slowly. This might be because some necessary low-order terms are missing.

In the above development, t and $r(t)$ are restricted to be real, and $f(t)$ and $\phi_i(t)$ can be complex. In many engineering problems, we may assume that $f(t)$ is real and $\phi_i(t)$ can be chosen to be real. Under these conditions we merely drop the conjugate notation in (2.37) and the C_i are real.

An unlimited number of sets of orthogonal functions can be found for any given weighting function $r(t)$ over the same range. For many practical problems, the most useful orthogonal functions associated with a given weighting function are the polynomials. Orthogonal polynomials associated with a given weighting function can usually be generated straightforwardly.

Starting with

$$\phi_0(t) = 1$$

set

$$\phi_1(t) = t + c_{10}\phi_0(t)$$

Orthogonality requires

$$\int_a^b \phi_0(t)\phi_1(t)r(t)\,dt = 0 \tag{2.39}$$

From (2.39), c_{10} can be evaluated for the given $r(t)$. Next let

$$\phi_2(t) = t^2 + c_{21}\phi_1(t) + c_{20}\phi_0(t)$$

Setting

$$\int_a^b \phi_0(t)\phi_2(t)r(t)\,dt = 0 \tag{2.40}$$

and

$$\int_a^b \phi_1(t)\phi_2(t)r(t)\,dt = 0 \tag{2.41}$$

and solving for c_{21} and c_{20} will determine $\phi_2(t)$.

Similarly,

$$\phi_3(t) = t^3 + c_{32}\phi_2(t) + c_{31}\phi_1(t) + c_{30}\phi_0(t)$$

can next be determined by its orthogonality with $\phi_0(t)$, $\phi_1(t)$, and $\phi_2(t)$.

For example, suppose we have $r(t) = \epsilon^{-t}$, $a = 0$, $b = \infty$. Application of (2.39) gives

$$c_{10} = -1$$

and

$$\phi_1(t) = t - 1$$

Application of (2.40) gives

$$c_{20} = -2$$

and application of (2.41) gives

$$c_{21} = -4$$

Hence

$$\phi_2(t) = t^2 - 4t + 2$$

Similarly, we shall have

$$\phi_3(t) = t^3 - 9t^2 + 18t - 6$$
$$\phi_4(t) = t^4 - 16t^3 + 72t^2 - 96t + 24$$

These polynomials are known as the *Laguerre polynomials* and usually are tabulated in normalized form.

Many other orthogonal polynomials are tabulated in mathematical tables and their explicit expressions are sometimes given in the form of recursive formulas [AB2].

2.5 PADÉ APPROXIMANTS

If a given function $H(s)$ is analytic at the origin, it can be represented by a power series in the neighborhood of the origin. This power series is the Taylor's expansion about the origin. A rational function in s can always be found such that its Taylor's expansion about the origin has the same leading terms as those of $H(s)$. Such a rational function is known as the Padé approximant of $H(s)$.

Generally, the coefficients of a Padé approximant depend on the degrees of both the numerator and the denominator polynomials [PA1].

Let the given Taylor's expansion be denoted by

$$H(s) = h_0 + h_1 s + h_2 s^2 + \ldots + h_k s^k + \sum_{i=k+1}^{\infty} h_i s^i \tag{2.42}$$

A corresponding approximant of the form

$$\frac{P_{mn}(s)}{Q_{mn}(s)} = \frac{a_0 + a_1 s + a_2 s^2 + \ldots + a_m s^m}{b_0 + b_1 s + b_2 s^2 + \ldots + b_n s^n} = (m, n) \tag{2.43}$$

is its Padé approximant of order m, n and denoted by (m, n). Although there are $m + n + 2$ coefficients in (m, n), only $m + n + 1$ of them are independent, because any of the $m + n + 2$ coefficients may be assigned an arbitrary value without any loss in generality. Thus, if $P_{mn}(s)/Q_{mn}(s)$ is to approximate $H(s)$ in the Taylor sense about the origin, the former must have a Taylor's expansion whose first $m + n + 1$ terms are identical to those of $H(s)$. Such a rational function is unique. In other words, if P_{mn} and Q_{mn} are such that

$$H(s)Q_{mn}(s) - P_{mn}(s) = \sum_{i=m+n+1}^{\infty} h'_i s^i \tag{2.44}$$

then

$$\frac{P_{mn}(s)}{Q_{mn}(s)}$$

is the Padé approximant (m, n). It does not matter how the a's and b's in P_{mn} and Q_{mn} are found. Several of these methods are described in this section.

The Padé table for a given function is simply a tabulation of successive Padé approximants in an orderly manner. An example of the Padé table is given in Table 2.4 for the function ϵ^s.

Although in the development of this section the origin is chosen as a reference point about which the Taylor's series is expanded, the reference point can be shifted to any other point in the s plane by a simple translation.

It should be noted that since the Padé approximant and the given function have the same leading terms in their Taylor's expansions, these two functions tend to approach each other in the neighborhood of the origin. Generally speaking, near the origin, the approximant approximates the given function very closely. The approximation at points far away from the origin can be very poor.

Another way to look at the manner a Padé approximant approximates a given function is to consider the behavior of $H(s)$ and its Taylor's series along the unit circle centered at the origin. The functions $1, s, s^2, \ldots, s^k, \ldots$, for $s = \epsilon^{j\theta}$, are orthogonal along the unit circle in the variable θ since

TABLE 2.4

Padé Table for the Function ϵ^s

m / n	0	1	2	3	4
0	$(0,0) = \dfrac{1}{1}$	$(1,0) = \dfrac{1+s}{1}$	$(2,0) = \dfrac{1+s+\frac{1}{2}s^2}{1}$	$(3,0) = \dfrac{1+s+\frac{1}{2}s^2+\frac{1}{6}s^3}{1}$	$(4,0) = \dfrac{1+s+\frac{1}{2}s^2+\frac{1}{6}s^3+\frac{1}{24}s^4}{1}$
1	$(0,1) = \dfrac{1}{1-s}$	$(1,1) = \dfrac{1+\frac{1}{2}s}{1-\frac{1}{2}s}$	$(2,1) = \dfrac{1+\frac{2}{3}s+\frac{1}{6}s^2}{1-\frac{1}{3}s}$	$(3,1) = \dfrac{1+\frac{3}{4}s+\frac{1}{4}s^2+\frac{1}{24}s^3}{1-\frac{1}{4}s}$	$(4,1) = \dfrac{1+\frac{4}{5}s+\frac{3}{10}s^2+\frac{1}{15}s^3+\frac{1}{120}s^4}{1-\frac{1}{5}s}$
2	$(0,3) = \dfrac{1}{1-s-\frac{1}{2}s^2}$	$(1,2) = \dfrac{1+\frac{1}{3}s}{1-\frac{2}{3}s+\frac{1}{6}s^2}$	$(2,2) = \dfrac{1+\frac{1}{2}s+\frac{1}{12}s^2}{1-\frac{1}{2}s+\frac{1}{12}s^2}$	$(3,2) = \dfrac{1+\frac{3}{5}s+\frac{3}{20}s^2+\frac{1}{60}s^3}{1-\frac{2}{5}s+\frac{1}{20}s^2}$	$(4,2) = \dfrac{1+\frac{2}{3}s+\frac{1}{5}s^2+\frac{1}{30}s^3+\frac{1}{360}s^4}{1-\frac{1}{3}s+\frac{1}{30}s^2}$
3	$(0,3) = \dfrac{1}{1-s+\frac{1}{2}s^2-\frac{1}{6}s^3}$	$(1,3) = \dfrac{1+\frac{1}{4}s}{1-\frac{3}{4}s+\frac{1}{4}s^2-\frac{1}{24}s^3}$	$(2,3) = \dfrac{1+\frac{2}{5}s+\frac{1}{20}s^2}{1-\frac{3}{5}s+\frac{3}{20}s^2-\frac{1}{60}s^3}$	$(3,3) = \dfrac{1+\frac{1}{2}s+\frac{1}{10}s^2+\frac{1}{120}s^3}{1-\frac{1}{2}s+\frac{1}{10}s^2-\frac{1}{120}s^3}$	$(4,3) = \dfrac{1+\frac{4}{7}s+\frac{1}{7}s^2+\frac{2}{105}s^3+\frac{1}{840}s^4}{1-\frac{3}{7}s+\frac{1}{14}s^2-\frac{1}{120}s^3}$
4	$(0,4) = \dfrac{1}{1-s+\frac{1}{2}s^2-\frac{1}{6}s^3+\frac{1}{24}s^4}$	$(1,4) = \dfrac{1+\frac{1}{5}s}{1-\frac{4}{5}s+\frac{3}{10}s^2-\frac{1}{15}s^3+\frac{1}{120}s^4}$	$(2,4) = \dfrac{1+\frac{1}{3}s+\frac{1}{30}s^2}{1-\frac{2}{3}s+\frac{1}{5}s^2-\frac{1}{30}s^3+\frac{1}{360}s^4}$	$(3,4) = \dfrac{1+\frac{3}{7}s+\frac{1}{14}s^2+\frac{1}{210}s^3}{1-\frac{4}{7}s+\frac{1}{7}s^2-\frac{2}{105}s^3+\frac{1}{840}s^4}$	$(4,4) = \dfrac{1+\frac{1}{2}s+\frac{3}{28}s^2+\frac{1}{84}s^3+\frac{1}{1680}s^4}{1-\frac{1}{2}s+\frac{3}{28}s^2-\frac{1}{84}s^3+\frac{1}{1680}s^4}$

$$\int_{-\pi}^{\pi} (s^{\alpha} \cdot \bar{s}^{\beta})\Big|_{s=\epsilon^{j\theta}} d\theta = \int_{-\pi}^{\pi} \epsilon^{j(\alpha-\beta)\theta}\, d\theta = 2\pi, \qquad \alpha = \beta$$
$$= 0, \qquad \alpha \neq \beta \tag{2.45}$$

Now if $H(s)$ is to be approximated along this contour by the series

$$\hat{H}(s) = d_0 + d_1 s + d_2 s^2 + \ldots + d_k s^k + \ldots \tag{2.46}$$

in the least-mean-square-error sense, from (2.37) for $r(t) = 1$, it is necessary to make

$$d_k = \frac{1}{2\pi}\int_{-\pi}^{\pi} \frac{H(s)}{s^k}\Big|_{s=\epsilon^{j\theta}} d\theta$$
$$= \frac{1}{2\pi j}\oint_C \frac{H(s)}{s^{k+1}}\, ds \tag{2.47}$$

where contour C is the unit circle. This expression is recognized to be that used to obtain the Taylor's expansion about the origin. Hence it can be said that a truncated Taylor's series about the origin gives the best approximation in the least-mean-square-error sense along the unit circle with respect to the set of orthogonal functions $1, s, s^2, \ldots, s^k, \ldots$.

Since $H(s)$, $\hat{H}(s)$, and (m, n) all have the same truncated Taylor's expansion about the origin, it can be said that the Padé approximant of a function tends to give a good approximation for the function in the least-mean-square-error sense along the unit circle centered at the origin.

1. Direct Method

Any particular approximant (m, n) can be found by first forming the explicit expression of the approximant with undetermined coefficients and then forcing the Taylor's expansion of this function about the origin to agree with that of the given function through the $(n + m + 1)$th term. Algorithmically, it is usually easier to multiply the denominator of the approximant into the power series of the function to be approximated and then match these coefficients to those of the numerator of the approximant. Using the notation of (2.43), we get

$$a_0 + a_1 s + a_2 s^2 + \ldots + a_m s^m$$
$$\cong (h_0 + h_1 s + h_2 s^2 + \ldots + h_{m+n} s^{m+n})(b_0 + b_1 s + b_2 s^2 + \ldots + b_n s^n) \tag{2.48}$$

and then equate coefficients on both sides of (2.48) through the s^{m+n} term.† In this consideration $a_i = 0$ for $m + 1 \leq i \leq m + n$.

†Any one of the coefficients of (m, n) may be assigned an arbitrary value.

This step of equating coefficients of (2.48) can be accomplished systematically by solving the following matrix equation.

$$\begin{bmatrix} 1 & 0 & 0 & \cdots & 0 & 0 & 0 & \cdots & 0 \\ 0 & 1 & 0 & \cdots & 0 & -h_0 & 0 & \cdots & 0 \\ 0 & 0 & 1 & \cdots & 0 & -h_1 & -h_0 & \cdots & 0 \\ \cdots & \cdots & \cdots & \cdots & \cdots & \cdots & \cdots & \cdots & \cdots \\ 0 & 0 & 0 & \cdots & 1 & -h_{m-1} & -h_{m-2} & \cdots & -h_{m-n} \\ 0 & 0 & 0 & \cdots & 0 & -h_m & -h_{m-1} & \cdots & -h_{m-n+1} \\ 0 & 0 & 0 & \cdots & 0 & -h_{m+1} & -h_m & \cdots & -h_{m-n+2} \\ \cdots & \cdots & \cdots & \cdots & \cdots & \cdots & \cdots & \cdots & \cdots \\ 0 & 0 & 0 & \cdots & 0 & -h_{m+n-1} & -h_{m+n-2} & \cdots & -h_m \end{bmatrix} \begin{bmatrix} a_0 \\ a_1 \\ a_2 \\ \vdots \\ a_m \\ b_1 \\ b_2 \\ \vdots \\ b_n \end{bmatrix} = \begin{bmatrix} h_0 b_0 \\ h_1 b_0 \\ h_2 b_0 \\ \vdots \\ h_m b_0 \\ h_{m+1} b_0 \\ h_{m+2} b_0 \\ \vdots \\ h_{m+n} b_0 \end{bmatrix} \tag{2.49}$$

In (2.49) it is assumed that $h_i = 0$ for $i < 0$.

As an example, suppose it is desired to find (2, 3) for

$$\frac{\tan^{-1} s}{s} = 1 - \frac{s^2}{3} + \frac{s^4}{5} - \frac{s^6}{7} + \cdots \tag{2.50}$$

We can solve the following matrix equation, assuming $b_0 = 1$.

$$\begin{bmatrix} 1 & 0 & 0 & 0 & 0 & 0 \\ 0 & 1 & 0 & -1 & 0 & 0 \\ 0 & 0 & 1 & 0 & -1 & 0 \\ 0 & 0 & 0 & \frac{1}{3} & 0 & -1 \\ 0 & 0 & 0 & 0 & \frac{1}{3} & 0 \\ 0 & 0 & 0 & -\frac{1}{5} & 0 & \frac{1}{3} \end{bmatrix} \begin{bmatrix} a_0 \\ a_1 \\ a_2 \\ b_1 \\ b_2 \\ b_3 \end{bmatrix} = \begin{bmatrix} 1 \\ 0 \\ -\frac{1}{3} \\ 0 \\ \frac{1}{5} \\ 0 \end{bmatrix}$$

which yields

$$a_0 = 1, \quad a_1 = 0, \quad a_2 = \tfrac{4}{15}, \quad b_1 = 0, \quad b_2 = \tfrac{3}{5}, \quad b_3 = 0$$

Hence†

$$(2, 3) = \frac{1 + \frac{4}{15} s^2}{1 + \frac{3}{5} s^2} \tag{2.51}$$

This method is straightforward and easy to follow. However, it has the obvious disadvantage that the number of equations is usually unnecessarily high.

†In this particular example, (2, 2) = (2, 3).

The next section gives an alternative procedure that somewhat alleviates this shortcoming.

2. Determinant Method

Although the equations that have to be solved in method 1 involve either $m + n + 1$ linear simultaneous equations or an $(m + n + 1) \times (m + n + 1)$ matrix equation, the solution of these equations can be simplified to a great extent. Specifically, those equations may be written in two separate groups.

$$
\begin{aligned}
a_0 &= h_0 b_0 \\
a_1 &= h_1 b_0 + h_0 b_1 \\
&\vdots \\
a_m &= h_m b_0 + h_{m-1} b_1 + \dots + h_0 b_m
\end{aligned}
\tag{2.52}
$$

$$
\begin{aligned}
0 &= h_{m+1} b_0 + h_m b_1 + \dots + h_{m-n+2} b_{n-1} + h_{m-n+1} b_n \\
0 &= h_{m+2} b_0 + h_{m+1} b_1 + \dots + h_{m-n+3} b_{n-1} + h_{m-n+2} b_n \\
&\vdots \\
0 &= h_{m+n} b_0 + h_{m+n-1} b_1 + \dots + h_{m+1} b_{n-1} + h_m b_n
\end{aligned}
\tag{2.53}
$$

Again, it is assumed that $h_i = 0$ if $i < 0$. Thus we can solve for $b_1, b_2, \dots, b_n$ from (2.53) in terms of h's and b_0. The value of b_0 may be arbitrarily assigned. Then these values of of b's may be substituted into (2.52) to obtain the values of a's. Solving (2.53) yields

$$
b_k = -\frac{b_0}{\Delta_{mn}}
\begin{vmatrix}
h_m & h_{m-1} & \cdots & h_{m-k+2} & h_{m+1} & h_{m-k} & \cdots & h_{m-n+1} \\
h_{m+1} & h_m & \cdots & h_{m-k+3} & h_{m+2} & h_{m-k+1} & \cdots & h_{m-n+2} \\
\cdots & \cdots & \cdots & \cdots & \cdots & \cdots & \cdots & \cdots \\
h_{m+n-1} & h_{m+n-2} & \cdots & h_{m-k+n+1} & h_{m+n} & h_{m-k+n-1} & \cdots & h_m
\end{vmatrix}
\quad (k\text{th column at the } h_{m+1} \text{ column})
\tag{2.54}
$$

(*k*th column) ↓ marks the column $h_{m+1}, h_{m+2}, \dots, h_{m+n}$.

where

$$
\Delta_{mn} =
\begin{vmatrix}
h_m & h_{m-1} & \cdots & h_{m-k+2} & h_{m-k+1} & h_{m-k} & \cdots & h_{m-n+1} \\
h_{m+1} & h_m & \cdots & h_{m-k+3} & h_{m-k+2} & h_{m-k+1} & \cdots & h_{m-n+2} \\
\cdots & \cdots & \cdots & \cdots & \cdots & \cdots & \cdots & \cdots \\
h_{m+n-2} & h_{m+n-3} & \cdots & h_{m-k+n} & h_{m-k+n-1} & h_{m-k+n-2} & \cdots & h_{m-1} \\
h_{m+n-1} & h_{m+n-2} & \cdots & h_{m-k+n+1} & h_{m-k+n} & h_{m-k+n-1} & \cdots & h_m
\end{vmatrix}
\tag{2.55}
$$

provided that $\Delta_{mn} \neq 0$. Now we may choose $b_0 = \Delta_{mn}$ to simplify the development. Next we write (2.54) as

$$b_k = \begin{vmatrix} -1 & 0 & 0 & \cdots & 0 & 0 & 0 & \cdots & 0 \\ h_{m-k+1} & h_m & h_{m-1} & \cdots & h_{m-k+2} & h_{m+1} & h_{m-k} & \cdots & h_{m-n+1} \\ h_{m-k+2} & h_{m+1} & h_m & \cdots & h_{m-k+3} & h_{m+2} & h_{m-k+1} & \cdots & h_{m-n+2} \\ \cdots & \cdots & \cdots & \cdots & \cdots & \cdots & \cdots & \cdots & \cdots \\ h_{m-k+n} & h_{m+n-1} & h_{m+n-2} & \cdots & h_{m-k+n+1} & h_{m+n} & h_{m-k+n-1} & \cdots & h_m \end{vmatrix} \tag{2.56}$$

Determinant (2.56) is $(n + 1) \times (n + 1)$ and its value is obviously identical to that of (2.54) with $b_0 = \Delta_{mn}$. The additional column (first) does not change the value of the bordered determinant. When the first and the $(k + 1)$st columns are interchanged and s^k is multiplied into both sides of (2.56), we get

$$b_k s^k = \begin{vmatrix} 0 & 0 & 0 & \cdots & 0 & s^k & 0 & \cdots & 0 \\ h_{m+1} & h_m & h_{m-1} & \cdots & h_{m-k+2} & h_{m-k+1} & h_{m-k} & \cdots & h_{m-n+1} \\ h_{m+2} & h_{m+1} & h_m & \cdots & h_{m-k+3} & h_{m-k+2} & h_{m-k+1} & \cdots & h_{m-n+2} \\ \cdots & \cdots & \cdots & \cdots & \cdots & \cdots & \cdots & \cdots & \cdots \\ h_{m+n} & h_{m+n-1} & h_{m+n-2} & \cdots & h_{m-k+n+1} & h_{m-k+n} & h_{m-k+n-1} & \cdots & h_m \end{vmatrix} \tag{2.57}$$

The elements of the determinant in (2.57) are now in their proper sequences and their subscripts can easily be followed. Now it is easy to see that

$$\begin{aligned} Q_{mn}(s) &= b_0 + b_1 s + b_2 s^2 + \ldots + b_n s^n \\ &= \begin{vmatrix} 1 & s & s^2 & \cdots & s^{n-1} & s^n \\ h_{m+1} & h_m & h_{m-1} & \cdots & h_{m-n+2} & h_{m-n+1} \\ h_{m+2} & h_{m+1} & h_m & \cdots & h_{m-n+3} & h_{m-n+2} \\ \cdots & \cdots & \cdots & \cdots & \cdots & \cdots \\ h_{m+n} & h_{m+n-1} & h_{m+n-2} & \cdots & h_{m+1} & h_m \end{vmatrix} \end{aligned} \tag{2.58}$$

The numerator polynomial $P_{mn}(s)$ can now be formed by evaluating the coefficients a's from (2.52). Also, since $Q_{mn}(s)$ is known, $P_{mn}(s)$ can be found by multiplying $Q_{mn}(s)$ into the Taylor's expansion of $H(s)$, retaining only the first $(m + 1)$ terms. This step can be accomplished systematically by the following procedure. Since

$$H(s) = \sum_{i=0}^{\infty} h_i s^i \tag{2.59}$$

$$H(s)Q_{mn}(s) = \begin{vmatrix} \sum_{i=0}^{\infty} h_i s^i & \sum_{i=0}^{\infty} h_i s^{i+1} & \cdots & \sum_{i=0}^{\infty} h_i s^{i+n-1} & \sum_{i=0}^{\infty} h_i s^{i+n} \\ h_{m+1} & h_m & \cdots & h_{m-n+2} & h_{m-n+1} \\ h_{m+2} & h_{m+1} & \cdots & h_{m-n+3} & h_{m-n+2} \\ \cdots & \cdots & \cdots & \cdots & \cdots \\ h_{m+n} & h_{m+n-1} & \cdots & h_{m+1} & h_m \end{vmatrix}$$

$$= P_{mn}(s) + \sum_{i=m+n+1}^{\infty} a_i s^i \qquad \textbf{(2.60)}$$

In (2.60) the s^{m+1} through s^{m+n} terms are missing by vertue of (2.53). Keeping only terms up to s^m in (2.60), we obtain

$$P_{mn}(s) = \begin{vmatrix} \sum_{i=0}^{m} h_i s^i & \sum_{i=0}^{m-1} h_i s^{i+1} & \cdots & \sum_{i=0}^{m-n+1} h_i s^{i+n-1} & \sum_{i=0}^{m-n} h_i s^{i+n} \\ h_{m+1} & h_m & \cdots & h_{m-n+2} & h_{m-n+1} \\ h_{m+2} & h_{m+1} & \cdots & h_{m-n+3} & h_{m-n+2} \\ \cdots & \cdots & \cdots & \cdots & \cdots \\ h_{m+n} & h_{m+n-1} & \cdots & h_{m+1} & h_m \end{vmatrix} \qquad \textbf{(2.61)}$$

In (2.61) a summation is zero if its upper index is negative.

As an example, let it be desired to find (2, 3) for the function $\frac{1}{2}(1 + \epsilon^s)$. We have

$$\tfrac{1}{2}(1 + \epsilon^s) = 1 + \tfrac{1}{2}s + \tfrac{1}{4}s^2 + \tfrac{1}{12}s^3 + \tfrac{1}{48}s^4 + \tfrac{1}{240}s^5 + \sum_{i=6}^{\infty} h_i s^i \qquad \textbf{(2.62)}$$

From (2.61),

$$P_{23}(s) = \begin{vmatrix} 1 + \frac{1}{2}s + \frac{1}{4}s^2 & s + \frac{1}{2}s^2 & s^2 & 0 \\ \frac{1}{12} & \frac{1}{4} & \frac{1}{2} & 1 \\ \frac{1}{48} & \frac{1}{12} & \frac{1}{4} & \frac{1}{2} \\ \frac{1}{240} & \frac{1}{48} & \frac{1}{12} & \frac{1}{4} \end{vmatrix}$$

$$= \frac{1}{48 \times 12}\left(1 + \frac{1}{10}s^2\right) \qquad \textbf{(2.63)}$$

From (2.58),

$$Q_{23}(s) = \begin{vmatrix} 1 & s & s^2 & s^3 \\ \frac{1}{12} & \frac{1}{4} & \frac{1}{2} & 1 \\ \frac{1}{48} & \frac{1}{12} & \frac{1}{4} & \frac{1}{2} \\ \frac{1}{240} & \frac{1}{48} & \frac{1}{12} & \frac{1}{4} \end{vmatrix} \qquad \textbf{(2.64)}$$

$$= \frac{1}{48 \times 12}\left(1 - \frac{1}{2}s + \frac{1}{10}s^2 - \frac{1}{120}s^3\right)$$

Thus

$$(2, 3) = \frac{1 + \frac{1}{10}s^2}{1 - \frac{1}{2}s + \frac{1}{10}s^2 - \frac{1}{120}s^3} \tag{2.65}$$

3. Recursion Fomulas

A set of recursion formulas have been derived by Frobenius [FR2, MA2]. These formulas enable us to calculate new approximants based on certain known ones. These formulas are included here without proof.

In this subsection the following notation will be used:

$$C_{mn} = \begin{vmatrix} h_{m-n+1} & h_{m-n+2} & \cdots & h_m \\ h_{m-n+2} & h_{m-n+3} & \cdots & h_{m+1} \\ \cdots & \cdots & \cdots & \cdots \\ h_{m-1} & h_m & \cdots & h_{m+n-2} \\ h_m & h_{m+1} & \cdots & h_{m+n-1} \end{vmatrix} = (-1)^{n(n-1)/2}\Delta_{mn} \tag{2.66}$$

which differs from Δ_{mn} of (2.55) only in that their column sequences are reversed.

$$D_{mn} = \begin{vmatrix} h_{m-n+1} & h_{m-n+2} & \cdots & h_{m-1} & h_m \\ h_{m-n+2} & h_{m-n+3} & \cdots & h_m & h_{m+1} \\ \cdots & \cdots & \cdots & \cdots & \cdots \\ h_{m-1} & h_m & \cdots & h_{m+n-3} & h_{m+n-2} \\ h_{m+1} & h_{m+2} & \cdots & h_{m+n-1} & h_{m+n} \end{vmatrix} \tag{2.67}$$

which differs from C_{mn} only in the last (nth) row, in which each subscript is increased by one. The letter R represents either P or Q, the numerator and the demoninator polynomials.

Before each formula is applied to a sequence of approximants, it is necessary that the coefficients in each approximant,

$$\frac{a_0 + a_1 s + a_2 s^2 + \dots + a_m s^m}{b_0 + b_1 s + b_2 s^2 + \dots + b_n s^n} = h_0 + h_1 s + h_2 s^2 + \dots + h_{m+n} s^{m+n} \tag{2.68}$$

be adjusted such that

$$C_{mn} = b_0 \tag{2.69}$$

For example, for

$$H(s) = 1 + \tfrac{1}{2}s + \tfrac{1}{4}s^2 + \tfrac{1}{12}s^3 + \tfrac{1}{48}s^4 + \dots$$

we have

$$(2, 1) = \frac{1 + \frac{1}{6}s + \frac{1}{12}s^2}{1 - \frac{1}{3}s} \tag{2.70}$$

for which $b_0 = 1$. But

$$C_{21} = h_2 = \tfrac{1}{4}$$

Hence we must use, instead,

$$rH(s) = 4 + 2s + s^2 + \tfrac{1}{3}s^3 + \tfrac{1}{12}s^4 + \dots,$$

for which

$$(2, 1) = \frac{4 + \frac{2}{3}s + \frac{1}{3}s^2}{1 - \frac{1}{3}s} \tag{2.71}$$

Furthermore, all approximants must have the same b_0.

(*a*) *Diagonal sequence* The following formulas relate entries that lie on the same diagonal in a Padé table:

$$\begin{aligned} C_{mn}^2 R_{m+1,\,n+1} &= \{C_{mn}C_{m+1,\,n+1} + (C_{m+1,\,n+1}D_{mn} - D_{mn}C_{m+1,\,n+1})s\}R_{mn} \\ &\quad - C_{m+1,n+1}^2 s^2 R_{m-1,n-1} \end{aligned} \tag{2.72}$$

$$\begin{aligned} C_{mn}^2 R_{m+1,n+1} &= \{C_{mn}C_{m+1,n+1} + (C_{m+1,n+2}C_{m,n-1} - C_{m+2,n+1}C_{m-1,n})s\}R_{mn} \\ &\quad - C_{m+1,n+1}^2 s^2 R_{m-1,n-1} \end{aligned} \tag{2.73}$$

Thus we may start with a pair of known $(a, 0)$ and $(a + 1, 1)$ or $(0, b)$ and $(1, b + 1)$ and proceed diagonally to obtain sucessive additional entries to the Padé table.

(*b*) *Horizontal sequence* The following formula relates entries that lie on the same horizontal row in a Padé table:

$$\begin{aligned} C_{m,n+1}C_{mn}R_{m+1,n} &= (C_{m+1,n}C_{m,n+1} - C_{m+1,n+1}C_{mn}s)R_{mn} \\ &\quad - C_{m+1,n}C_{m+1,n+1}sR_{m-1,n} \end{aligned} \tag{2.74}$$

Thus we may start with any pair of adjacent entries on the same row and proceed to obtain other successive entries on the same row.

(*c*) *Vertical sequence* The following formula relates entries that lie on the same vertical column in a Padé table:

$$\begin{aligned} C_{m+1,n}C_{mn}R_{m,n+1} &= (C_{m+1,n}C_{m,n+1} - C_{m+1,n+1}C_{mn}s)R_{mn} \\ &\quad - C_{m,n+1}C_{m+1,n+1}sR_{m,n-1} \end{aligned} \tag{2.75}$$

Thus we may start with any two adjacent entries on the same column and proceed to obtain other successive entries on the same column.

4. Successive Modification Method

When the approximant $(m, 0)$ of a given function is known, the other approximant $(m - k, k)$ can always be found by solving a set of k linear simultaneous equations. Since $(m, 0)$ for a given function is the first $(m + 1)$ terms of its Taylor's expansion about the origin, it is possible to obtain the series of entries along the diagonal that runs from the top right toward the bottom left of the Padé table by solving sets of simultaneous equations of no more than m equations in each set.

Let

$$(m, 0) = \frac{a_0 + a_1 s + a_2 s^2 + \dots + a_m s^m}{1} = h_0 + h_1 s + h_2 s^2 + \dots + h_m s^m \tag{2.76}$$

Multiplying both the denominator and the numerator of $(m, 0)$ by the same polynomial $(1 + c_1 s + c_2 s^2 + \dots + c_k s^k)$ obviously does not alter the function. Thus

$$\frac{(a_0 + a_1 s + a_2 s^2 + \dots + a_m s^m)(1 + c_1 s + c_2 s^2 + \dots + c_k s^k)}{1 + c_1 s + c_2 s^2 + \dots + c_k s^k}$$
$$= h_0 + h_1 s + h_2 s^2 + \dots + h_m s^m \tag{2.77}$$

Now we can choose the c's such that all coefficients of the s^{m-k+1} through the s^m terms in the numerator product are zero. Since all coefficients of the expansion through the s^m term are independent of the coefficients of the s^{m+1} and higher-order terms, the approximation is still valid in the Taylor's sense through the s^m term if these high-degree terms are dropped. When this is done, what remains should be identical to $(m - k, k)$.

For example, for ϵ^{-s} we have

$$(4, 0) = 1 - s + \frac{s^2}{2} - \frac{s^3}{6} + \frac{s^4}{24} \tag{2.78}$$

$$(4, 0)(1 + as + bs^2) = 1 + (a - 1)s + (b - a + \tfrac{1}{2})s^2 + (\tfrac{1}{2}a - b - \tfrac{1}{6})s^3$$
$$+ \left(\frac{1}{2}b - \frac{1}{6}a + \frac{1}{24}\right)s^4 + \left(\frac{a}{24} - \frac{b}{6}\right)s^5 + \frac{b}{24}s^6$$

Setting the coefficients of s^3 and s^4 terms to zero yields

$$a = \tfrac{1}{2}, \qquad b = \tfrac{1}{12}$$

Hence

$$(2, 2) = \frac{1 - \frac{1}{2}s + \frac{1}{12}s^2}{1 + \frac{1}{2}s + \frac{1}{12}s^2} \tag{2.79}$$

Generally the step of setting the s^{m-k+1} through s^m terms to zero can be implemented in a systematic way by solving the matrix equation

$$\begin{bmatrix} h_{m-1} & h_{m-2} & \cdots & h_{m-k} \\ h_{m-2} & h_{m-3} & \cdots & h_{m-k-1} \\ \vdots & \vdots & & \vdots \\ h_{m-k} & h_{m-k-1} & \cdots & h_{m-2k} \end{bmatrix} \begin{bmatrix} c_1 \\ c_2 \\ \vdots \\ c_k \end{bmatrix} = \begin{bmatrix} -h_m \\ -h_{m-1} \\ \vdots \\ -h_{m-k} \end{bmatrix} \tag{2.80}$$

in which it is understood that $h_i = 0$ if $i < 0$.

After the c's are found, the numerator polynomial $d_0 + d_1 s + d_2 s^2 + \ldots + d_{m-k}s^{m-k}$ can be formed by evaluating its coefficients through the following formulas:

$$\begin{aligned} d_0 &= h_0 \\ d_i &= \sum_{j=0}^{i} h_j c_{i-j}, \qquad i = 1, 2, \ldots, m-k \end{aligned} \tag{2.81}$$

For example, if the (1, 3) for ϵ^{-s} is desired, it is only necessary to solve

$$\begin{bmatrix} -\frac{1}{6} & \frac{1}{2} & -1 \\ \frac{1}{2} & -1 & 1 \\ -1 & 1 & 0 \end{bmatrix} \begin{bmatrix} c_1 \\ c_2 \\ c_3 \end{bmatrix} = \begin{bmatrix} -\frac{1}{24} \\ \frac{1}{6} \\ -\frac{1}{2} \end{bmatrix}$$

which gives

$$c_1 = \tfrac{3}{4}, \qquad c_2 = \tfrac{1}{4}, \qquad c_3 = \tfrac{1}{24}$$

and

$$d_1 = h_0 c_1 + h_1 c_0 = -\tfrac{1}{4}$$

Hence

$$(1, 3) = \frac{1 - \frac{1}{4}s}{1 + \frac{3}{4}s + \frac{1}{4}s^2 + \frac{1}{24}s^3}$$

This procedure can also be used in the opposite direction—starting with a particular $(0, m)$ and work toward $(m, 0)$. The steps are exactly the same as described above except the roles of the numerator and the denominator are reversed. The basic principal is to multiply the numerator and the denominator of (0, m) by the polynomial $1 + c_1 s + c_2 s^2 + \ldots + c_k s^k$ and then force the necessary coefficients to go to zero and drop the higher-order terms.

The corresponding (m, 0) for a known (0, m) of the same function can be found easily by the following procedure.

Let

$$(m, 0) = H_0(1 + f_1 s + f_2 s^2 + \ldots + f_m s^m) = H_0(1 + \psi) \tag{2.82}$$

and apply the binomial expansion of $1/(1 + \psi)$. It is clear that

$$(m, 0) = \frac{H_0}{1 - \psi + \psi^2 - \psi^3 + \ldots} \tag{2.83}$$

Now if the like terms of ψ^k are collected in the denominator, and only s^m and lower-order terms are retained, the first ($m + 1$) terms of the Taylor's expansion of the new function must obviously agree with those of (m, 0). Hence the new function must be (0, m).

If for a given (m, 0), (0, m) is obtained first, the intermediate entries of the Padé table can be generated without ever solving linear simultaneous equations of order more then $m/2$ or $(m + 1)/2$, whichever is an integer.

5. Continued-Fraction Method

The Padé approximant of a function is closely related to the continued-fraction expansion of the function. In fact, if $H(s)$ is expressed in the form [KH1]

$$H(s) = h_0 + h_1 s + h_2 s^2 + \ldots + h_k s^k + \cfrac{g_0^{(k)} s^{k+1}}{1 - \cfrac{g_1^{(k)} s}{1 - \cfrac{g_2^{(k)} s}{1 - \ddots}}} \tag{2.84}$$

and the continued fraction is truncated after successively starting with dropping the g_0 terms, the truncated continued fractions are the following sequence of Padé approximants:

$$\begin{array}{ccccc}
(k, 0) & \rightarrow & (k + 1, 0) & & \\
& & \downarrow & & \\
& & (k + 1, 1) & \rightarrow & (k + 2, 1) \\
& & & & \downarrow \\
& & & & (k + 2, 2) \rightarrow \\
& & & & \quad \ddots
\end{array}$$

Alternatively, if the reciprocal of $H(s)$ is expressed in the form of (2.84), or

$$H(s) = \cfrac{1}{f_0 + f_1 s + f_2 s^2 + \dots + f_k s^k + \cfrac{\gamma_0^{(k)} s^{k+1}}{1 - \cfrac{\gamma_1^{(k)} s}{1 - \cfrac{\gamma_2^{(k)} s}{1 - \ddots}}}} \tag{2.85}$$

then the successive truncated expansions are the following Padé approximants:

$$\begin{array}{lll}
(0, k) & & \\
\downarrow & & \\
(0, k+1) \rightarrow (1, k+1) & & \\
\qquad\qquad\quad\ \downarrow & & \\
\qquad\qquad (1, k+2) \rightarrow (2, k+2) & & \\
\qquad\qquad\qquad\qquad\quad\ \ \downarrow & & \\
\qquad\qquad\qquad\qquad\quad\ \ \ddots & &
\end{array}$$

For example, since

$$\epsilon^s = 1 + \cfrac{s}{1 - \cfrac{\frac{1}{2}s}{1 + \cfrac{\frac{1}{1 \cdot 2}s}{1 - \cfrac{\frac{1}{2 \cdot 3}s}{1 + \cfrac{\frac{1}{2 \cdot 5}s}{1 - \cfrac{\frac{1}{2 \cdot 5}s}{1 + \cfrac{\frac{1}{2 \cdot 7}s}{1 - \ddots}}}}}}}$$

we have

$$(0, 0) = 1$$

$$(1, 0) = 1 + s$$

$$(1, 1) = 1 + \frac{s}{1 - \frac{1}{2}s} = \frac{2 + s}{2 - s}$$

$$(2, 1) = 1 + \cfrac{s}{1 - \cfrac{\frac{1}{2}s}{1 + \frac{1}{6}s}} = \frac{6 + 4s + s^2}{6 - 2s}$$

and so on.

Alternatively, since

$$e^s = \cfrac{1}{1 - \cfrac{s}{1 + \cfrac{\frac{1}{2}s}{1 - \cfrac{\frac{1}{2\cdot 3}s}{1 + \cfrac{\frac{1}{2\cdot 3}s}{1 - \cfrac{\frac{1}{2\cdot 5}s}{1 + \cfrac{\frac{1}{2\cdot 5}s}{1 - \cfrac{\frac{1}{2\cdot 7}s}{1 + \ddots}}}}}}}}$$

we have

$$(0, 0) = 1$$

$$(0, 1) = \frac{1}{1 - s}$$

$$(1, 1) = \cfrac{1}{1 - \cfrac{s}{1 + \frac{1}{2}s}} = \frac{2 + s}{2 - s}$$

$$(1, 2) = \cfrac{1}{1 - \cfrac{s}{1 + \cfrac{\frac{1}{2}s}{1 - \frac{1}{6}s}}} = \frac{6 + 2s}{6 - 4s + s^2}$$

and so on.

For more thorough discussions of the theory on continued fractions and methods for finding the continued-fraction expansion of a given function, the reader is referred to the references [KH1, WA1].

PROBLEMS

2.1. Derive (2.26), (2.27), (2.28), and (2.29).

2.2. Derive (2.30).

2.3. Find the first five polynomials in the orthogonal set for the range $|t| \leq 1$ with respect to the weighting function $r(t) = 1$. [*Note*: These polynomials are known as the Legendre polynomials and usually denoted by $P_n(t)$.]

2.4. Show that the Chebyshev polynomials of the second kind,

$$U_n(t) = \frac{\sin\,[(n + 1)\cos^{-1} t]}{\sqrt{1 - t^2}}$$

are orthogonal over the range $-1 \leq t \leq 1$ with respect to the weighting function

$$r(t) = \sqrt{1 - t^2}$$

2.5. Generate a set of polynomials, orthogonal over the range $-1 \leq t \leq 1$ with respect to the weighting function

$$r(t) = \frac{1}{\sqrt{1 - t^2}}$$

In particular, write out explicit expressions for $n = 0, 1, 2, 3, 4$. (*Hint*: To integrate, let $t = \cos x$.) [*Note*: This set of orthogonal polynomials is known as the Chebyshev polynomials (of the first kind), usually denoted by $T_n(t)$.]

2.6. Use (2.37), find the orthogonal-polynomial approximation of $f(t)$ in Fig. P2.1 by using Chebyshev polynomials of the first kind for (a) $n = 2$, (b) $n = 3$, and (c) $n = 4$. Plot the approximate functions for each degree in the range $-1 \leq t \leq 1$.

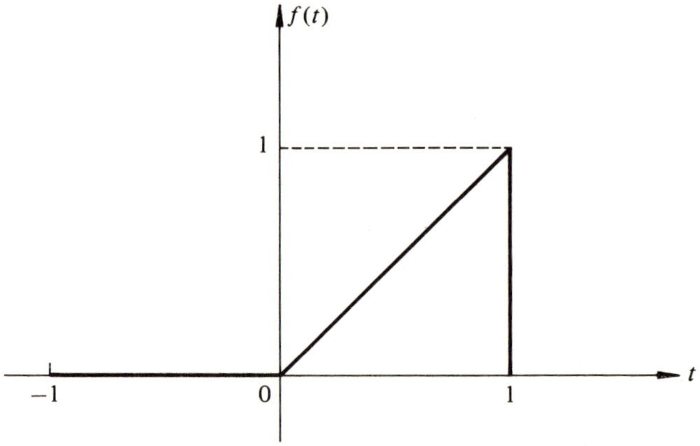

Figure P2-1

2.7. Legendre's polynomials, which are orthogonal in the range $|t| \leq 1$ and $r(t) = 1$, may be expressed as

$$P_n(t) = \frac{1}{2^n n!} \frac{d^n}{dt^n} (t^2 - 1)^n$$

Use these polynomials to approximate the function

$$f(t) = 1 - |t|$$

in the range $|t| \leq 1$. Find the approximating function $f_n(t)$ for $n = 0, 2, 4$. Also compute the total error,

$$E_2 = \int_{-1}^{1} |f(t) - f_n(t)|^2 \, dt$$

for each n.

2.8. Generate a Padé table for $m, n \leq 4$ for the function $\frac{1}{2}(1 + \epsilon^s)$ in the following sequence:

(a) Fill in (0, 0), (1, 0), (2, 0), (3, 0), and (4, 0) by inspection.

(b) Obtain (0, 1) from (1, 0), and (1, 1) and (0, 2) from (2, 0) by the successive modification method.
(c) Starting with (0, 2) obtain (0, 3), (1, 3), and (1, 4), successively using the continued fraction of the form of (2.85).
(d) Starting with (2, 0) and (3, 0), obtain successively (3, 1) and (4, 1) using the continued fraction of the form of (2.84).
(e) Find (2, 1) by the direct method.
(f) Starting with (2, 0) and (2, 1), find (2, 2), (2, 3), and (2, 4) using (2.75).
(g) Starting with (1, 0) and (2, 1), find (3, 2) and (4, 3) using (2.72) or (2.73).
(h) Starting with (2, 2) and (3, 2), find (4, 2) using (2.51).
(i) Find (1, 2) and (3, 3) by the determinant method.
(j) Complete the table by any method you desire.

2.9. Find the Padé approximants (4, 0), (6, 0), (8, 0), (0, 4), (1, 4), (4, 4), and (5, 5) for the function

$$f(t) = \frac{1 + t - t^3}{1 - t^3}$$

3

The Impulse Response of a Network

In the frequency-domain synthesis, we are accustomed to formulating our problem in terms of a desired network function $H(s)$. Then methods are devised to obtain networks that realize this network function.

Analogously it would also be convenient to formulate our time-domain problems in terms of an impulse response $h(t)$, which is the inverse Laplace transform of $H(s)$. However, in many practical situations, instead of the impulse response of the network, excitation, $f_1(t)$, and response, $f_2(t)$, are given. In these situations it is necessary to first determine the impulse response of the network.

The problem of determining the impulse response for a given pair of $f_1(t)$ and $f_2(t)$ involves the solution of (1.4) for $h(t)$.† A procedure that enables us to find $h(t)$ for a pair of $f_1(t)$ and $f_2(t)$ is sometimes known as deconvolution or convolution inversion.† Convolution is analogous to the multiplication in frequency domain; deconvolution is analogous to the division (1.1).

After the deconvolution procedure has been applied to a pair of $f_1(t)$ and $f_2(t)$, we are then interested to know whether the resulting $h(t)$ is realizable by the permitted class of networks or not. If so, will the network be stable? If not, how can an approximate impulse response $h^*(t)$ be determined such that the corresponding network will render a response $f_2^*(t) = f_1(t) \circledast h^*(t)$ that is an acceptable approximation of $f_2(t)$? Some answers to these questions are provided in this chapter.

†Since the convolution is commutative ($f_1 \circledast h = h \circledast f_1$), this problem is equivalent to that of finding $f_1(t)$ for a given pair of $f_2(t)$ and $h(t)$.

3.1 CAUSALITY, STABILITY, AND REALIZABILITY

When a pair of $f_1(t)$ and $f_2(t)$ are specified, it would be advantageous to be able to ascertain in advance whether or not the response $f_2(t)$ can be produced by a stable, linear, lumped network when it is excited by the accompanying $f_1(t)$.

Since the network is assumed to be initially idle, $f_2(t)$ cannot be nonzero before $f_1(t)$ becomes nonzero. This is simply the causal property we require of our networks.

In order for the network to be stable, it should not have any finite response when no excitation is applied to it [SU2]. That is,

$$f_2(t) \equiv 0, \qquad \text{for } f_1(t) \equiv 0 \tag{3.1}$$

In the frequency domain the necessary and sufficient condition for a network to be stable is that its network function has no j-axis† or right-halfplane poles. This condition is equivalent to

$$\int_0^\infty |h(t)|\, dt < \infty \tag{3.2}$$

in the time domain. Hence, for a pair of $f_1(t)$ and $f_2(t)$ to be meaningful, their corresponding $h(t)$ must be absolutely integrable.

If a network whose $h(t)$ is absolutely integrable, the next point to consider is its realizability. By "realizability" we mean whether an absolutely integrable $h(t)$ can be realized by a certain class of linear networks or not. The requirement for an $h(t)$ to be realizable cannot be stated easily in general terms. It will depend on the nature (voltage or current) and localities of the excitation and response. Also it depends on what components or elements are available as the building blocks of the network. For example, an $h(t)$ may be realizable by an *RLC* network but not by an *RC* one.

Since our plan in handling a time-domain synthesis problem is to transform the $h(t)$ in question to the frequency domain, the realizability consideration is most conveniently done in the frequency domain. In the frequency domain, the realizability criteria for various classes of linear systems are well formulated. One common necessary requirement for all classes of stable and realizable linear systems in the frequency domain is that $H(s)$ be a rational function with no right-halfplane pole or multiple j-axis pole. Hence $h(t)$ can contain only terms of following type:

†In passive networks, simple j-axis poles are permitted. However, we prefer to view these poles as left-halfplane poles that lie infinitesimally close of the j axis.

$$(A_0 + A_1 t + A_2 t^2 + \cdots + A_n t^n)\epsilon^{s_v t} \tag{3.3}$$

where

$$s_v = \sigma_v + j\omega_v, \quad \sigma_v \leq 0; \qquad n = 0 \text{ if } \sigma_v = 0$$

If an $h(t)$ contains terms that do not meet this restriction, then a stable network that will realize this $h(t)$ cannot be found. If the nonrealizability is due to the fact that its Laplace transform, $H(s)$, is not rational, then the only possible recourse is to find another impulse response $h^*(t)$ that approximates this $h(t)$ in some fashion. Its corresponding Laplace transform, $H^*(s)$, will then be used as a substituting network function.

If the nonrealizability is due to the presence of right-halfplane poles, either another approximate impulse response must be found or, in many practical problems, these poles can simply be deleted. The latter situation is, of course, theoretically impossible, because $h(t)$ must be absolutely integrable. However, in many practical solutions of a time-domain problem, the integral of (3.2) is taken only up to certain large, but finite, time. In some approximation techniques, therefore, right-halfplane poles could creep in without showing up in the finite time interval used. When this happens, the right-halfplane poles will usually have very small residues. The omission of these poles will not have a noticeable effect on the accuracy of the approximations in the time interval.

3.2 STRAIGHTFORWARD METHOD OF COMPUTING $h^*(t)$

When a pair of $f_1(t)$ and $f_2(t)$ is given, the following basic method may be used to determine an approximate impulse response for the pair. The method is simple in theory but may not always be easy to implement. However, it serves to illustrate the nature of a time-domain approximation problem and the form its solution may typically entail.

The convolution formula (1.4) can be used to compute directly an approximate impulse response $h^*(t)$ from a given pair of $f_1(t)$ and $f_2(t)$ if the natural frequencies of the system are somehow known. These frequencies may be known because of certain a priori knowledge of the system. They are also determinable by certain experimental means. Other methods by which the natural frequencies corresponding to an $h(t)$ can be determined are given in Chapter 4. In other situations, these frequencies can merely be estimated, since they are not unique, to give the same accuracy of approximation.

Once the natural frequencies of a system have been determined or estimated and with the pair of $f_1(t)$ and $f_2(t)$ given, the approximating $h^*(t)$ can be evaluated directly by one of the possible criteria. For simplicity, assume that all poles are simple and let

$$\begin{aligned} h^*(t) &= \sum_{\nu=1}^{M} (A_\nu + jB_\nu)\epsilon^{-(\alpha_\nu + j\beta_\nu)t} \\ &\quad + \sum_{\nu=1}^{M} (A_\nu - jB_\nu)\epsilon^{-(\alpha_\nu - j\beta_\nu)t} + \sum_{\mu=1}^{N} C_\mu \epsilon^{-\sigma_\mu t} \\ &= \sum_{\nu=1}^{M} 2A_\nu \epsilon^{-\alpha_\nu t} \cos \beta_\nu t - \sum_{\nu=1}^{M} 2B_\nu \epsilon^{-\alpha_\nu} \sin \beta_\nu t \\ &\quad + \sum_{\mu=1}^{N} C_\mu \epsilon^{-\sigma_\mu t} \end{aligned} \tag{3.4}$$

where α_ν, β_ν, σ_μ, A_ν, B_ν, and C_μ are real. The first two summations in (3.4) include all complex poles. The third summation includes all real poles. In (3.4) it is assumed that α_ν, β_ν, and σ_μ are all known.

The convolution of $f_1(t)$ and $h^*(t)$ gives

$$\begin{aligned} f_2^*(t) &= \sum_{\nu=1}^{M} 2A_\nu \int_0^t f_1(t-\tau)\epsilon^{-\alpha_\nu \tau} \cos \beta_\nu \tau \, d\tau \\ &\quad - \sum_{\nu=1}^{M} 2B_\nu \int_0^t f_1(t-\tau)\epsilon^{-\alpha_\nu \tau} \sin \beta_\nu \tau \, d\tau \\ &\quad + \sum_{\mu=1}^{N} C_\mu \int_0^t f_1(t-\tau)\epsilon^{-\sigma_\mu \tau} \, d\tau \end{aligned} \tag{3.5}$$

Now let

$$a_\nu(t) = \int_0^t f_1(t-\tau)\epsilon^{-\alpha_\nu \tau} \cos \beta_\nu \, d\tau \tag{3.6}$$

$$b_\nu(t) = -\int_0^t f_1(t-\tau)\epsilon^{-\alpha_\nu \tau} \sin \beta_\nu \tau \, d\tau \tag{3.7}$$

$$c_\mu(t) = \int_0^t f_1(t-\tau)\epsilon^{-\sigma_\mu \tau} \, d\tau \tag{3.8}$$

Then (3.5) may be written as

$$f_2^*(t) = \sum_{\nu=1}^{M} 2[A_\nu a_\nu(t) + B_\nu b_\nu(t)] + \sum_{\mu=1}^{N} C_\mu c_\mu(t) \tag{3.9}$$

Since $a_\nu(t)$, $b_\nu(t)$, and $c_\mu(t)$ are dependent only on $f_1(t)$, the problem is that of finding a set of numbers A_ν, B_ν, and C_μ such that $f_2^*(t)$ approximates $f_2(t)$ in some fashion.

1. Series of Matched Points

One scheme of finding a set of A_ν, B_ν, and C_μ for an $f_2^*(t)$ is to judiciously select a series of points along the t axis and force $f_2^*(t)$ and $f_2(t)$ to equal each other at these points. Since there are $N + 2M$ undetermined coefficients in (3.4), as many points of t may be chosen. Let these points be t_k; $k = 1, 2, \ldots, N + 2M$. For each t_k, an equation

$$\sum_{\nu=1}^{M} 2[a_\nu(t_k)A_\nu - b_\nu(t_k)B_\nu] + \sum_{\mu=1}^{N} c_\mu(t_k)C_\mu = f_2(t_k) \tag{3.10}$$

is obtained. There are $N + 2M$ such equations. The solution of these equations for A_ν, B_ν, and C_μ gives the desired $h^*(t)$.

Example. Find the $h^*(t)$ with natural frequencies assumed at $s = -1$, $-0.5 \pm j0.5$ for the excitation

$$f_1(t) = u_{-1}(t) - u_{-1}(t - 2.5) \tag{3.11}$$

such that $f_1(t) \circledast h^*(t)$ is equal to

$$f_2(t) = \mathscr{L}^{-1}\left(\frac{1 - \epsilon^{-2.5s}}{2s^2} - \frac{1 - 2\epsilon^{-2.5s} + \epsilon^{-5s}}{5s^3}\right) \tag{3.12}$$

at $t = 0.5$, 1.5, and 3.5. This $f_2(t)$ is given explicitly by

$$\begin{aligned} f_2(t) &= \tfrac{1}{2}t - \tfrac{1}{10}t^2, & 0 < t < 2.5 \\ &= \tfrac{5}{4} - \tfrac{1}{10}t^2 + \tfrac{1}{5}(t - 2.5)^2, & 2.5 < t < 5 \\ &= 0, & 5 < t \end{aligned}$$

For this pair of $f_1(t)$ and $f_2(t)$, the impulse response can be found analytically to be

$$h(t) = \mathscr{L}^{-1}\left[\frac{1}{25} - \frac{1 - 2.5\epsilon^{-2.5s}}{5s}\right] \tag{3.13}$$

The functions $f_1(t)$ and $f_2(t)$ are shown in Fig. 3.1, together with the exact $h(t)$ associated with these functions.

Solution. Assume, as in (3.4), that

$$h^*(t) = (A + jB)\epsilon^{-(0.5+j0.5)t} + (A - jB)\epsilon^{-(0.5-j0.5)t} + C\epsilon^{-t}$$

For the given $f_1(t)$, (3.6), (3.7), and (3.8) give

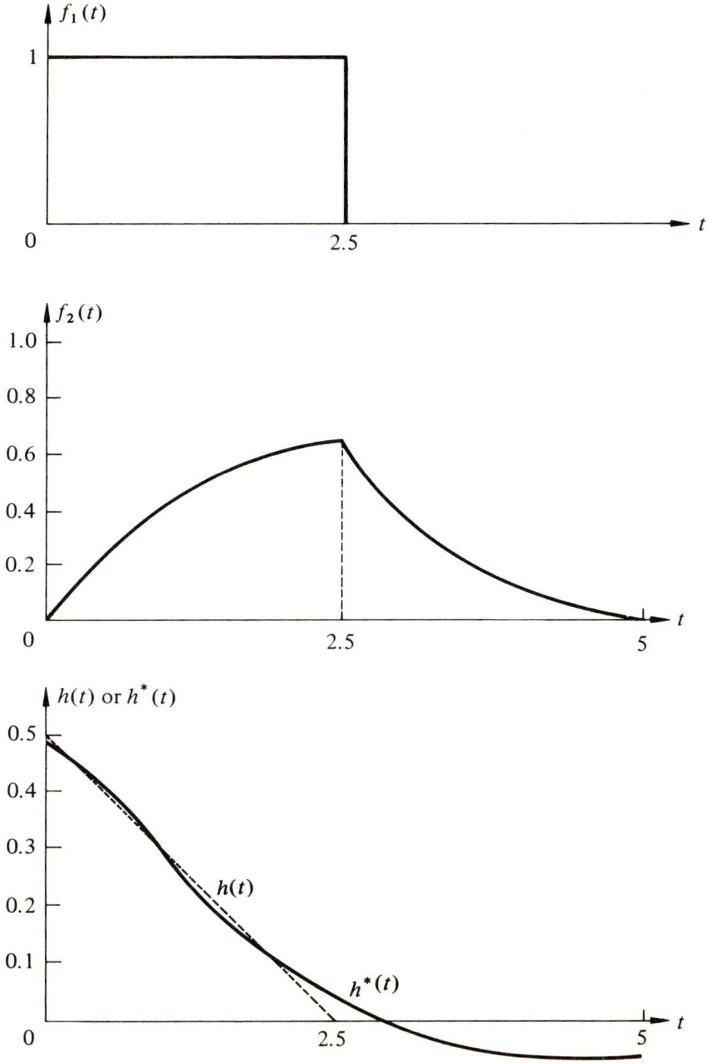

Fig. 3.1 Functions in Equations (3.11) to (3.14).

$$a(0.5) = \int_0^{0.5} \epsilon^{-0.5\tau} \cos 0.5\tau \, d\tau = 0.4381$$

$$b(0.5) = -\int_0^{0.5} \epsilon^{-0.5\tau} \sin 0.5\tau \, d\tau = -0.05273$$

$$c(0.5) = \int_0^{0.5} \epsilon^{-\tau} \, d\tau = 0.3935$$

$$a(1.5) = \int_0^{1.5} \epsilon^{-0.5\tau} \cos 0.5\tau \, d\tau = 0.9764$$

$$b(1.5) = -\int_0^{1.5} \epsilon^{-0.5\tau} \sin 0.5\tau \, d\tau = -0.3324$$

$$c(1.5) = \int_0^{1.5} \epsilon^{-\tau} \, d\tau = 0.7769$$

$$a(3.5) = \int_1^{3.5} \epsilon^{-0.5\tau} \cos 0.5\tau \, d\tau = 0.4435$$

$$b(3.5) = -\int_1^{3.5} \epsilon^{-0.5\tau} \sin 0.5\tau \, d\tau = -0.6831$$

$$c(3.5) = \int_1^{3.5} \epsilon^{-\tau} \, d\tau = 0.3377$$

(3.10) now gives

$$0.4381(2A) + 0.05273(2B) + 0.3935(C) = f_2(0.5) = 0.225$$
$$0.9764(2A) + 0.3324(2B) + 0.7769(C) = f_2(1.5) = 0.525$$
$$0.4435(2A) + 0.6831(2B) + 0.3377(C) = f_2(3.5) = 0.225$$

Solution of these three simultaneous equations yields

$$2A = 0.7918, \qquad 2B = -0.0340, \qquad C = -0.3047$$

Thus the desired approximate impulse response is

$$h^*(t) = 0.7918\epsilon^{-0.5t} \cos 0.5t - 0.0340\epsilon^{-0.5t} \sin 0.5t - 0.3047\epsilon^{-t} \tag{3.14}$$

This function is plotted in Fig. 3.1.

2. Least-Mean-Square-Error Approximation

Another scheme of finding a set of A_ν, B_ν, and C_μ is to force $f_2^*(t)$ to approximate $f_2(t)$ in the least-mean-square-error sense. In this case, any number of points along the t axis may be chosen. The squared error

$$\int_0^\infty [f_2(t) - f_2^*(t)]^2 \, dt \tag{3.15}$$

is then respresented by

$$E_2 = \sum_k [f_2(t_k) - f_2^*(t_k)]^2 \tag{3.16}$$

This E_2 may then be minimized with respect to all A_ν, B_ν, and C_μ by setting

$$\begin{aligned} \frac{\partial E_2}{\partial A_\nu} &= 0 \\ \frac{\partial E_2}{\partial B_\nu} &= 0 \\ \frac{\partial E_2}{\partial C_\mu} &= 0 \end{aligned} \tag{3.17}$$

for all ν's and μ's. The solution of the $N + 2M$ equations in (3.17) fixes the desired $h^*(t)$.

As a practical matter, the number of t_k should be larger than $N + 2M$ for this method to make sense. If this number is equal to $N + 2M$, the method in Section 3.1 gives $E_2 = 0$, and no further reduction in E_2 is possible. If this number is less than $N + 2M$, the requirements are overspecified.

3. Remarks

The basic principle of the foregoing methods is simple. However, these methods have one major drawback. Their success depends on too many factors that are arbitrary. These factors are the

1. Number of poles used
2. Positions of these poles
3. Number and values of t_k

Although for any choice of these three factors, an answer always exists by either the point-matching or the least-squares scheme, there is no assurance that the approximation will be of any predeterminable quality.

Fortunately, some methods of determining the pole positions exist (Chapter 4). Hence we do not have to work completely in the dark initially. The above method can be used as one step in a series of cut-and-try procedures until an acceptable $h^*(t)$ is obtained. If an initial $h^*(t)$ is not acceptable, any one of the three factors may be altered with the hope that the next $h^*(t)$ might be an improvement over the previous one. Section 3.3 offers an analytic scheme by which a set of pole positions can be modified. The locations of matching points (t_k) can usually be modified by inspection by plotting both the $h^*(t)$ and the necessary $h(t)$ determined by other methods offered in this chapter.

Although the function assumed in this development is restricted to the form of (3.4) which has only simple poles, the reader should be able to see that the same procedure applies to functions with multiple poles as well. In fact, there is no reason why the same procedure cannot be applied to any arbitrarily assumed function $\hat{h}(t)$, such as the polynomial, any orthogonal functions, and so on. The advantage of using the $h^*(t)$ of (3.4) is, of course, that its Laplace transform $H^*(s)$ is in the correct form to be realized in the frequency domain. If another $\hat{h}(t)$ is used instead of this $h^*(t)$, another approximation step remains to be carried out before network realization can be accomplished.

3.3 ADJUSTMENT OF POLE LOCATIONS

The success of the foregoing method depends largely on the closeness of the estimated or arbitrarily chosen locations of the poles of $H^*(s)$. If these poles are exactly coinciding with the necessary poles, the resulting $h^*(t)$ will be exact. If these poles are only approximations of their true locations, the resulting $h^*(t)$ will very likely be only an approximation of the true $h(t)$. If these poles are not close to their necessary locations, the method does not guarantee a close approximation for $h^*(t)$.

If the $h^*(t)$ is not a sufficiently good approximation of $h(t)$, there are two recourses. One may either increase the number of poles and repeat the process or relocate the poles. The following scheme is a systematic method of improving the pole locations.

For each term corresponding to a real pole

$$h_\mu(t) = C_\mu \epsilon^{-\sigma_\mu t}$$

we replace it with

$$h'_\mu(t) = C'_\mu \epsilon^{-\sigma_\mu t} + C''_\mu t \epsilon^{-\sigma_\mu t} \tag{3.18}$$

This amounts to increasing the order of each pole by one. In addition to (3.8), we may define

$$c'_\mu(t) = \int_0^t f_1(t-\tau)\tau\epsilon^{-\sigma_\mu \tau}\, d\tau \tag{3.19}$$

For each pair of complex-conjugate terms corresponding to a pair of complex-conjugate poles,

$$\begin{aligned} h_\nu(t) &= (A_\nu + jB_\nu)\epsilon^{-(\alpha_\nu + j\beta_\nu)t} + (A_\nu - jB_\nu)\epsilon^{-(\alpha_\nu - j\beta_\nu)t} \\ &= 2A_\nu \epsilon^{-\alpha_\nu t}\cos\beta_\nu t + 2B_\nu \epsilon^{-\alpha_\nu t}\sin\beta_\nu t \end{aligned} \tag{3.20}$$

we replace it with

$$h''_\nu(t) = 2(A'_\nu + A''_\nu t)\epsilon^{-\alpha_\nu t}\cos\beta_\nu t + 2(B'_\nu + B''_\nu t)\epsilon^{-\alpha_\nu t}\sin\beta_\nu t \tag{3.21}$$

This step also amounts to doubling the order of both poles. In addition to (3.6), we may define

$$a'_\nu(t) = \int_0^t f_1(t - \tau)\tau\epsilon^{-\alpha_\nu\tau} \cos \beta_\nu \tau \, d\tau \tag{3.22}$$

and, in addition to (3.7),

$$b'_\nu(t) = -\int_0^t f_1(t - \tau)\tau\epsilon^{-\alpha_\nu\tau} \sin \beta_\nu \tau \, d\tau \tag{3.23}$$

Now the procedure of Section 3.2 may be repeated to determine the unknowns A'_ν, A''_ν, B'_ν, B''_ν, C'_μ, and C''_μ. The number of unknowns involved now is twice that of the first application of the procedure—$2N + 4M$. The results obtained here may then be used to modify the locations of the original $N + 2M$ poles as follows.

The following procedure is only feasible if the pole locations are already fairly close to the required exact locations. Hence it is only intended to effect a fine secondary adjustment rather than a gross correction of pole positions. Hence we shall assume that

$$A_\nu \cong A'_\nu, \qquad B_\nu \cong B'_\nu, \qquad C_\mu \cong C'_\mu$$

Consider a slight shift of a real pole from $-\sigma_\mu$ to $-(\sigma_\mu + \Delta\sigma_\mu)$. Its corresponding time-domain term is varied from $C_\mu\epsilon^{-\sigma_\mu t}$ to

$$\begin{aligned} &C_\mu\epsilon^{-(\sigma_\mu+\Delta\sigma_\mu)t} \\ &\quad = C_\mu\epsilon^{-\sigma_\mu t}\epsilon^{-\Delta\sigma_\mu t} \\ &\quad = C_\mu\epsilon^{-\sigma_\mu t}\left[1 - \Delta\sigma_\mu t + \frac{(\Delta\sigma_\mu)^2}{2!} - \dots\right] \\ &\quad \cong C_\mu\epsilon^{-\sigma_\mu t}(1 - \Delta\sigma_\mu t) \end{aligned} \tag{3.24}$$

for small $\Delta\sigma_\mu$. A comparison of (3.24) with (3.18) shows that

$$\Delta\sigma_\mu = -\frac{C''_\mu}{C_\mu} \tag{3.25}$$

Next consider the shift of a pair of complex poles from $\alpha_\nu + jB_\nu$ to $\alpha_\nu + \Delta\alpha_\nu \pm j(\beta_\nu + \Delta\beta_\nu)$. The new $h_\nu(t)$ of (3.20) becomes

$$\begin{aligned} 2A_\nu\epsilon^{-(\alpha_\nu+\Delta\alpha_\nu)t} &\cos(\beta_\nu + \Delta\beta_\nu)t + 2B_\nu\epsilon^{-(\alpha_\nu+\Delta_\nu)t} \sin(\beta_\nu + \Delta\alpha_\nu)t \\ &= 2A_\nu\epsilon^{-\alpha_\nu t}\epsilon^{-\Delta\alpha_\nu t}(\cos\beta_\nu t \cos\Delta\beta_\nu t - \sin\beta_\nu t \sin\Delta\beta_\nu t) \\ &\quad + 2B_\nu\epsilon^{-\alpha_\nu t}\epsilon^{-\Delta\alpha_\nu t}(\sin\beta_\nu t \cos\Delta\beta_\nu t + \cos\beta_\nu t \sin\Delta\beta_\nu t) \end{aligned} \tag{3.26}$$

Now for small $\Delta\alpha_\nu$ and $\Delta\beta_\nu$ we may make the approximation

$$\begin{aligned} \epsilon^{-\Delta\alpha_\nu t} &\cong 1 - \Delta\alpha_\nu t \\ \cos \Delta\beta_\nu t &\cong 1 \\ \sin \Delta\beta_\nu t &\cong \Delta\beta_\nu t \end{aligned} \tag{3.27}$$

The quantity in (3.26) becomes approximately

$$\begin{aligned} &2A_\nu \epsilon^{-\alpha_\nu t}(1 - \Delta\alpha_\nu t)(\cos \beta_\nu t - \Delta\beta_\nu t \sin \beta_\nu t) \\ &\quad + 2B_\nu \epsilon^{-\alpha_\nu t}(1 - \Delta\alpha_\nu t)(\sin \beta_\nu t + \Delta\beta_\nu t \cos \beta_\nu t) \\ &= 2[A_\nu + (-A_\nu \Delta\alpha_\nu + B_\nu \Delta\beta_\nu)t]\epsilon^{-\alpha_\nu t} \cos \beta_\nu t \\ &\quad + 2[B_\nu + (-B_\nu \Delta\alpha_\nu - A_\nu \Delta\beta_\nu)t]\epsilon^{-\alpha_\nu t} \sin \beta_\nu t \end{aligned} \tag{3.28}$$

Comparing (3.28) with (3.21) yields the following identification:

$$A''_\nu = -A_\nu \Delta\alpha_\nu + B_\nu \Delta\beta_\nu \tag{3.29}$$

$$B''_\nu = -B_\nu \Delta\alpha_\nu - A_\nu \Delta\beta_\nu \tag{3.30}$$

Solution of $\Delta\alpha_\nu$ and $\Delta\beta_\nu$ from (3.29) and (3.30) gives

$$\Delta\alpha_\nu = -\frac{A_\nu A''_\nu + B_\nu B''_\nu}{A_\nu^2 + B_\nu^2} \tag{3.31}$$

$$\Delta\beta_\nu = -\frac{A_\nu B''_\nu - A''_\nu B_\nu}{A_\nu^2 + B_\nu^2} \tag{3.32}$$

The accuracy of the approximation of

$$\begin{aligned} h^*(t) &= \sum_{\nu=1}^{M} 2A_\nu \epsilon^{-(\alpha_\nu + \Delta\alpha_\nu)t} \cos(\beta_\nu + \Delta\beta_\nu)t \\ &\quad + \sum_{\nu=1}^{M} 2B_\nu \epsilon^{-(\alpha_\nu + \Delta\alpha_\nu)t} \sin(\beta_\nu + \Delta\beta_\nu)t \\ &\quad + \sum_{\mu=1}^{M} C_\mu \epsilon^{-(\sigma_\mu + \Delta\sigma_\mu)t} \end{aligned} \tag{3.33}$$

should be comparable to that of

$$\sum_{\nu=1}^{M} h''_\nu(t) + \sum_{\mu=1}^{N} h'_\mu(t) \tag{3.34}$$

where $h''_\nu(t)$ and $h'_\mu(t)$ are defined in (3.18) and (3.21).

In summary, this method depends on the intermediate step of increasing the degree of each simple pole to two, thereby doubling the number of adjustable parameters. Because of the increase in the number of parameters, generally a better approximation can be expected of the augmented function.

After this better approximation has been obtained, another set of simple poles are found such that this new set of simple poles accomplishes approximately what the double poles do. This last approximation is done on parts of the functions in the Taylor's sense about the origin as in (3.24) and (3.27). Thus for this scheme to be convergent, the initial pole locations must be sufficiently close to what are necessary to give a good approximation. Otherwise, the repeated applications of this procedure may not converge. In that case the original locations of the poles will need be reestimated.

3.4 COMPUTATION OF *h(t)* BY TRAINS OF SINGULARITY FUNCTIONS—THE DIVISION METHOD

This method of computing $h(t)$ is based on the assumption that both $f_1(t)$ and $f_2(t)$ are specified as sequences of singularity functions uniformly spaced along the time axis [BA1, LE2]. These singularity functions may either be the only available representation of the actual functions or the approximate representations† of the actual $f_2(t)$ and $f_1(t)$ specified in one of the three forms described in Chapter 1. For practical reasons it is convenient to choose the singularity functions from Table 3.1.

TABLE 3.1

Useful Singularity Functions

Singularity function	Notation	Analytical description (function vanishes for $t < 0$)	Laplace transform
Impulse	$u_0(t)$	$\delta(t)$	1
Step	$u_{-1}(t)$	1 or $u(t)$	$1/s$
Ramp	$u_{-2}(t)$	t	$1/s^2$
Parabolic	$u_{-3}(t)$	$t^2/2$	$1/s^3$

If a given response $f_2(t)$ is a sequence of singular functions of the same type τ seconds apart, say

$$f_2(t) = b_0 u_{-b}(t) + b_1 u_{-b}(t-\tau) + b_2 u_{-b}(t-2\tau) + \dots + b_n u_{-b}(t-n\tau) + \dots \tag{3.35}$$

then its Laplace transform is

†When approximations are present, $f_1(t)$, $f_2(t)$, and $h(t)$ should be replaced by $\hat{f}_1(t)$, $\hat{f}_2(t)$, and $\hat{h}(t)$ in this discussion.

$$F_2(s) = \frac{1}{s^b}(b_0 + b_1\epsilon^{-\tau s} + b_2\epsilon^{-2\tau s} + \ldots + b_n\tau^{-n\tau s} + \ldots) \tag{3.36}$$

Similarly, if the corresponding excitation $f_1(t)$ is given by another sequence of singularity functions, say

$$f_1(t) = a_0 u_{-a}(t) + a_1 u_{-a}(t - \tau) + \ldots + a_n u_{-a}(t - n\tau) + \ldots \tag{3.37}$$

where a may or may not be equal to b, then the Laplace transform of $f_1(t)$ is

$$F_1(s) = \frac{1}{s^a}(a_0 + a_1\epsilon^{-\tau s} + a_2\epsilon^{-2\tau s} + \ldots + a_n\epsilon^{-n\tau s} + \ldots) \tag{3.38}$$

Both $F_1(s)$ and $F_2(s)$ expressed this way are Dirichlet series in $\epsilon^{-\tau s}$. Hence

$$\begin{aligned} H(s) = \frac{F_2(s)}{F_1(s)} &= \frac{1}{s^{b-a}} \frac{b_0 + b_1\epsilon^{-\tau s} + b_2\epsilon^{-2\tau s} + \ldots}{a_0 + a_1\epsilon^{-\tau s} + a_2\epsilon^{-2\tau s} + \ldots} \\ &= \frac{1}{s^c}(c_0 + c_1\epsilon^{-\tau s} + c_2\epsilon^{-2\tau s} + \ldots + c_n\epsilon^{-n\tau s} + \ldots) \end{aligned} \tag{3.39}$$

is another Dirichlet series. The index $c = b - a$ depends only on the type of functions used for $f_1(t)$ and $f_2(t)$. The coefficients in the new series, c's, can be obtained by dividing the two series [RE1, p. 157]. Let $\epsilon^{-\tau s} = x$. For example, if

$$F_1(s) = \frac{1}{s}(1 + 3x + 2x^2 + 5x^3 + \ldots)$$

$$F_2(s) = \frac{1}{s^2}(1 + 4x + 3x^2 + 6x^3 + \ldots)$$

then we perform the long division

$$\begin{array}{r} 1 + x - 2x^2 + 5x^3 + \ldots \\ 1 + 3x + 2x^2 + 5x^3 + \ldots \,/\, \overline{1 + 4x + 3x^2 + 6x^3 + \ldots} \\ \underline{1 + 3x + 2x^2 + 5x^3 + \ldots} \\ x + x^2 + x^3 + \ldots \\ \underline{x + 3x^2 + 2x^3 + \ldots} \\ -2x^2 - x^3 + \ldots \\ \underline{-2x^2 - 6x^3 + \ldots} \\ 5x^3 + \ldots \end{array} \tag{3.40}$$

Thus

$$H(s) = \frac{1}{s}(1 + \epsilon^{-\tau s} - 2\epsilon^{-\tau s} + 5\epsilon^{-3\tau s} + \ldots)$$

The power terms in such a long-division algorithm may be omitted and (3.40) abbreviated as

$$
\begin{array}{r|llll}
 & 1, & 1, & -2, & 5, \dots \\
\hline
1, 3, 2, 5, \dots & 1, & 4, & 3, & 6, \dots \\
 & 1, & 3, & 2, & 5, \dots \\
\cline{2-5}
 & & 1, & 1, & 1, \dots \\
 & & 1, & 3, & 2, \dots \\
\cline{3-5}
 & & & -2, & -1, \dots \\
 & & & -2, & -6, \dots \\
\cline{4-5}
 & & & & 5, \dots
\end{array}
\tag{3.41}
$$

Henceforth the division of (3.41) shall be written as

$$\frac{\{1, 4, 3, 6, \dots\}}{\{1, 3, 2, 5, \dots\}} = \{1, 2, -2, 5, \dots\} \tag{3.42}$$

and more generally, for series (3.36), (3.38), and (3.39),

$$\{c_0, c_1, c_2, \dots\} = \frac{\{b_0, b_1, b_2, \dots\}}{\{a_0, a_1, a_2, \dots\}} \tag{3.43}$$

After the coefficients of (3.39) have been obtained, its inverse Laplace transform, $h(t)$, is another train of singularity functions which is readily recognizable and the character of $h(t)$ is immediately revealed.

The type of singularity train which $h(t)$ turns out to be depends entirely, of course, on the type of singularity functions used to describe $f_1(t)$ and $f_2(t)$. Table 3.2 lists the types of singularity functions among $f_1(t)$, $f_2(t)$, and $h(t)$.

TABLE 3.2

Relations of Types of Singularity Functions Among $f_1(t)$, $f_2(t)$, and $h(t)$

$f_1(t)$ \ $f_2(t)$ → $h(t)$	Impulse $b = 0$	Step $b = 1$	Ramp $b = 2$	Parabolic $b = 3$
Impulse $a = 0$	Impulse $c = 0$	Step $c = 1$	Ramp $c = 2$	Parabolic $c = 3$
Step $a = 1$		Impulse $c = 0$	Step $c = 1$	Ramp $c = 2$
Ramp $a = 2$			Impulse $c = 0$	Step $c = 1$
Parabolic $a = 3$				Impulse $c = 0$

Example 1. Suppose we have

$$f_1(t) = 10u_{-2}(t) - 10u_{-2}(t-1) \tag{3.44}$$

$$f_2(t) = u_{-3}(t) - 2u_{-3}(t-1) + u_{-3}(t-2) \tag{3.45}$$

$$\tau = 1$$

Then

$$F_1(s) = \frac{10}{s^2}(1 - \epsilon^{-s})$$

$$F_2(s) = \frac{1}{s^3}(1 - 2\epsilon^{-s} + \epsilon^{-2s})$$

and

$$H(s) = \frac{1}{10s}\frac{1 - \epsilon^{-s} + \epsilon^{-2s}}{1 - \epsilon^{-s}} = \frac{1}{10s}(1 - \epsilon^{-s})$$

Thus

$$h(t) = 0.1[u_{-1}(t) - u_{-1}(t-1)] \tag{3.46}$$

This $h(t)$ is a rectangular pulse of 1 unit in duration and 0.1 unit in strength. That this is the correct and exact answer for the given pair of $f_1(t)$ and $f_2(t)$ can easily be verified by performing the convolution between $f_1(t)$ and $h(t)$. These functions are depicted in Fig. 3.2.

Example 2. Suppose $f_1(t)$ and $f_2(t)$ are given as in Fig. 3.3(a) and (b). It is easy to show that

$$F_1(s) = \frac{1}{s}(2 - 5\epsilon^{-\tau s} + 2\epsilon^{-2\tau s}) \tag{3.47}$$

$$F_2(s) = \frac{1}{s^2}(1 - 4\epsilon^{-\tau s} + 5\epsilon^{-2\tau s} - 2\epsilon^{-3\tau s}) \tag{3.48}$$

whence

$$\{c_0, c_1, c_2, \ldots\} = \frac{\{1, -4, 5, -2\}}{\{2, -5, 2\}} = \left\{\frac{1}{2}, -\frac{3}{4}, \frac{1}{8}, \frac{1}{16}, \frac{1}{32}, \ldots\right\}$$

and

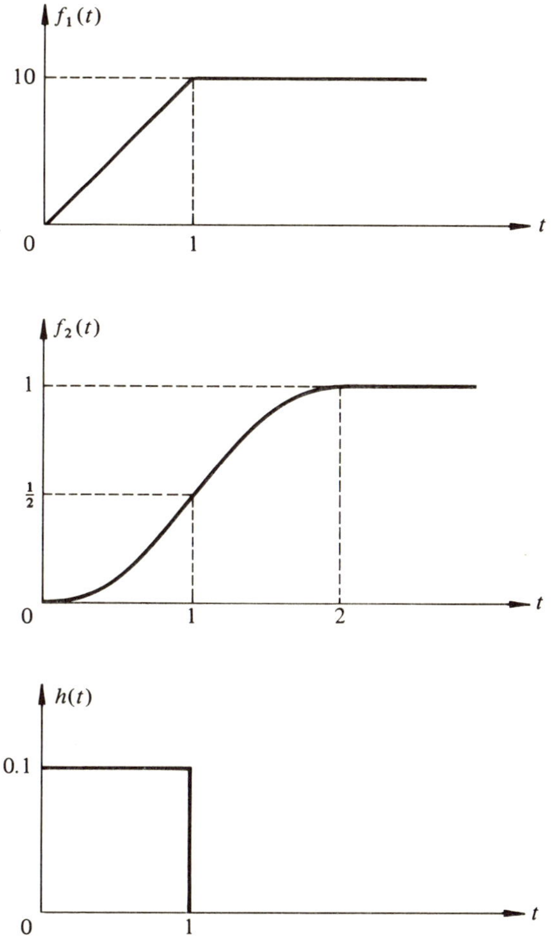

Fig. 3.2 Functions (3.44), (3.45), and (3.46) in Example 1.

$$H(s) = \frac{1}{s}\left(\frac{1}{2} - \frac{3}{4}\epsilon^{-\tau s} + \frac{1}{8}s^{-2\tau s} + \frac{1}{16}\epsilon^{-3\tau s} + \frac{1}{32}\epsilon^{-4\tau s} + \dots + \frac{1}{2^n}\epsilon^{-n\tau s} + \dots\right) \tag{3.49}$$

The inverse Laplace transform of (3.49) is shown in Fig. 3.3(c)

Example 3. Suppose now the $f_2(t)$ of Example 2 is changed to that shown in Fig. 3.4, which has a Laplace transform

$$F_2(s) = \frac{1}{s^2}(2 - 6\epsilon^{-\tau s} + 6\epsilon^{-2\tau s} - 2\epsilon^{-3\tau s}) \tag{3.50}$$

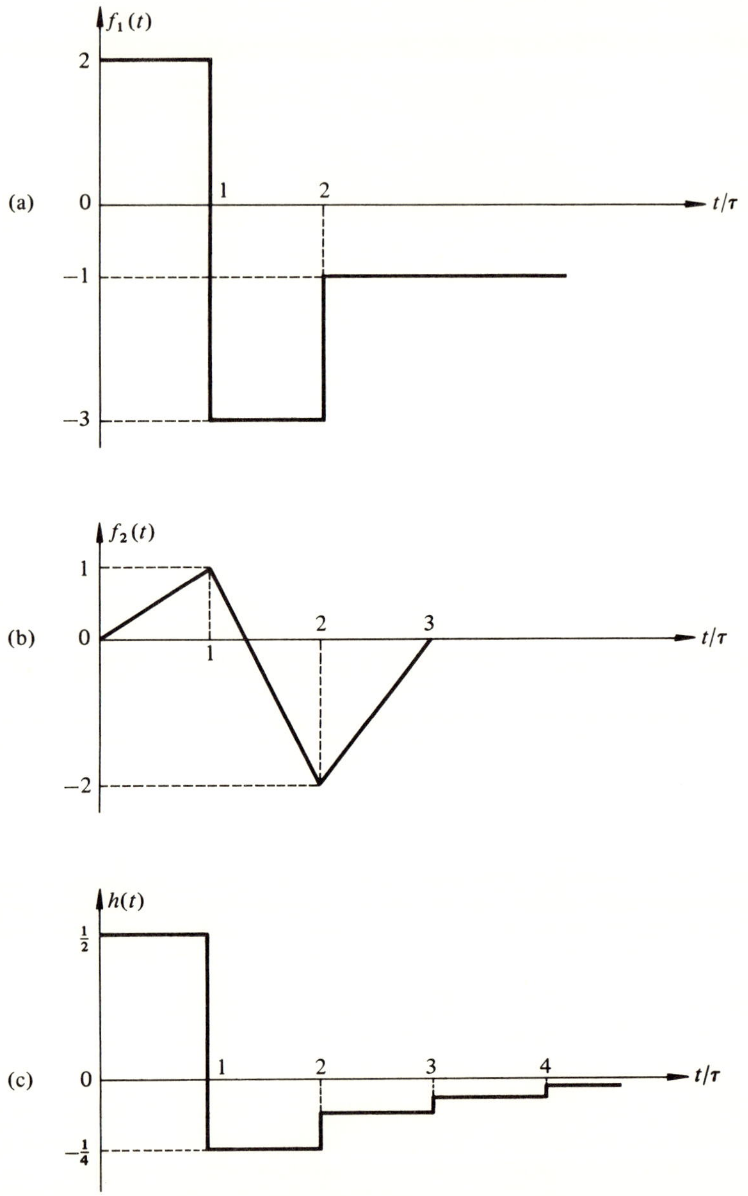

Fig. 3.3 Time functions corresponding to (3.47), (3.48), and (3.49) of Example 2.

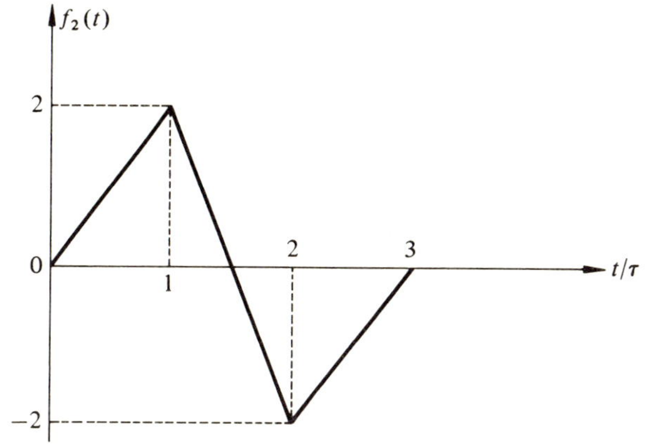

Fig. 3.4 Time function corresponding to (3.50) of Example 3.

Equation (3.43) gives

$$\{c_0, c_1, c_2, \ldots\} = \frac{\{2, -6, 6, -2\}}{\{2, -5, 2\}}$$

$$= \left\{1, -\frac{1}{2}, \frac{3}{4}, \frac{11}{8}, \frac{43}{16}, \frac{171}{32}, \ldots\right\}$$

This number sequence is obviously not convergent. This conclusion may be quite surprising at first in view of the fact that $f_2(t)$ of this example is really very similar to the $f_2(t)$ of Example 2. Yet this slight change in $f_2(t)$ causes the resulting $h(t)$ to become unbounded and certainly not absolutely integrable.

This phenomenon can easily be explained if we once again resort to the frequency-domain transform of these time functions. For the $f_1(t)$ of both Example 2 and Example 3,

$$F_1(s) = \frac{1}{s}(1 - 2\epsilon^{-\tau s})(2 - \epsilon^{-\tau s}) \tag{3.51}$$

Hence the zeros of $F_1(s)$ are given by the equations

$$\epsilon^{-\tau s} = \frac{1}{2}$$

and

$$\epsilon^{-\tau s} = 2$$

or

$$\tau s = 0.69315 - j2n\pi$$

and

$$\tau s = -0.69315 - j2n\pi$$

respectively. The zeros of $F_1(s)$ become poles of $H(s)$. Since the first group of zeros are in the right halfplane, the resulting $H(s)$ will have right-halfplane poles (which cause instability) unless these poles are canceled out by the same zeros contained in $F_2(s)$.

If we examine the zeros of $F_2(s)$ of (3.48), we see that

$$F_2(s) = \frac{1}{s}(1 - 2\epsilon^{-\tau s})(1 - \epsilon^{-\tau s})^2 \tag{3.52}$$

The first of these two factors of this $F_2(s)$ is the one that cancels out the like factor in $F_1(s)$.

On the other hand, the $F_2(s)$ of (3.50) is

$$F_2(s) = \frac{2}{s^2}(1 - \epsilon^{-\tau s})^3 \tag{3.53}$$

which does not have the factor $(1 - 2\epsilon^{-\tau s})$. Hence $H(s)$ obtained with this $F_2(s)$ has right-halfplane poles at

$$s = \frac{1}{\tau}(0.69315 - j2n\pi)$$

and the corresponding $h(t)$ is not bounded. Hence we cannot produce the $f_2(t)$ of Fig. 3.4 for the $f_1(t)$ of Fig. 3.3(a) with a stable, linear, lumped network.

Example 4. Examples 1 through 3 illustrate the simple algorithm that is necessary to be performed when $f_1(t)$ and $f_2(t)$ are each a uniformly spaced train of singularity functions. The same principle can be used when $f_1(t)$ and $f_2(t)$ can be approximately represented by these trains. To illustrate this application, let

$$f_1(t) = \epsilon^{-4t} \tag{3.54}$$

$$f_2(t) = t\epsilon^{-4t} - (t - 1)\epsilon^{-4(t-1)}u(t - 1) \tag{3.55}$$

Since

$$F_1(s) = \frac{1}{s + 4}$$

$$F_2(s) = \frac{1}{(s + 4)^2} - \frac{\epsilon^{-s}}{(s + 4)^2}$$

we have

$$H(s) = \frac{1}{s+4} - \frac{\epsilon^{-s}}{s+4}$$

and

$$h(t) = \epsilon^{-4t} - \epsilon^{-4(t-1)}u(t-1) \tag{3.56}$$

These time functions are shown in Fig. 3.5. Now suppose the t axis is divided into equal intervals of τ each and let $\tau = 0.1$. Further, suppose each of $f_1(t)$ and $f_2(t)$ is approximated by a series of impulse functions situated at the centers of these intervals. For simplicity the area of each impulse is set equal to the functional value at the center of the interval times τ. With these assumptions, the coefficients in the two series

$$\mathscr{L}[\hat{f}_1(t)] = \tau\epsilon^{-0.05s}(a_0 + a_1 s^{-0.1s} + a_2 s^{-0.2s} + \dots) \tag{3.57}$$

$$\mathscr{L}[\hat{f}_2(t)] = \tau\epsilon^{-0.05s}(b_0 + b_1 s^{-0.1s} + b_2 s^{-0.2s} + \dots) \tag{3.58}$$

are simply the values of $f_1(t)$ and $f_2(t)$ at $t = 0.05, 0.15, \dots$. These values are tabulated in Table 3.3. The approximate impulse response $\hat{h}(t)$ is given by

$$\mathscr{L}[\hat{h}(t)] = c_0 + c_1 s^{-0.1s} + c_2 s^{-0.2s} + \dots \tag{3.59}$$

where

$$\{c_0, c_1, c_2, \dots\} = \frac{\{b_0, b_1, b_2, \dots\}}{\{a_0, a_1, a_2, \dots\}} \tag{3.60}$$

These c's are also tabulated in Table 3.3. It is significant to note that the factors $\tau\epsilon^{-0.05s}$ in (3.57) and (3.58) cancel out during the division implied by (3.60) and do not appear in (3.59). Hence the durations of the intervals used for obtaining $\hat{f}_1(t)$ and $\hat{f}_2(t)$ are not materials so long as they are the same.

Now the $\hat{h}(t)$ given by (3.59) can be interpreted as another impulse train with the first impulse located at $t = 0$, not $t = 0.05$. Note should be made that because $\hat{h}(t) = 0$ for $t < 0$, the first impulse represents the area under $h(t)$ for the interval $0 < t < 0.05$, whereas the subsequent impulses represent areas over intervals 0.1 in width. If the values of c's are interpreted by the reverse of the way a's and b's are obtained, the values of $\hat{h}(t)$ can be obtained and are tabulated in Table 3.4, which also shows the actual values of $h(t)$. At $t = 1$, where $h(t)$ has a discontinuity, $\hat{h}(t)$ gives the average of the two approximate values of $h(t)$ at that point. The close agreement between the values of $\hat{h}(t)$ obtained by such a crude approximation of $f_1(t)$ and $f_2(t)$ and the actual $h(t)$ is startling indeed.

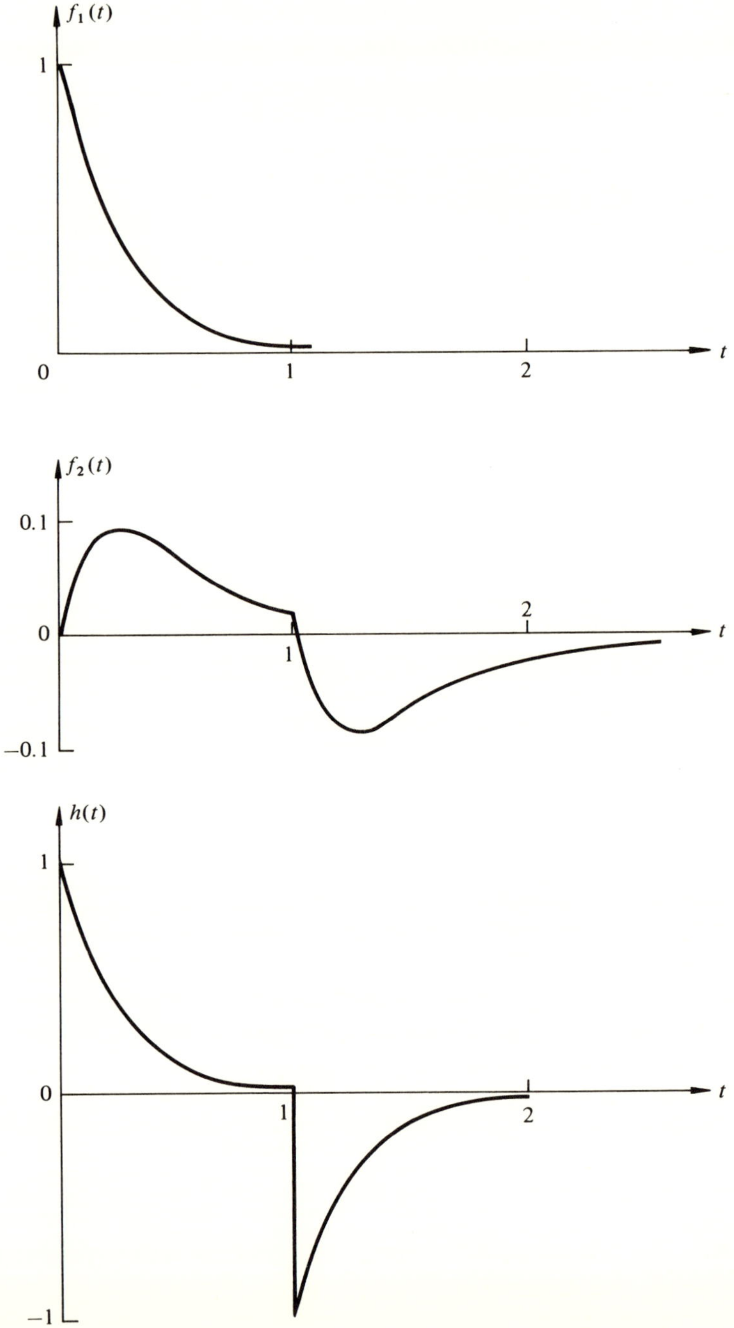

Fig. 3.5 Functions (3.54), (3.55), and (3.56) in Example 4.

TABLE 3.3

Number Sequences Representing $f_1(t)$, $f_2(t)$, and $h(t)$ of Example 4

t	$a_i = f_1(t)$	$b_i = f_2(t)$	c_i
0.05	0.818731	0.040937	0.050000
0.15	0.548812	0.082322	0.067032
0.25	0.367879	0.091970	0.044933
0.35	0.246597	0.086309	0.030119
0.45	0.165299	0.074384	0.020190
0.55	0.110803	0.060942	0.013534
0.65	0.074274	0.048278	0.009072
0.75	0.049787	0.037340	0.006081
0.85	0.033373	0.028367	0.004076
0.95	0.022371	0.021252	0.002732
1.05	0.014996	−0.025191	−0.048168
1.15	0.010052	−0.070762	−0.065804
1.25	0.006738	−0.083547	−0.044110
1.35	0.004517	−0.080212	−0.029568
1.45	0.003028	−0.069995	−0.019820
1.55	0.002029	−0.057796	−0.013286
1.65	0.001360	−0.046033	−0.008906
1.75	0.000912	−0.035745	−0.005970
1.85	0.000611	−0.027236	−0.004002
1.95	0.000410	−0.020453	−0.002682
2.05	0.000275	−0.015182	−0.001798
2.15	0.000184	−0.011164	−0.001205
2.25	0.000123	−0.008145	−0.000808
2.35	0.000083	−0.005903	−0.000542
2.45	0.000055	−0.004254	−0.000363
2.55	0.000037	−0.003051	−0.000243
2.65	0.000025	−0.002179	−0.000163
2.75	0.000017	−0.001550	−0.000109
2.85	0.000011	−0.001099	−0.000073
2.95	0.000008	−0.000777	−0.000049

Although the error of this method at any one point tends to be retained at all succeeding points, the accuracy of each problem can always be improved by reducing the value of the basic interval, τ. Also, singularity functions other than the impulse may also be used for approximating these time functions. The choice of what singularity functions to use will depend somewhat on the wave shape to be approximated.

Another very important feature of this method is the fact that the value of $\hat{h}(t)$ at any instant depends only on the values of $f_1(t)$ and $f_2(t)$ at and before that instant. The values of $\hat{h}(t)$ may be found as $f_1(t)$ and $f_2(t)$ are still unfolding themselves. Hence this method may be viewed as a real-time scheme of finding $\hat{h}(t)$.

TABLE 3.4

Comparison of $\hat{h}(t)$ and $h(t)$ of Example 4

t	$\hat{h}(t)$	$h(t)$
0	1.00000	1.00000
0.1	0.67032	0.67032
0.2	0.44933	0.44933
0.3	0.30119	0.30119
0.4	0.20190	0.20190
0.5	0.13534	0.13534
0.6	0.09072	0.09072
0.7	0.06081	0.06081
0.8	0.04076	0.04076
0.9	0.02732	0.02732
1.0^-	−0.48168	0.01832
1.0^+	−0.48168	−0.98168
1.1	−0.65804	−0.65804
1.2	−0.44110	−0.44110
1.3	−0.29568	−0.29567
1.4	−0.19820	−0.19820
1.5	−0.13286	−0.13286
1.6	−0.08906	−0.08906
1.7	−0.05970	−0.05970
1.8	−0.04002	−0.04001
1.9	−0.02682	−0.02682
2.0	−0.01798	−0.01798
2.1	−0.01205	−0.01206
2.2	−0.00808	−0.00808
2.3	−0.00542	−0.00537
2.4	−0.00363	−0.00360
2.5	−0.00243	−0.00241
2.6	−0.00163	−0.00159
2.7	−0.00109	−0.00106
2.8	−0.00073	−0.00061
2.9	−0.00049	−0.00041

3.5 ABSOLUTE INTEGRABILITY OF THE IMPULSE RESPONSE

Examples 2 and 3 of Section 3.4 suggest that whether a given set of $f_2(t)$ and $f_1(t)$ corresponds to an absolutely integrable $h(t)$ or not depends largely on the locations of the zeros of $F_1(s)$, since they are also the poles of $H(s)$. An absolutely integrable $h(t)$ is, of course, necessary for the corresponding network to be stable. When right-halfplane zeros are present in $F_1(s)$, the only way the corresponding $h(t)$ will be absolutely integrable is for $F_2(s)$ to have the same zeros. The occurrence of this coincidence is a rather rare matter in a randomly chosen pair of $f_2(t)$ and $f_1(t)$. In fact, it is rather difficult to specify a pair of $f_2(t)$ and $f_1(t)$ each with a number of singularity terms to have exactly the same right-halfplane zeros in their transforms. This is especially true when $f_2(t)$ and $f_1(t)$ are only some form of approximation of the real response to an actual excitation in a network.

On the other hand, if there is some means by which the presence or the absence of right-halfplane zeros in $F_1(s)$ can be determined in advance, then the stability problem of the synthesis task would have been answered barring the rare coincidence just mentioned. Specifically, if we can ascertain that the given train of singularity functions for $f_1(t)$ does not have right-halfplane zeros, then we may proceed with the long division with the knowledge that the resulting $h(t)$ will be absolutely integrable. While if right-halfplane zeros are present in $F_1(s)$, long division need not be carried too far if evidence does not point to a convergent $h(t)$ to encourage the acquisition of additional points.

It is seldom practical for us to perform the type of factorization as was done in (3.51), (3.52), and (3.53) when the number of points involved is large. Hence some means of determining whether right-halfplane zeros are present in $F_1(s)$ or not, when $f_1(t)$ is a train of singularity functions, is very helpful. Since $F_1(s)$ is a polynomial in $\epsilon^{-\tau s}$, it may be written as

$$F_1(s) = A \prod_i (\sigma_i - \epsilon^{-\tau s}) \prod_j (\alpha_j^2 + \beta_j^2 - 2\alpha_j \epsilon^{-\tau s} + \epsilon^{-2\tau s}) \tag{3.61}$$

where the first group of factors represents real zeros in the $\epsilon^{-\tau s}$ plane and the second group of factors represents complex-conjugate zeros.

Hence the real zeros are given by

$$\epsilon^{-\tau s_i} = \sigma_i$$

or

$$s_i = \frac{\ln|\sigma_i| \pm jn\pi}{-\tau} \tag{3.62}$$

where n is odd if σ_i is negative and even if σ_i is positive. Hence we may be say that

if $\lvert\sigma_i\rvert > 1$,	s_i is in the left halfplane
if $\lvert\sigma_i\rvert = 1$.	s_i is on the j axis
if $\lvert\sigma_i\rvert < 1$,	s_i is in the right halfplane

For each complex pole

$$\epsilon^{-\tau s_i} = \alpha_j + j\beta_j = A_j \underline{/\theta_j}$$

$$s_i = \frac{\ln|A_j| + j(\theta_j + 2n\pi)}{-\tau} \tag{3.63}$$

where n is any integer. Here we may say that

$$\begin{aligned} &\text{if } |A_i| > 1, && s_i \text{ is in the left halfplane} \\ &\text{if } |A_i| = 1, && s_i \text{ is on the } j \text{ axis} \\ &\text{if } |A_i| < 1, && s_i \text{ is in the right halfplane} \end{aligned}$$

It is seen that to determine whether $F_1(s)$ has right-halfplane zero or not is equivalent to determining whether the polynomial in p obtained by replacing every $\epsilon^{-\tau s}$ in $F_1(s)$ by p [call it $G(p)$] has zeros in the interior of the unit circle centered at the origin in the p plane or not. To ascertain whether any zeros are present in $G(p)$ inside the unit circle, the following bilinear transformation is helpful:

$$\lambda = \frac{1-p}{1+p}, \qquad p = \frac{1-\lambda}{1+\lambda} \tag{3.64}$$

This transformation transforms the interior of a unit circle centered at the origin onto the right halfplane, the circle to the j axis, and the exterior to the left halfplane. Hence the transformation

$$\lambda = \frac{1-\epsilon^{-\tau s}}{1+\epsilon^{-\tau s}} \tag{3.65}$$

maps the right halfplane of s onto the right halfplane of λ, the j axis onto the j axis, and the left halfplane onto the left halfplane. But $G[(1-\lambda)/(1+\lambda)]$ is a rational function in λ. Hence we merely need to apply the Hurwitz test to the numerator polynomial of $G[(1-\lambda)/(1+\lambda)]$ in the variable λ. A failure of this test will indicate the presence of right-halfplane zeros in $F_1(s)$.

For example, for the $f_1(t)$ given in Fig. 3.6, we have

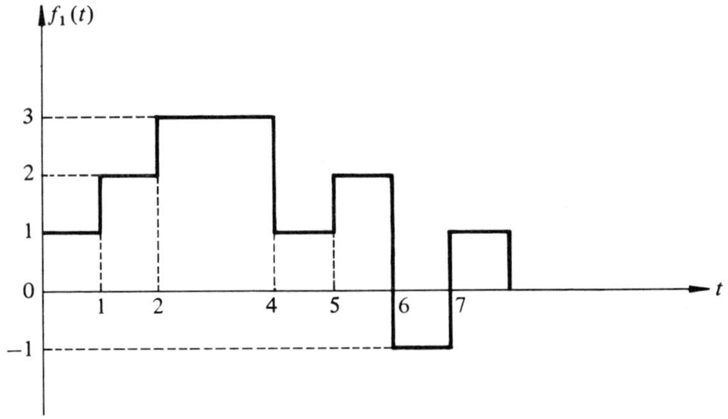

Fig. 3.6 Time function corresponding to $F_1(s)$ of (3.66).

$$
\begin{aligned}
F_1(s) &= \frac{1}{s}(1 + p + p^2 - 2p^4 + p^5 - 3p^6 + 2p^7 - p^8) \\
&= \frac{1}{s(1+\lambda)^8}[(1+\lambda)^8 + (1+\lambda)^7(1-\lambda) + (1+\lambda)^6(1-\lambda)^2 \\
&\quad - 2(1+\lambda)^4(1-\lambda)^4 \\
&\quad + (1+\lambda)^3(1-\lambda)^5 - 3(1+\lambda)^2(1-\lambda)^6 \\
&\quad + 2(1+\lambda)(1-\lambda)^7 - (1-\lambda)^8]
\end{aligned} \tag{3.66}
$$

Simplifying we obtain

$$
F_1(s) = \frac{8\lambda(3 + 5\lambda + 11\lambda^2 + \lambda^3 + 13\lambda^4 - 5\lambda^5 + 5\lambda^6 - \lambda^7)}{s(1+\lambda)^8} \tag{3.67}
$$

The factor λ in the numerator of (3.67) indicates that zeros are present on the j axis in $F_1(s)$ on the s plane. The other factor is obviously not Hurwitz, because it has negative coefficients. Hence $F_1(s)$ does have zeros in the right half of the s plane.

To facilitate algebra of the type used in going from (3.66) to (3.67), tables of expansions of $(1 + \lambda)^m(1 - \lambda)^n$ for $m + n$ up to 15 are given in Appendix A.

The foregoing method soon gets out of hand as the number of singularity functions in the sequence gets larger and larger. As this number reaches beyond, say 15, one may as well go ahead and perform the long division to get the number sequence $\{c_0, c_1, c_2, \ldots\}$ instead of checking the Hurwitzness of a high-degree polynomial. The matter of absolute integrability of $h(t)$ car usually be ascertained after a definite pattern of c's begins to emerge.

The advantage of using the transform method is, of course, that the inference drawn that way is conclusive. Once the Hurwitzness of the polynomial is ascertained, there is no question that the $h(t)$ associated with any finite $f_2(t)$ will be absolutely integrable. On the other hand, when the division of $\{b_0, b_1, b_2, \ldots\}/\{a_0, a_1, a_2, \ldots\}$ is performed and the pattern of the resultant coefficients is observed, it cannot be assumed with certainty in most cases that the conclusion is correct. For example, the value of $h(t)$ may display violent variation of long duration before it dies down. On the other hand, another $h(t)$ may appear to taper off before it eventually goes haywire again.

Except for the situation when right-halfplane zeros in both $F_1(s)$ and $F_2(s)$ coincide, it is usually sufficient in determining the suitability of a given $F_1(s)$ to assume that $F_2(s) = 1$. For example, for the $f_1(t)$ of Fig. 3.6, we may investigate the absolute integrability of $h(t)$ for $f_2(t) = \delta(t)$ by performing the division and obtaining

$$
\begin{aligned}
&\frac{\{1\}}{\{1, 1, 1, 0, -2, 1, -3, 2, -1\}} \\
&= \{1, -1, 0, 1, 1, -5, 8, -6, 2, -5, 25, -56, -76, -71, \ldots\}
\end{aligned} \tag{3.68}
$$

Looking over the quotient of this synthetic-division divergence might be suspected after the coefficient 8 was obtained. Divergence was fairly apparent when the number 25 appeared. But we really have no absolute assurance that by some strange sequence of events, convergence of coefficients may not take place in future steps. The negative answer of the Hurwitz test of the numerator of (3.67), however, enables us to draw the conclusion, with not a trace of doubt, that $H_1(s)$ has right-halfplane zeros.

3.6 ALTERNATIVE DERIVATION OF THE DIVISION METHOD

The method of obtaining the impulse response when an excitation-and-response pair is given (Section 3.4) can be arrived at by a different approach. This alternative approach makes use of a number of theorems on convolution. Suppose

$$f_2(t) = \int_0^t f_1(\tau)h(t - \tau)\, d\tau \tag{3.69}$$

where $f_1(t)$ is piecewise linear and continuous. The following statements and formulas are true:

1. $f_2^{(j+k)}(t) = \int_0^t f_1^{(k)}(\tau)h^{(j)}(t - \tau)\, d\tau$ **(3.70)**

2. If $f_1(t) = \delta(t)$, then $f_2(t) = h(t)$.
3. If $f_1(t) = \delta(t)$ and $h(t) = \delta(t)$, then $f_2(t) = \delta(t)$.
4. If $f_1(t)$ is replaced by $f_1(t - t_1)$ and $h(t)$ is replaced by $h(t - t_2)$, then $f_2(t)$ becomes $f_2(t - t_1 - t_2)$.

Application of several of these theorems leads to the conclusion that if $f_1^{(n)} = \delta(t)$, then $f_2^{(n)}(t) = h(t)$. In other words, if the nth derivative of $f_1(t)$ is a unit impulse, then the corresponding nth derivation of $f_2(t)$ is the impulse response of the system.

On the other hand, if the nth derivative of $f_1(t)$ is a unit impulse, and the mth derivative of $f_2(t)$ is also a unit impulse, then the $(m - n)$th derivative of $h(t)$ is a unit impulse. In that case, $h(t)$ is the $(m - n)$th integral of the unit impulse.

These statements can easily be modified when the impulses in the above theorems and statements are not all unit in strength. The algebra of convolution, of course, follows the law of linearity.

When any of the impulses is delayed, the effect can be seen by item 4. When either $f_1(t)$ or $h(t)$ is delayed, $f_2(t)$ is delayed by the same amount.

When both $f_1(t)$ and $h(t)$ are delayed by the same or different amounts of time, $f_2(t)$ is delayed by the sum of the two delays. On the other hand, to delay the response, either $f_1(t)$ or $h(t)$ or both may be delayed as long as the total amount of delay of both functions is the same as the desired delay for $f_2(t)$.

When $f_1(t)$ and $h(t)$ are sequences of impulses, their convolution will produce another sequence of impulses. For example, if

$$f_1(t) = a_1 u_0(t - t_1) + a_2 u_0(t - t_2) \tag{3.71}$$

$$h(t) = c_1 u_0(t - t_3) + c_2 u_0(t - t_4) \tag{3.72}$$

then

$$f_2(t) = a_1 c_1 u_0(t - t_1 - t_3) + a_1 c_2 u_0(t - t_1 - t_4) + a_2 c_1 u_0(t - t_2 - t_3) + a_2 c_2 u_0(t - t_2 - t_4) \tag{3.73}$$

This convolution process of impulse trains is seen to be analogous to the multiplication of a sequence of complex numbers when each complex number is written in polar form; magnitudes multiply while phase angles add.

Conversely, when the convolution of two impulse sequences is given in the form of another impulse sequence, and if one of the two sequences which convolve is known, the other convolving sequence should be obtainable by division.

This is precisely what was done in Section 3.4, where we wrote, say,

$$F_1(s) = \frac{1}{s^a}(a_0 + a_1 \epsilon^{-\tau s} + \dots + a_k \epsilon^{-k\tau s} + \dots) \tag{3.74}$$

the factor $1/s^a$ indicates that if $f_1(t)$ is differentiated a times [$F_1(s)$ is multiplied by s^a], the resultant function, $f_1^{(a)}(t)$, will be what is contained in the parentheses of (3.74). The series of terms inside the parentheses is the transform of a sequence of impulses spaced τ seconds apart. When the operation

$$H(s) = \frac{F_2(s)}{F_1(s)} = \frac{1}{s^{b-a}} \frac{b_0 + b_1 \epsilon^{-\tau s} + b_2 \epsilon^{-2\tau s} + \dots}{a_0 + a_1 \epsilon^{-\tau s} + a_2 \epsilon^{-2\tau s} + \dots} \tag{3.75}$$

is performed, $H(s)$ is obtained. In (3.75), the term $1/s^{b-a}$ may be interpreted as the $(b - a)$th integral in the time domain. The second fraction may be expanded into a power series in $\epsilon^{-\tau s}$, the inverse Laplace transform of which represents another impulse train. The coefficients of this power series may, of course, be obtained by dividing the coefficients of the numerator polynomial by those of the denominator polynomial.

3.7 OTHER ELEMENTARY FUNCTIONS

The long-division methods described in the previous sections are based on breaking up both $f_1(t)$ and $f_2(t)$ into smaller elementary parts, all of which have the same shape but with varying magnitudes and commencing at different moments. The wave shape of these small elementary parts may be different between $f_1(t)$ and $f_2(t)$, but within $f_1(t)$ or $f_2(t)$ they are uniform. Although the elementary functions used there are limited to the few singularity functions, this principle can be extended to wave forms other than the singularity functions.

Let us say first that $f_1(t)$ is composed of a time sequence of functions all of the type $v_1(t)$. Referring to Fig. 3.7, we are defining

$$f_1(t) = a_0 v_1(t) + a_1 v_1(t - \tau) + \dots + a_k v_1(t - k\tau) + \dots \tag{3.76}$$

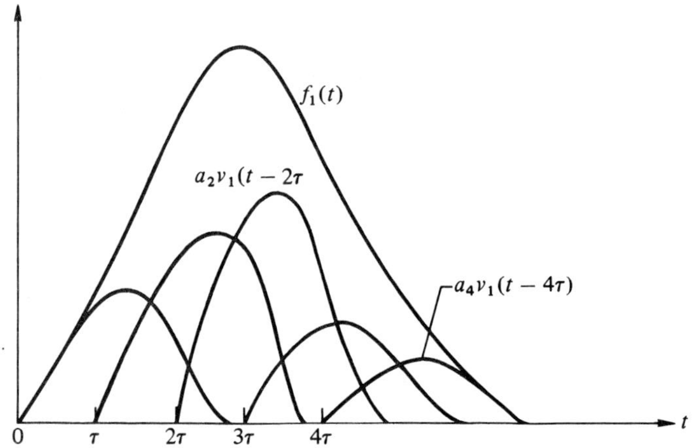

Fig. 3.7 Representation of a time function by a sequence of elementary functions.

Similarly, $f_2(t)$ is represented by another time sequence of elementary functions, all of type $v_2(t)$,

$$f_2(t) = b_0 v_2(t) + b_1 v_2(t - \tau) + \dots + b_k v_2(t - k\tau) + \dots \tag{3.77}$$

In the frequency domain

$$F_1(s) = V_1(s)(a_0 + a_1 \epsilon^{-\tau s} + \dots + a_k \epsilon^{-k\tau s} + \dots) \tag{3.78}$$

$$F_2(s) = V_2(s)(b_0 + b_1 \epsilon^{-\tau s} + \dots + b_k \epsilon^{-k\tau s} + \dots) \tag{3.79}$$

$$H(s) = \frac{F_2(s)}{F_1(s)} = \frac{V_2(s)}{V_1(s)}(c_0 + c_1 \epsilon^{-\tau s} + \dots + c_k \epsilon^{-k\tau s} + \dots) \tag{3.80}$$

In (3.80) the coefficients in the series may be obtained by long division as in (3.60). In fact, this series denotes another time sequence of similar functions spaced τ apart. If we denote

$$\bar{h}(t) = \mathscr{L}^{-1}\left(\frac{V_2(s)}{V_1(s)}\right) \tag{3.81}$$

then

$$h(t) = c_0\bar{h}(t) + c_1\bar{h}(t-\tau) + \ldots + c_k\bar{h}(t-k\tau) + \ldots \tag{3.82}$$

Function $\bar{h}(t)$ of (3.81) may be interpreted as the impulse response of a (fictitious) network which when excited by $v_1(t)$ will produce $v_2(t)$ as its response. It is easy to see that if $v_1(t)$ and $v_2(t)$ are arbitrary functions, the determination of $\bar{h}(t)$ can be just as difficult as the original problem of determining $h(t)$. To take advantage of this method, $v_1(t)$ and $v_2(t)$ must be related to each other in some simple fashion.

One easy choice to facilitate the determination of $\bar{h}(t)$ is to make

$$v_1(t) \equiv v_2(t) \tag{3.83}$$

In that case, we have simply

$$\bar{h}(t) = \delta(t) \tag{3.84}$$

and

$$h(t) = c_0\delta(t) + c_1\delta(t-\tau) + \ldots + c_k\delta(t-k\tau) + \ldots \tag{3.85}$$

For example, one of the most useful elementary functions is the triangular wave form, as shown in Fig. 3.8 [TU1]. Let us call it $\Delta(t)$. When a sequence of these functions are staggered τ apart, the resultant function is a series

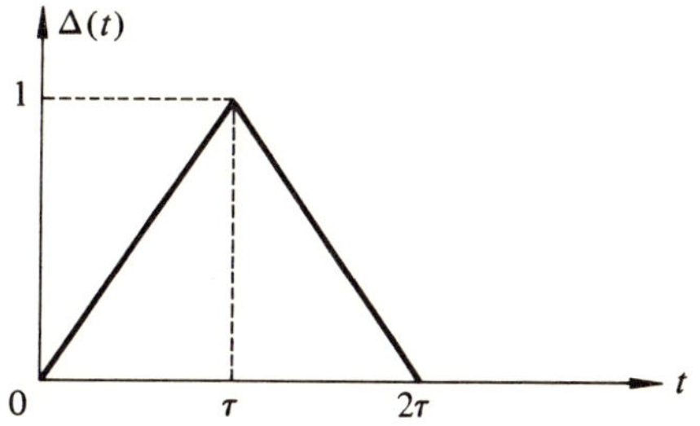

Fig. 3.8 The unit triangular wave, $\Delta(t)$.

of straight-line segments obtained by joining the peaks of these triangles. An example of a particular sequence of these triangular pulses in shown in Fig. 3.9.

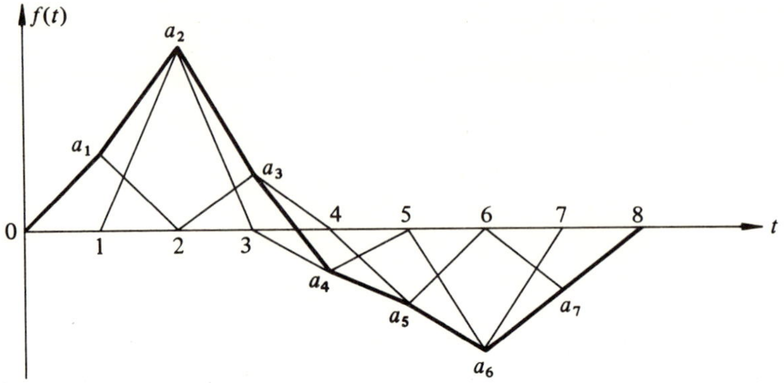

Fig. 3.9 A sequence of triangular waves and its resultant time function.

Hence, when a given function $f_1(t)$ or $f_2(t)$ is given in the form of a sequence of triangular waves, the function can be plotted very easily. Conversely, if a function is given or is to be approximated by straight-line segments, the magnitudes of the successive triangular waves can readily be obtained by reading off the ordinates of the function at regular intervals.

For example, let $f_1(t)$ and $f_2(t)$ be as shown in Fig. 3.10. These functions may be represented exactly by ramp functions, $u_{-2}(t)$. Using the notation of Section 3.4, we have

$$\{a_0, \ldots, a_4\} = \{1, -4, 3, 2, -2\}$$
$$\{b_0, \ldots, b_7\} = \{-1, 6, -10, -2, 17, -8, -6, 4\}$$

whence

$$\{c_0, \ldots, c_3\} = \{-1, 2, 1, -2\}$$

and

$$h(t) = -\delta(t) + 2\delta(t-1) + \delta(t-2) - 2\delta(t-3)$$

If, instead, the triangular elementary function is used, the number sequences corresponding to $f_1(t)$ and $f_2(t)$ are, respectively,

$$\{a_0, a_1, a_2\} = \{1, -2, -2\}$$
$$\{b_0, \ldots, b_5\} = \{-1, 4, -1, -8, 2, 4\}$$

with the same resulting $h(t)$. These numbers sequences are much easier to obtain than the ramp functions.

The reader may have noticed the resemblance between this method and the procedure used in Example 4 of Section 3.4.

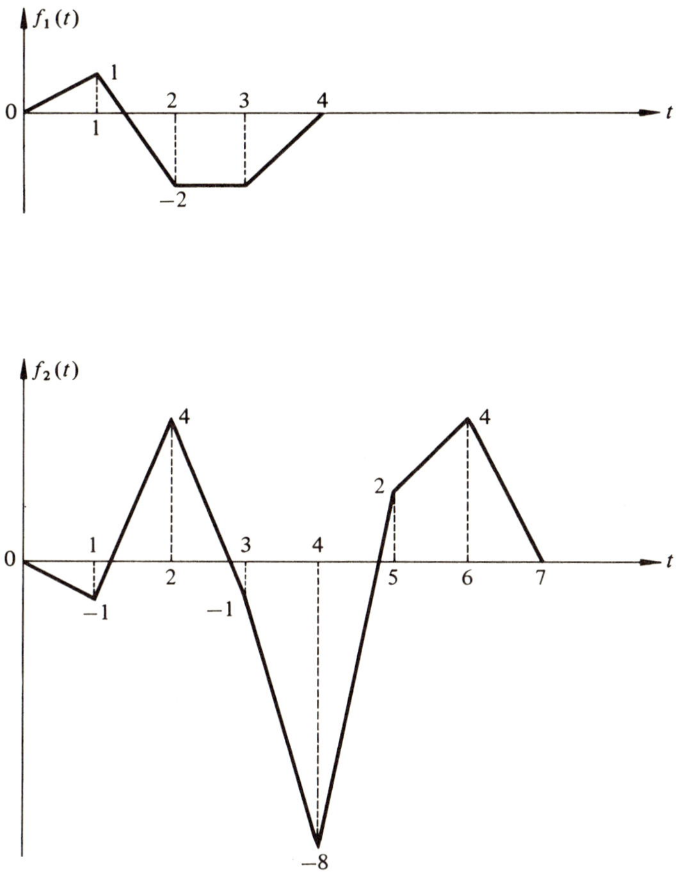

Fig. 3.10 A pair of time functions expressible in terms of triangular waves.

Another example in the choice of the elementary functions might be to choose $v_1(t)$ and $v_2(t)$ so that one is the derivative of the other. For instance, if we let

$$v_1(t) = \frac{d}{dt} v_2(t) \tag{3.86}$$

then

$$\bar{h}(t) = u(t)$$

and $h(t)$ is a sequence of step functions. This may be illustrated by referring to Example 2 of Section 3.4, where $v_1(t) = u_{-1}(t)$, $v_2(t) = u_{-2}(t)$, and

$$h(t) = \tfrac{1}{2}u_{-1}(t) - \tfrac{3}{4}u_{-1}(t-\tau) + \tfrac{1}{8}u_{-1}(t-2\tau) + \tfrac{1}{16}u_{-1}(t-3\tau) + \tfrac{1}{32}u_{-1}(t-4\tau) + \ldots$$

Another choice of $v_1(t)$ and $v_2(t)$ according to (3.86) is to use the $\Delta(t)$ of Fig. 3.8 for $v_2(t)$; the corresponding $v_1(t)$ is then given in Fig. 3.11. Now for the $f_1(t)$ and $f_2(t)$ of Fig. 3.12, the division involved is

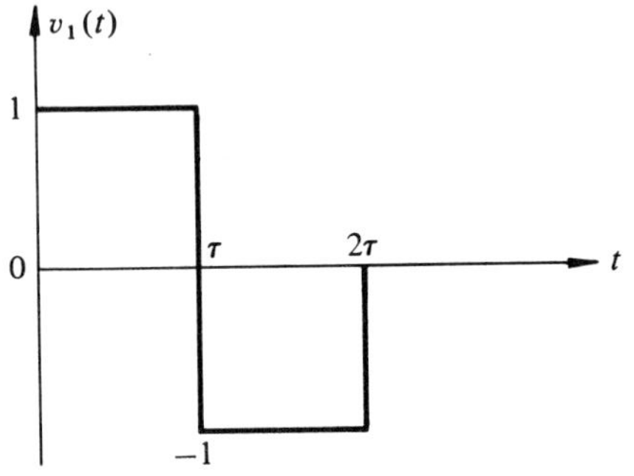

Fig. 3.11 The derivative of the triangular wave of Fig. 3.8.

$$\{c_0, \ldots\} = \frac{\{2, 3, 3, 4, 4, 4\}}{\{3, 2, -1\}} = \left\{\frac{2}{3}, \frac{5}{9}, \frac{11}{27}, \ldots\right\}$$

and

$$h(t) = \tfrac{2}{3}u_{-1}(t-\tau) + \tfrac{5}{9}u_{-1}(t-2\tau) + \tfrac{11}{27}u_{-1}(t-3\tau) + \ldots$$

It is generally difficult to get a given $f_1(t)$ in terms of $v_1(t)$ of the type given by Fig. 3.11. There are two shortcuts.

1. Since $\Delta(t)$ is the integral of the $v_1(t)$ of Fig. 3.11, we can integrate the given $f_1(t)$, and then get the corresponding $f_1^{(-1)}(t)$ in terms of $\Delta(t)$.
2. Since $v_1(t) = u_{-1}(t) - 2u_{-1}(t-1) + u_{-1}(t-2)$, we may first write $f_1(t)$ in terms of the step functions. Then we can divide the coefficients of these step functions by $\{1, -2, 1\}$. For example, for $f_1(t)$ of Fig. 3.12 we perform

$$\frac{\{3, -4, -2, 4, -1\}}{\{1, -2, 1\}} = \{3, 2, -1\}$$

Still another possibility in the convenient choice of the elementary functions is to form $v_1(t)$ and $v_2(t)$ such that

$$v_2(t) = v_1(t) \circledast v_1(t) \tag{3.87}$$

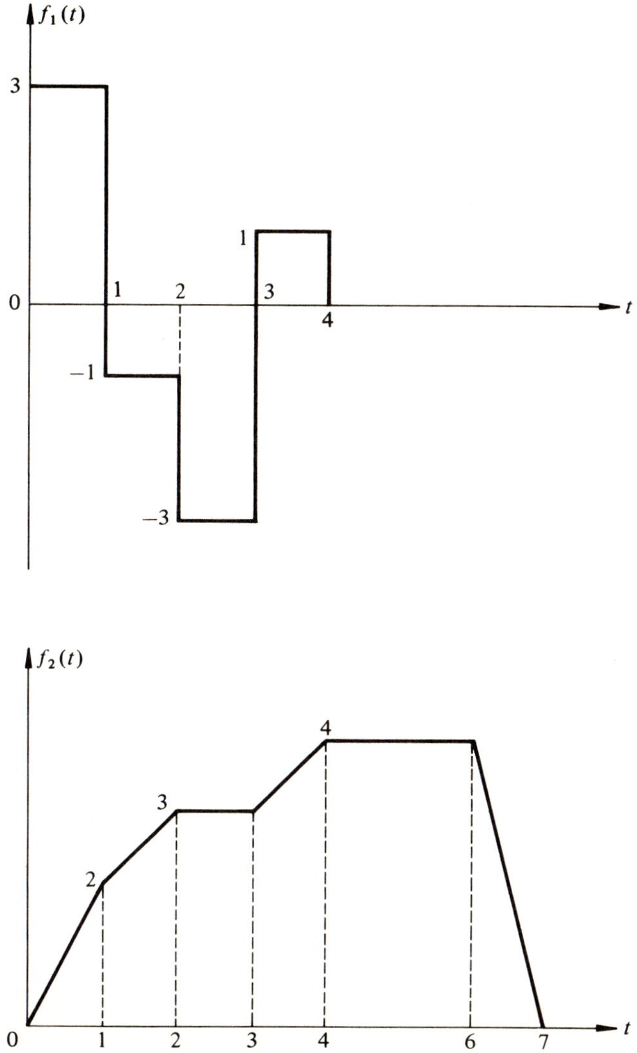

Fig. 3.12 Excitation and response corresponding to the impulse response of (3.86).

or, in the frequency domain,

$$V_2(s) = V_1^2(s) \tag{3.88}$$

Then

$$\bar{h}(t) = v_1(t) \tag{3.89}$$

An example of such a pair of functions is

$$v_1(t) = u_{-1}(t) - u_{-1}(t - \tau) \tag{3.90}$$

This $v_1(t)$ is a rectangular pulse which is convenient to use in approximating a given time function by a series of pulses. The corresponding $v_2(t)$ is the $\Delta(t)$ of Fig. 3.8, which is also very convenient to use. The resulting $h(t)$ is then in the form of another train of rectangular pulses.

3.8 CONCLUSIONS

This chapter is largely concerned with the problem of determining either the impulse response $h(t)$ or an approximate impulse response $\hat{h}(t)$ or $h^*(t)$ of a network when a corresponding excitation-and-response pair is prescribed. Referring to Fig. 1.2, we have been concerned in going from the boxes in the first column to those in the second column, not necessary on the same row. Thus we have taken the first step in going from a given pair of $f_1(t)$ and $f_2(t)$ toward a realizable $H^*(s)$.

All procedures in this chapter may be regarded as methods of deconvolution. Of course, when an approximate $\hat{h}(t)$ or $h^*(t)$ is obtained, the deconvolution procedure only gives an approximate answer. Also many numerical deconvolution procedures exist [PA2]. We have only covered those techniques that are especially useful to a time-domain network problem.

Collaterally, some of the methods developed in subsequent chapters that will enable us to find an $\hat{H}(s)$ or $H^*(s)$ for a given pair of $f_1(t)$ and $f_2(t)$ may also be regarded as deconvolution procedures, because the corresponding $\hat{h}(t)$ or $h^*(t)$ may be found readily, either by tables or numerically, by the inverse Laplace transform.

Some of the necessary conditions for a pair of $f_1(t)$ and $f_2(t)$ to be realizable by a linear lumped network were discussed in Section 3.1. A method for testing the absolute integrability of an $h(t)$ or its approximation is given in Section 3.5.

PROBLEMS

3.1 For the input and output functions shown in Fig. P3.1, assume the poles of $H^*(s)$ to be located at -6 and $-5 \pm j5$. For the given $f_1(t)$, compute an $H^*(s)$ such that its corresponding $h^*(t)$ renders an output time function equal to the given $f_2(t)$ at $t = 0.2$, 0.8, and 1.5.

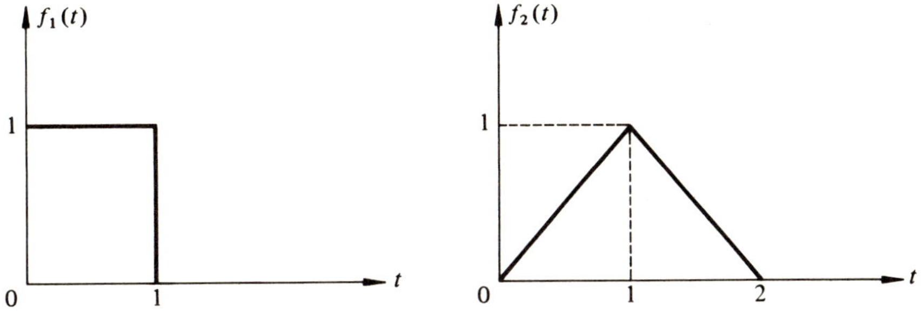

Fig. P3.1

3.2 Repeat Problem 3.1 for the set of assumed poles at -6, $-5 \pm j3$, and $-4 \pm j6$ and matched $f_2(t)$ points at $t = 0.2, 0.5, 0.9, 1.25$, and 1.8.

3.3 Repeat Problem 3.1 by using the least-mean-square-error criterion instead of point matching. Choose $t_k = 0.05, 0.15, 0.25, \ldots, 1.95$.

3.4 Repeat Problem 3.2 by using the least-mean-square-error criterion at $t_k = 0.05, 0.15, 0.25, \ldots, 1.95$.

3.5 Modify the locations of the poles given in Problem 3.1 by the method outlined in Section 3.3.

3.6 Repeat Problem 3.5 for the poles of Problem 3.2.

3.7 For the $f_1(t)$ and $f_2(t)$ of Problem 3.1, use two impulse trains to approximate these functions. Obtain an approximate impulse response by long division. Let $\tau = 0.2$. Plot your answer and compare it with the exact $h(t)$.

3.8 Let

$$f_1(t) = u_{-1}(t) - u_{-1}(t-1)$$
$$f_2(t) = (1 - \epsilon^{-2t})u_{-1}(t) - [1 - \epsilon^{-2(t-1)}]u_{-1}(t-1)$$

Using a series of step functions to represent $f_1(t)$ and a series of ramp functions to represent $f_2(t)$, obtain approximate $h(t)$ for (a) $\tau = 0.1$, (b) $\tau = 0.2$, and (c) $\tau = 0.5$. Plot all answers and compare them with the exact $h(t)$.

3.9 Convolve the $h(t)$ of Fig. 3.2 and the $f_1(t)$ of Fig. 3.3 by interpreting the product of two number sequences, each representing a sequence of singularity functions.

3.10 Using $u_{-1}(t)$ for $f_1(t)$ and $\Delta(t)$ for $f_2(t)$ of Fig. 3.3, obtain a number sequence for each function. Divide the number sequence of $f_2(t)$ by that of $f_1(t)$ to obtain the number sequences for $h(t)$. What elementary function does the number sequence of $h(t)$ imply? How can the coefficients of (3.49) be obtained from this sequence?

3.11 For each of the time functions in Fig. P3-2, determine whether its Laplace transform has right-halfplane zeros.

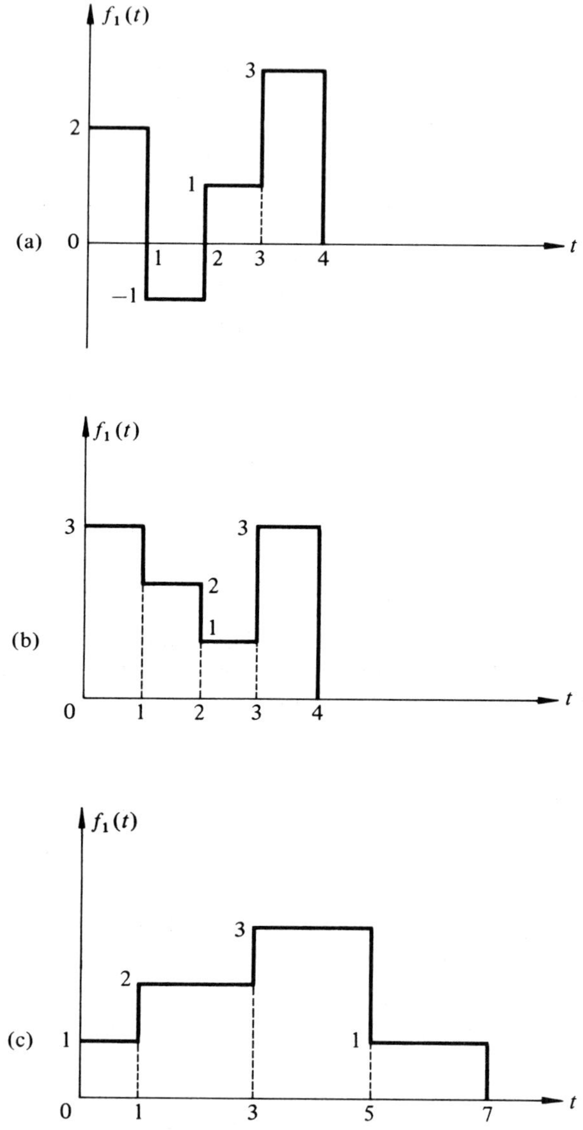

Fig. P3.2

4

Determination or Estimation of Pole Locations of a Network Function

In a synthesis problem in which either $h(t)$ is given or an $\hat{h}(t)$ has been determined by one of the methods given in Chapter 3, it would be very helpful if the number and the locations of poles of a network function can be determined or estimated. Once this is done, the balance of the network function can usually be found much more easily than when this information is not available.

In this chapter several methods are presented by which a set of poles for $H^*(s)$ can be determined for a known $h(t)$ or $\hat{h}(t)$. Because of the different nature of different methods, the necessary information on $h(t)$ may vary from one method to the other. The required information may be either that $h(t)$ is known analytically or at distinct points $t = t_k$, $k = 0, 1, 2, \ldots$. If the necessary form is $h(t_k)$ and the given information is the analytic expression of either $h(t)$ or $\hat{h}(t)$, these values can simply be calculated. Conversely, if only the $\hat{h}(t_k)$ are known and an analytic expression is required, then an application of the analytic substitution may be necessary to determine an $\hat{h}(t)$ as an approximate representation of $h(t)$ before preceeding with the pole determination.

4.1 ESTIMATION OF POLE LOCATIONS BY FREQUENCY-DOMAIN APPROXIMATION

For a given $h(t)$, a quick and straightforward way of determining the poles of its corresponding $H^*(s)$ is to first obtain an analytical approximation of $h(t)$. We can denote this approximating function by $\hat{h}(t)$, which is Laplace transformable but need not be realizable. Then the Laplace transform of $\hat{h}(t)$ can be found [call it $\hat{H}(s)$]. Once this transform has been obtained, any of the several frequency-domain approximation methods can be used to find a rational function $\hat{H}^*(s)$ to approximate $\hat{H}(s)$. The poles of $\hat{H}^*(s)$ can now be used as a first approximation of the poles of $H^*(s)$.

For instance, we may take the expression of $\hat{H}(s)$ and expand it into a Taylor's series about the origin or infinity,

$$\hat{H}(s) = a_0 + a_1 s + a_2 s^2 + \dots \tag{4.1}$$

$$= \frac{b_1}{s} + \frac{b_2}{s^2} + \frac{b_3}{s^3} + \dots \tag{4.2}$$

Then a rational function $\hat{H}^*(s)$ can be found to approximate this $\hat{H}(s)$. One way to accomplish this approximation is to use a Padé approximant in the variable s or $1/s$.

As an example, suppose it is desired to find a set of poles for $H^*(s)$ to approximately realize the impulse response of Fig. 4.1, for which we have

$$h(t) = 0.5u_{-1}(t) - 0.2u_{-2}(t) + 0.2u_{-2}(t - 2.5) \tag{4.3}$$

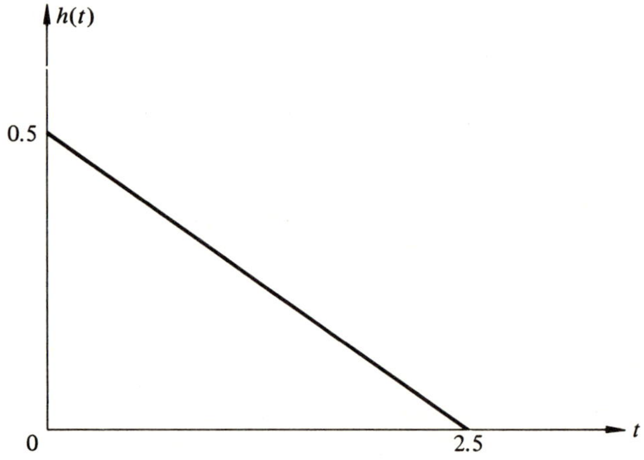

Fig. 4.1 Impulse response (4.3).

This $h(t)$ is Laplace transformable, so finding an $\hat{h}(t)$ is not necessary. Here

$$H(s) = \frac{0.5}{s} - \frac{0.2}{s^2} + \frac{0.2}{s^2}\epsilon^{-2.5s} \tag{4.4}$$

Replacing $\epsilon^{-2.5s}$ by its Taylor's expansion about the origin, we obtain

$$H(s) = 0.625 - 0.52083s + 0.32552s^2 - 0.16276s^3 + 0.06782s^4 - 0.02422s^5 + \dots \tag{4.5}$$

Now suppose a 3-pole $\hat{H}^*(s)$ is desired; we may use the Padé approximant (2, 3) of the series in (4.5). Since only poles of $\hat{H}^*(s)$ are of interest, we only need compute $Q_{23}(s)$. From (2.58),

$$Q_{23}(s) = \begin{vmatrix} 1 & s & s^2 & s^3 \\ -0.16276 & 0.32552 & -0.52083 & 0.625 \\ 0.06782 & -0.16276 & 0.32552 & -0.52083 \\ -0.02422 & 0.06782 & -0.16276 & 0.32552 \end{vmatrix}$$

$$= (s^3 + 6s^2 + 14.4s + 13.44) \times 3.4219 \times 10^{-6} \tag{4.6}$$

the zeros of which yield the approximate pole locations at

$$s_1 = -2.2595$$
$$s_{2,3} = -1.8703 \pm j1.5653$$

A question may be raised at this point. Why did we use the (2, 3) approximant when a three-pole $H^*(s)$ is assumed? The reason is that only this approximant will give a finite $h^*(t)$ at $t = 0$. Since

$$h^*(0) = \lim_{s\to\infty} sH^*(s)$$

it is seen that we must have $m = n - 1$ for $h^*(0)$ to be finite. Since the series in (4.5) for $H(s)$ is the Taylor expansion about the origin, the truncated series is only a good approximation in the vicinity of $s = 0$. Since the behavior of $\hat{H}^*(s)$ near $s = 0$ determines largely the behavior of $\hat{h}^*(t)$ for large values of t, it is not surprising that we cannot use just any arbitrary approximant to approximate $h(t)$ by $\hat{h}^*(t)$ for small values of t through this series.

Collaterally, if $h(0) = 0$ we may say that

$$h'(0) = \lim_{s\to\infty} s^2 H(s) \tag{4.7}$$

Hence, if $h'(0) \neq 0$, we must make $m = n - 2$. Similarly, if $h(0) = 0$, $h'(0) = 0, \ldots, h^{(k-1)}(0) = 0$, but $h^{(k)}(0) \neq 0$, we have

$$h^{(k)}(0) = \lim_{s\to\infty} s^{(k+1)} H(s) \tag{4.8}$$

and we must choose $m = n - k - 1$.

Probably the easiest way to obtain a Laplace-transformable $\hat{h}(t)$ for a given $h(t)$ is to use a series of singularity functions similar to the scheme used in Section 3.4. For example, if the function of Fig. 4.1 is approximated by a series of step functions as shown in Fig. 4.2, we have

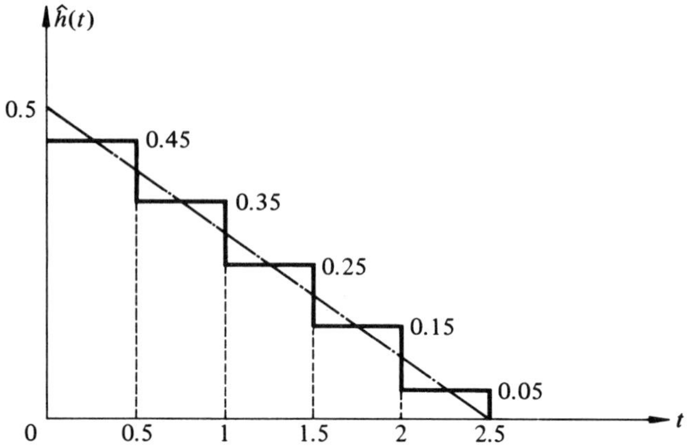

Fig. 4.2 The staircase approximation of the $h(t)$ of Fig. 4.1.

$$\begin{aligned}\hat{h}(t) = 0.45u(t) - 0.1u(t-0.5) - 0.1u(t-1) - 0.1u(t-1.5)\\ - 0.1u(t-2) - 0.05u(t-2.5)\end{aligned} \tag{4.9}$$

and

$$\hat{H}(s) = \frac{1}{s}(0.45 - 0.1\epsilon^{-0.5s} - 0.1\epsilon^{-s} - 0.1\epsilon^{-1.5s} - 0.1\epsilon^{-2s} - 0.05\epsilon^{-2.5s}) \tag{4.10}$$

Now each exponential term may be represented by its Taylor's series. When the same power terms are grouped, we get

$$\hat{H}(s) = 0.525 - 0.5313s + 0.3385s^2 - 0.1736s^3 + \dots \tag{4.11}$$

This series is seen to be very close to the one obtained from the exact $h(t)$, (4.5).

From the algebra of the foregoing example, it is easily seen that the same scheme will work equally well if the spacing and types of singularities used are not uniform throughout the interval during which $h(t)$ is not negligible. For example, suppose we are given

$$h(t) = \frac{1}{1 + \epsilon^t} \tag{4.12}$$

This function is plotted in Fig. 4.3 as the dashed curve. We then may approximate this curve by an $\hat{h}(t)$ represented by the series of ramp and step functions as shown in the solid curve of Fig. 4.3. This $\hat{h}(t)$ may then be written as

$$\hat{h}(t) = 0.5u_{-1}(t) - 0.2167u_{-2}(t) + 0.1310u_{-2}(t - 1.5) + 0.08571u_{-2}(t - 3.25)$$
$$- 0.015u_{-1}(t - 4.25) - 0.0075u_{-1}(t - 5.5)$$
$$- 0.0025u_{-1}(t - 7) \tag{4.13}$$

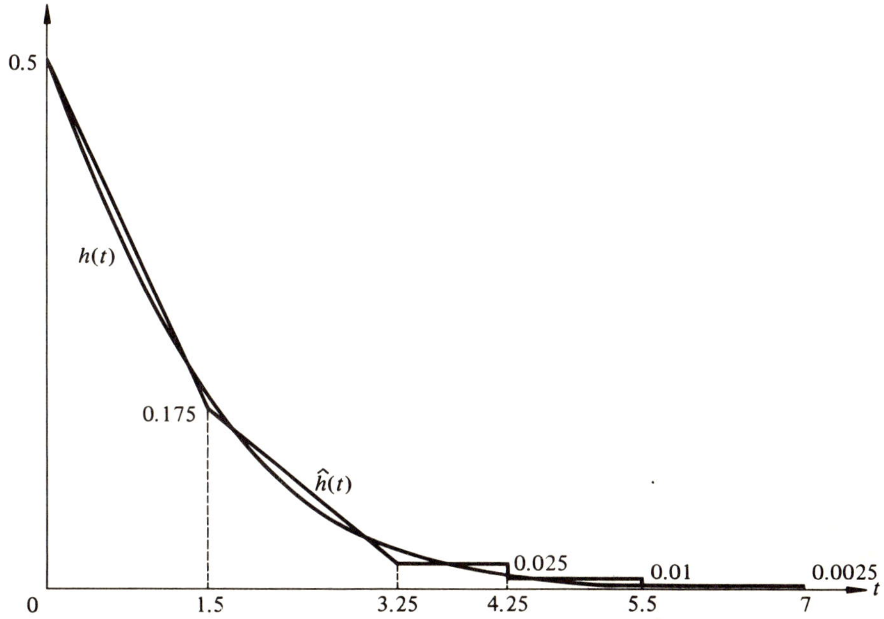

Fig. 4.3 The $h(t)$ of (4.12) and the corresponding $\hat{h}(t)$ of (4.13).

for which

$$\hat{H}(s) = \frac{0.5}{s} - \frac{0.2167}{s^2} + \frac{0.1310}{s^2}\epsilon^{-1.5s} + \frac{0.08571}{s^2}\epsilon^{-3.25s}$$
$$- \frac{0.015}{s}\epsilon^{-4.25s} - \frac{0.0075}{s}\epsilon^{-5.5s} - \frac{0.0025}{s}\epsilon^{-7s} \tag{4.14}$$

After substituting the Taylor's series for each exponential function, we obtain

$$\hat{H}(s) = 0.7225 - 0.8742s + 0.9689s^2 - 1.0073s^3 + 0.9804s^4 - \ldots \tag{4.15}$$

The appropriate Padé approximant for this series can then be found and its poles calculated.

Although the Laplace transform $H(s)$ of (4.12) is not known in closed form, its power series can be found. Since

$$H(s) = \int_0^\infty \frac{\epsilon^{-st}}{1+\epsilon^t}\,dt \tag{4.16}$$

we may replace ϵ^{-st} by its series expansion and then perform the integration term by term. Thus

$$\begin{aligned} H(s) &= \int_0^\infty \frac{dt}{1+\epsilon^t} - s\int_0^\infty \frac{t\,dt}{1+\epsilon^t} + \frac{s^2}{2!}\int_0^\infty \frac{t^2\,dt}{1+\epsilon^t} - \frac{s^3}{3!}\int_0^\infty \frac{t^3\,dt}{1+\epsilon^t} + \dots \\ &= 0.6932 - 0.8225s + 0.9016s^2 - 0.9470s^3 + \dots \end{aligned} \tag{4.17}$$

which agrees approximately with the series obtained from the singularity function approximation, (4.15).

4.2 DETERMINATION OF POLE LOCATIONS BY TIME-DOMAIN APPROXIMATION

The frequency-domain approximation method of determining the approximate locations of poles of $H^*(s)$ for a given $\hat{h}(t)$ provides a quick and simple means of obtaining a first-order estimation of these poles. Usually, poles obtained this way cannot be expected to be very accurate. This is basically because a good frequency-domain approximation does not always lead to a good time-domain approximation.

When more accurate pole locations are desired, the determination must be done in the time domain. In the time domain, the problem becomes one of finding the best group of numbers s_ν such that the time function

$$h^*(t) = \sum_{\nu=1}^{N} A_\nu \epsilon^{s_\nu t} \tag{4.18}$$

best fits a given $h(t)$ or $\hat{h}(t)$ under certain criteria, where s_ν may be either real or in conjugate pairs. This problem is basically the same one investigated by Prony in the nineteenth century [HA1].

The relationship between the poles of a network function and the characteristic differential equation of the network should be familiar to electrical and systems engineers. That is, the transient response $f(t)$ of a network with a characteristic equation

$$a_N s^N + a_{N-1}s^{N-1} + \dots + a_1 s + a_0 s = 0 \tag{4.19}$$

must satisfy the differential equation

$$\sum_{k=0}^{N} a_k \frac{d^k}{dt^k} f(t) = 0 \tag{4.20}$$

Since the impulse response of a network is, in reality, a transient response, $h(t)$ must satisfy,

$$\sum_{k=0}^{N} a_k \frac{d^k}{dt^k} h(t) = 0 \tag{4.21}$$

for $t > 0$. Of course, (4.21) can only be satisfied if $h(t)$ is the correct, exact representation of an impulse response. When $h^*(t)$ is used in place of $h(t)$, (4.21) is only approximately satisfied. The different methods in the succeeding sections are based on different schemes of approximating (4.21).

4.3 POINT-MATCHING TECHNIQUE FOR THE DETERMINATION OF POLES

The first method by which (4.21) can be approximately satisfied by an approximate impulse response $h^*(t)$ is to force (4.21) to be satisfied at equally spaced points. Before devising a procedure to accomplish this, it is expedient to delineate a particular relationship between the poles of $H(s)$ [or coefficients in the exponents of $h^*(t)$] and the derivatives of $h^*(t)$ at the origin.

Let us suppose the impulse response is

$$h^*(t) = \sum_{\nu=1}^{N} A_\nu \epsilon^{s_\nu t} \tag{4.22}$$

and its Laplace transform is

$$\begin{aligned} H^*(s) &= \frac{B(s)}{s^N + a_{N-1}s^{N-1} + a_{N-2}s^{N-2} + \dots + a_2 s^2 + a_1 s + a_0} \\ &= \frac{B(s)}{(s - s_1)(s - s_2)\dots(s - s_N)} \end{aligned} \tag{4.23}$$

in which

$$\begin{aligned} a_0 &= -(s_1 + s_2 + \dots + s_N) \\ a_1 &= s_1 s_2 + s_1 s_3 + \dots + s_{N-1} s_N \\ &\vdots \\ a_{N-1} &= (-1)^N s_1 s_2 \dots s_N \end{aligned} \tag{4.24}$$

From (4.21), with $a_N = 1$, it is known that $h(t)$ must satisfy

$$h^{(N)}(t) + a_{N-1} h^{(N-1)}(t) + \dots + a_1 h'(t) + a_0 h(t) = 0 \tag{4.25}$$

Differentiating successively and substituting $t = 0$ we obtain

$$h^{(N)}(0) + a_{N-1}h^{(N-1)}(0) + \dots + a_1 h'(0) + a_0 h(0) = 0$$
$$h^{(N+1)}(0) + a_{N-1}h^{(N)}(0) + \dots + a_1 h''(0) + a_0 h'(0) = 0 \quad \textbf{(4.26)}$$
$$\cdots\cdots\cdots\cdots\cdots\cdots\cdots\cdots$$
$$h^{(2N-1)}(0) + a_{N-1}h^{(2N-2)}(0) + \dots + a_1 h^{(N)}(0) + a_0 h^{(N-1)}(0) = 0$$

Thus if all $h^{(k)}(0)$ $[k = 0, 1, 2, \dots, 2N - 1]$ are known of a function, $a_0, a_1, \dots, a_{N-1}$ can be found by solving the set of simultaneous equations (4.26). Hence (4.26) may be regarded as the link between the poles of $H(s)$ (which are determinable once a's are known) and the initial derivatives of $h(t)$.

Although equations (4.26) are theoretically sound, they are difficult to implement in practice. The high derivatives of a given function is rarely known for a given $h(t)$ except in situations in which $h(t)$ is given analytically by a single function for the entire time region $0 < t < \infty$. However, since

$$h^*(t) = \sum_{\nu=1}^{N} A_\nu \epsilon^{s_\nu t}$$
$$h^{*\prime}(t) = \sum_{\nu=1}^{N} A_\nu s_\nu \epsilon^{s_\nu t} \quad \textbf{(4.27)}$$
$$\cdots\cdots\cdots\cdots$$
$$h^{*(k)}(t) = \sum_{\nu=1}^{N} A_\nu s_\nu^k \epsilon^{s_\nu t}$$
$$\vdots$$

we have

$$h^*(0) = \sum_{\nu=1}^{N} A_\nu$$
$$h^{*\prime}(0) = \sum_{\nu=1}^{N} A_\nu s_\nu \quad \textbf{(4.28)}$$
$$\cdots\cdots\cdots\cdots$$
$$h^{*(k)}(0) = \sum_{\nu=1}^{N} A_\nu s_\nu^k$$
$$\vdots$$

Hence coefficients in (4.26) may be replaced by the summations given by (4.28). These relationships are restricted to the case in which only simple poles are present. If multiple poles are present, appropriate modifications must be made to obtain a similar set of relationships.

Now let us return to the problem of finding a set of s_ν such that the $h^*(t)$ of (4.22) may be forced to pass through a set of points of the known $h(t)$. Suppose the given $h(t)$ is known at $2N$ equally spaced points starting with the origin along the t axis. Let the spacing of the points be Δt. Thus $h(0)$, $h(\Delta t)$,

$h(2\,\Delta t), \ldots, h([2N-1]\,\Delta t)$ are given for the particular impulse response. We let

$$w_\nu = \epsilon^{s_\nu \Delta t} \tag{4.29}$$

Then we have

$$\begin{aligned} h^*(0) &= \sum_{\nu=1}^{N} A_\nu \\ h^*(\Delta t) &= \sum_{\nu=1}^{N} A_\nu w_\nu \\ h^*(2\,\Delta t) &= \sum_{\nu=1}^{N} A_\nu w_\nu^2 \\ &\;\vdots \\ h^*([2N-1]\,\Delta t) &= \sum_{\nu=1}^{N} A_\nu w_\nu^{2N-1} \end{aligned} \tag{4.30}$$

A comparison between (4.28) and (4.30) shows the parallelism of the roles of w_ν's and s_ν's in these equations. Since s_ν's satisfy

$$s^N + a_{N-1}s^{N-1} + a_{N-2}s^{N-2} + \ldots + a_1 s + a_0 = 0 \tag{4.31}$$

where a's satisfy (4.26), we may, by analogy, conclude that w_ν's must satisfy

$$w^N + a_{N-1}w^{N-1} + a_{N-2}w^{N-2} + \ldots + a_1 w + a_0 = 0 \tag{4.32}$$

in which a_i's satisfy

$$\begin{aligned} h^*(N\,\Delta t) + a_{N-1}h^*([N-1]\,\Delta t) + \ldots + a_1 h^*(\Delta t) + a_0 h^*(0) &= 0 \\ h^*([N+1]\,\Delta t) + a_{N-1}h^*(N\,\Delta t) + \ldots + a_1 h^*(2\,\Delta t) + a_0 h^*(\Delta t) &= 0 \\ \cdots\cdots\cdots\cdots\cdots\cdots\cdots\cdots\cdots\cdots\cdots\cdots \\ h^*([2N-1]\Delta t) + a_{N-1}h^*([2N-2]\Delta t) + \ldots + a_1 h^*(N\Delta t) + a_0 h^*([N-1]\Delta t) &= 0 \end{aligned} \tag{4.33}$$

which is parallel to (4.26). Now if we let $h^*(k\,\Delta t) = h(k\,\Delta t)$, $k = 0, 1, 2, \ldots, 2N-1$, (4.33) may be solved for $a_{N-1}, a_{N-2}, \ldots, a_2, a_1$, and a_0. Then the values of w_ν may be found by solving (4.32). Once w_ν are known, the poles of $H^*(s)$ are given by

$$s_\nu = \frac{1}{\Delta t} \ln w_\nu \tag{4.34}$$

It should be noted that the analogy between (4.31) and (4.32), and between (4.26) and (4.33), is only in mathematical forms. Generally, for the same $h(t)$; $h'(0), h''(0), \ldots, h^{(2N-1)}(0)$ will be different from $h(\Delta t), h(2\,\Delta t), \ldots,$

$h([2N-1]\,\Delta t)$. Hence the a's of (4.31) and those of (4.32) are different. However, as a practical matter, we may expect the two sets of poles to be fairly close to each other. On the other hand, if $h(0), h(\Delta t), h(2\,\Delta t), \ldots, h([2N-1]\,\Delta t)$, of one problem and $h(0), h'(0), h''(0), \ldots, h^{(2N-1)}(0)$ of another problem happen to be identical, then their a's will necessarily be the same. In this event, their poles are different but related by (4.34).

Example. Suppose it is desired to find a set of three poles for an $H^*(s)$ such that its inverse Laplace transform approximates

$$h(t) = \frac{1}{t+1} \tag{4.35}$$

We arbitrarily chose $\Delta t = 0.8$. Then (4.33) gives

$$0.29412 + 0.38462a_2 + 0.55556a_1 + 1.00000a_0 = 0$$
$$0.23810 + 0.29412a_2 + 0.38462a_1 + 0.55556a_0 = 0$$
$$0.20000 + 0.23810a_2 + 0.29412a_1 + 0.38462a_0 = 0$$

Solution of these equations gives

$$a_2 = -1.56080, \qquad a_1 = 0.66857, \qquad a_0 = -0.06555$$

Hence w_ν are given by the roots of

$$w^3 - 1.56000w^2 + 0.66857w - 0.06555 = 0$$

whence

$$w_1 = 0.89448$$
$$w_2 = 0.52628$$
$$w_3 = 0.13924$$

and

$$s_1 = -0.13940$$
$$s_2 = -0.80239$$
$$s_3 = -2.46445$$

Thus it is possible to make

$$h^*(t) = A_1\epsilon^{-0.13940t} + A_2\epsilon^{-0.80239t} + A_3\epsilon^{-2.46445} \tag{4.36}$$

coincide with $h(t)$ at $t = 0, 0.8, 1.6, 2.4, 3.2, 4.0$.

It is seen from this numerical example that the point-matching technique usually entails the solution of a set of simultaneous equations. When good accuracy and, therefore, a large number of points are involved, the solution of these equations might be quite laborious. However, with the wide availability of electronic computers, this need not be a major difficulty.

The selection of the number of poles, N, and the spacing Δt is a matter of engineering judgment. Usually, there is no way of predicting the accuracy of these poles in advance. In some sense, the selection of these parameters is a compromise between a good fit with large N and/or a small Δt and a poor fit with a small N and/or a large Δt.

The foregoing development can be viewed equally well through the principles of the difference equation. Equation (4.33) can be considered as a partial statement of a difference equation. If we let

$$h_k = h(k\,\Delta t) \tag{4.37}$$

(4.33) becomes

$$h_{i+N} + a_{N-1}h_{i+N-1} + \ldots + a_2 h_{i+2} + a_1 h_{i+1} + a_0 h_i = 0 \tag{4.38}$$

where $i = 0, 1, 2, \ldots, N - 1$. Leaving out the qualification on i makes (4.38) a homogeneous linear difference equation of order N.

If the difference operator E is used,

$$Eh_k = h_{k+1} \tag{4.39}$$

(4.38) becomes

$$(E^N + a_{N-1}E^{N-1} + \ldots + a_2 E^2 + a_1 E + a_0)h_i = 0 \tag{4.40}$$

A usual technique to solve a difference equation of the type of (4.40) is to assume that

$$h_\nu = w^\nu = (\epsilon^{s\Delta t})^\nu \tag{4.41}$$

Then any w that satisfies

$$w^N + a_{N-1}w^{N-1} + \ldots + a_2 w^2 + a_1 w + a_0 = 0 \tag{4.42}$$

is a solution to (4.38). Each value of w_ν, in turn, gives a corresponding

$$s_\nu = \frac{1}{\Delta t}\ln(w_\nu) \tag{4.43}$$

Equation (4.43), of course, is identical to (4.34). The general solution of the difference equation is any linear combination of all possible terms of the type of (4.41) which satisfy (4.42). Hence

$$h_i = h(i\,\Delta t) = \sum_{\nu=1}^{N} A_\nu \epsilon^{s_\nu i \Delta t} \tag{4.44}$$

Replacing $i\Delta t$ by t in (4.44) converts the discrete form of $h(t)$ to its continuous version.

4.4 POLE DETERMINATION BY THE LEAST-MEAN-SQUARE-ERROR CRITERION

The method in Section 4.3 enables us to find a set of N poles such that its corresponding $h^*(t)$ *can be* made to pass through the $2N$ points on $h(t)$ exactly. The result of this strategy is that unless the number of points used is very large, which is uneconomical from engineering standpoint, the final approximation of $h^*(t)$ to $h(t)$ cannot be expected to be very good except either over a very narrow region or at isolated points. In many occasions it is better to obtain a fairly good approximation over a broad range of time so that the general wave-shape of $h^*(t)$ will resemble $h(t)$ in the entire range $0 < t < \infty$.

The following method, which is based on the least-mean-square-error criterion, will accommodate this consideration and minimize the total error over as many equally spaced points as desired. Let the number of poles be N and the number of equally spaced points chosen be M, $M > 2N$.

Suppose the M points of $h(t)$ are denoted by

$$h_i = h(i\,\Delta t), \qquad i = 0, 1, 2, \ldots, M \tag{4.45}$$

Since $M > 2N$, the equation

$$h_{i+N} + a_{N-1}h_{i+N-1} + a_{N-2}h_{i+N-2} + \ldots + a_2 h_{i+2} + a_1 h_{i+1} + a_0 h_i = 0 \tag{4.46}$$

cannot always be satisfied exactly by all values of i. One way to satisfy (4.46) approximately is to require that a set of coefficient a's be found such that the total deviation between all possible h_{i+N} computed from (4.46) from the set of given h_i and the corresponding given h_{i+N} is minimum. In other words, we denote the computed values based on (4.46) by

$$\hat{h}_j = -a_{N-1}h_{j-1} - a_{N-2}h_{j-2} - \ldots - a_1 h_{j-N-1} - a_0 h_{j-N}, \quad j = N, N+1, \ldots, M \tag{4.47}$$

For each j the error is

$$\epsilon_j = |h_j - \hat{h}_j| \tag{4.48}$$

The total squared error is

$$E_2 = \sum_{j=N}^{M} \epsilon_j^2$$
$$= \sum_{j=N}^{M} (h_j + a_{N-1}h_{j-1} + a_{N-2}h_{j-2} + \ldots + a_k h_{k+j-N} + \ldots + a_1 h_{j-N+1} + a_0 h_{j-N})^2 \tag{4.49}$$

To minimize E_2, we set

$$\frac{\partial}{\partial a_k}\left[\sum_{j=N}^{M} \epsilon_j^2\right] = 0, \qquad k = 0, 1, \ldots, N-1 \tag{4.50}$$

or

$$\sum_{j=N}^{M} [h_{k+j-N}(h_j + a_{N-1}h_{j-1} + a_{N-2}h_{j-2} + \ldots + a_1 h_{i-N+1} + a_0 h_{j-N})] = 0,$$
$$k = 0, 1, \ldots, N-1 \tag{4.51}$$

Equation (4.51) may be alternatively expressed in matrix form

$$\begin{bmatrix} \sum_{j=N}^{M} h_{j-1}^2 & \sum_{j=N}^{M} h_{j-2}h_{j-1} & \cdots \sum_{j=N}^{M} h_{j-N}h_{j-1} \\ \sum_{j=N}^{M} h_{j-1}h_{j-2} & \sum_{j=N}^{M} h_{j-2}^2 & \cdots \sum_{j=N}^{M} h_{j-N}h_{j-2} \\ \cdots & \cdots & \cdots \\ \sum_{j=N}^{M} h_{j-1}h_{j-N+1} & \sum_{j=N}^{M} h_{j-2}h_{j-N+1} & \cdots \sum_{j=N}^{M} h_{j-N}h_{j-N+1} \\ \sum_{j=N}^{M} h_{j-1}h_{j-N} & \sum_{j=N}^{M} h_{j-2}h_{j-N} & \cdots \sum_{j=N}^{M} h_{j-N}^2 \end{bmatrix} \begin{bmatrix} a_{N-1} \\ a_{N-2} \\ \vdots \\ a_1 \\ a_0 \end{bmatrix} = \begin{bmatrix} -\sum_{j=N}^{M} h_j h_{j-1} \\ -\sum_{j=N}^{M} h_j h_{j-2} \\ \vdots \\ -\sum_{j=N}^{M} h_j h_{j-N+1} \\ -\sum_{j=N}^{M} h_j h_{j-N} \end{bmatrix} \tag{4.52}$$

Solution of (4.52) for a's gives the coefficients of an approximate difference equation. When these coefficients are entered into (4.42), the roots of that equation and (4.34) then give the locations of the estimated poles.

When $M = 2N - 1$, this method should render $E_2 = 0$ and the resulting estimated poles should be identical to those obtained by point-matching scheme of Section 4.3. In other words, if $M = 2N - 1$, then the solutions of (4.33) and (4.52) can be expected to be identical. In that case, (4.33) is much preferred, because it is much simpler.

Example. Using the six points of $h(i\,\Delta t)$ of the example in Section 4.3, find the three approximate poles of $H^*(s)$ according to the least-squares criterion.

Solution. We have $N = 3$, $M = 5$, and

$$h_0 = 1.00000$$
$$h_1 = 0.55556$$
$$h_2 = 0.38462$$
$$h_3 = 0.29412$$
$$h_4 = 0.23810$$
$$h_5 = 0.20000$$

Substituting these values into (4.52) gives

$$\begin{bmatrix} 0.29113 & 0.39683 & 0.63960 \\ 0.39683 & 0.54309 & 0.88236 \\ 0.63960 & 0.88236 & 1.45658 \end{bmatrix} \begin{bmatrix} a_2 \\ a_1 \\ a_0 \end{bmatrix} = \begin{bmatrix} -0.23077 \\ -0.31380 \\ -0.50332 \end{bmatrix}$$

Solution of this matrix equation yields

$$a_2 = -1.5608, \qquad a_1 = 0.6686, \qquad a_0 = -0.0656$$

which are identical to those values obtained in Section 4.3.

4.5 DIFFERENTIAL-EQUATION FORMULATION OF THE LEAST-SQUARES CRITERION

The development in Section 4.4 is based on applying the least-squares criterion to a series of equally spaced points related by a difference equation. Analogous developments that are of considerable theoretical interest are the applications of the same criterion to the functions $h(t)$ and $h^*(t)$ when they are given or assumed to be in analytical form and are therefore continuous.

The first development is based on the assumption that sufficient number of derivatives of the given $h(t)$ are known at a point which may or may not be the origin. As was stated in Section 4.2, if

$$h^*(t) = \sum_{\nu=1}^{N} A_\nu \epsilon^{s_\nu t} \tag{4.53}$$

is the impulse response of a network, then $h^*(t)$ is a solution to the differential equation

$$\begin{aligned} h^{*(i+N)}(t) + a_{N-1}h^{*(i+N-1)}(t) + a_{N-2}h^{*(i+N-2)}(t) + \dots \\ + a_1 h^{*(i+1)}(t) + a_0 h^{*(i)}(t) = 0 \end{aligned} \tag{4.54}$$

and s_ν are the solution to the equation

$$s^N + a_{N-1}s^{N-1} + a_{N-2}s^{N-2} + \dots + a_1 s + a_0 = 0 \tag{4.55}$$

Now suppose the derivatives of $h(t)$ are known through the Mth one at $t = t_0$ and $M > 2N - 1$. The errors based on the repeated application of (4.54) and the known derivatives can first be computed. That is,

$$h^{*(j)}(t_0) = -a_{N-1}h^{(j-1)}(t_0) - a_{N-2}h^{(j-2)}(t_0) - \dots - a_1 h^{(j-N+1)}(t_0) - a_0 h^{(j-N)}(t_0) \tag{4.56}$$

and

$$\epsilon_j = h^{(j)}(t_0) - h^{*(j)}(t_0) \qquad j = N, N+1, \dots, M \tag{4.57}$$

The total squared error is

$$\begin{aligned} E_2 &= \sum_{j=N}^{M} \epsilon_j^2 \\ &= \sum_{j=N}^{M} [h^{(j)}(t_0) + a_{N-1}h^{(j-1)}(t_0) + a_{N-2}h^{(j-2)}(t_0) + \dots \\ &\qquad + a_1 h^{(j-N-1)}(t_0) + a_0 h^{(j-N)}(t_0)]^2 \end{aligned} \tag{4.58}$$

By setting

$$\frac{\partial E_2}{\partial a_k} = 0, \qquad k = 0, 1, \dots, N-1 \tag{4.59}$$

we may obtain an equation that is analogous to (4.52):

$$\begin{bmatrix} \sum\limits_{j=N}^{M} h^{(j-1)}(t_0)h^{(j-1)}(t_0) & \sum\limits_{j=N}^{M} h^{(j-2)}(t_0)h^{(j-1)}(t_0) & \cdots & \sum\limits_{j=N}^{M} h^{(j-N)}(t_0)h^{(j-1)}(t_0) \\ \sum\limits_{j=N}^{M} h^{(j-1)}(t_0)h^{(j-2)}(t_0) & \sum\limits_{j=N}^{M} h^{(j-2)}(t_0)h^{(j-2)}(t_0) & \cdots & \sum\limits_{j=N}^{M} h^{(j-N)}(t_0)h^{(j-2)}(t_0) \\ \cdots & \cdots & \cdots & \cdots \\ \sum\limits_{j=N}^{M} h^{(j-1)}(t_0)h^{(j-N)}(t_0) & \sum\limits_{j=N}^{M} h^{(j-2)}(t_0)h^{(j-N)}(t_0) & \cdots & \sum\limits_{j=N}^{M} h^{(j-N)}(t_0)h^{(j-N)}(t_0) \end{bmatrix} \begin{bmatrix} a_{N-1} \\ a_{N-2} \\ \vdots \\ a_1 \\ a_0 \end{bmatrix} = \begin{bmatrix} -\sum\limits_{j=N}^{M} h^{(j)}(t_0)h^{(j-1)}(t_0) \\ -\sum\limits_{j=N}^{M} h^{(j)}(t_0)h^{(j-2)}(t_0) \\ \vdots \\ -\sum\limits_{j=N}^{M} h^{(j)}(t_0)h^{(j-N)}(t_0) \end{bmatrix} \tag{4.60}$$

The a's may be obtained by solving (4.60). If $M = 2N - 1$, the solution becomes identical to the method given in Section 4.3, except for a shift of origin to $t = t_0$. In that case it would be easier to use (4.26) instead of (4.60).

The second development uses N derivatives over an arbitrary range $t_1 < t < t_2$, instead of M derivatives at $t = t_0$. The deviation between the computed Nth derivative is then minimized in the least-squares sense. Let $j = N$ in (4.56) and (4.57) to give

$$\begin{aligned}\epsilon(t) &= h^{(N)}(t) - h^{*(N)}(t) \\ &= h^{(N)}(t) + a_{N-1}h^{(N-1)}(t) + \dots + a_1 h^{(1)}(t) + a_0 h(t)\end{aligned} \tag{4.61}$$

and then minimize

$$\begin{aligned}E_2 &= \int_{t_1}^{t_2} \epsilon^2(t)\,dt \\ &= \int_{t_1}^{t_2} [h^{(N)}(t) + a_{N-1}h^{(N-1)}(t) + \dots + a_1 h^{(1)}(t) + a_0 h(t)]^2\,dt\end{aligned} \tag{4.62}$$

with respect to a's. Since

$$\begin{aligned}\frac{\partial E_2}{\partial a_k} &= 2\int_{t_1}^{t_2} [h^{(N)}(t) + a_{N-1}h^{(N-1)}(t) + \dots + a_1 h^{(1)}(t) + a_0 h(t)]h^{(k)}(t)\,dt, \\ k &= 0, 1, \dots, N-1\end{aligned} \tag{4.63}$$

Setting $\partial E_2/\partial a_k = 0$, $k = 0, 1, \dots, N-1$, and rewriting the equations in matrix form, we have

$$\begin{bmatrix} \int_{t_1}^{t_2} h^{(N-1)}(t)h^{(N-1)}(t)\,dt & \int_{t_1}^{t_2} h^{(N-2)}(t)h^{(N-1)}(t)\,dt \dots & \int_{t_1}^{t_2} h(t)h^{(N-1)}(t)\,dt \\ \int_{t_1}^{t_2} h^{(N-1)}(t)h^{(N-2)}(t)\,dt & \int_{t_1}^{t_2} h^{(N-2)}(t)h^{(N-2)}(t)\,dt \dots & \int_{t_1}^{t_2} h(t)h^{(N-2)}(t)\,dt \\ \cdots & \cdots & \cdots \\ \int_{t_1}^{t_2} h^{(N-1)}(t)h^{(1)}(t)\,dt & \int_{t_1}^{t_2} h^{(N-2)}(t)h^{(1)}(t)\,dt & \dots \int_{t_1}^{t_2} h(t)h^{(1)}(t)\,dt \\ \int_{t_1}^{t_2} h^{(N-1)}(t)h(t)\,dt & \int_{t_1}^{t_2} h^{(N-2)}(t)h(t)\,dt & \dots \int_{t_1}^{t_2} h(t)h(t)\,dt \end{bmatrix} \begin{bmatrix} a_{N-1} \\ a_{N-2} \\ \vdots \\ a_1 \\ a_0 \end{bmatrix} = \begin{bmatrix} -\int_{t_1}^{t_2} h^{(N)}(t)h^{(N-1)}(t)\,dt \\ -\int_{t_1}^{t_2} h^{(N)}(t)h^{(N-2)}(t)\,dt \\ \vdots \\ -\int_{t_1}^{t_2} h^{(N)}(t)h^{(1)}(t)\,dt \\ -\int_{t_1}^{t_2} h^{(N)}(t)h(t)\,dt \end{bmatrix} \tag{4.64}$$

The solution of (4.64) gives the a's, which, in turn, will readily yield the poles of $H^*(s)$.

The following mathematical artifice, which makes use of the orthogonal functions, provides an alternative to solving (4.64) and enables us to obtain all sets of poles from $N = 1$ up to any number in a systematic fashion [KA4].

We first assume a series of functions

$$\begin{aligned}
\phi_0(t) &= h(t) \\
\phi_1(t) &= B_{10}\phi_0(t) + h'(t) \\
\phi_2(t) &= B_{20}\phi_0(t) + B_{21}\phi_1(t) + h''(t) \\
&\;\vdots \\
\phi_i(t) &= B_{i0}\phi_0(t) + B_{i1}\phi_1(t) + B_{i2}\phi_2(t) + \ldots + B_{i,i-1}\phi_{i-1}(t) + h^{(i)}(t)
\end{aligned} \tag{4.65}$$

Now we require that these functions be orthogonal in the range $t_1 < t < t_2$. This demands that

$$\begin{aligned}
\int_{t_1}^{t_2} \phi_j(t)\phi_k(t) &= I_k \neq 0, && \text{for } k = j \\
&= 0, && \text{for } k \neq j
\end{aligned} \tag{4.66}$$

For instance,

$$\int_{t_1}^{t_2} \phi_1(t)\phi_0(t)\,dt = B_{10}\int_{t_1}^{t_2} \phi_0^2(t)\,dt + \int_{t_1}^{t_2} \phi_0(t)h'(t)\,dt = 0$$

leads to

$$B_{10} = -\frac{\int_{t_1}^{t_2} \phi_0(t)h'(t)\,dt}{\int_{t_1}^{t_2} \phi_0^2(t)\,dt} = -\frac{\int_{t_1}^{t_2} h(t)h'(t)\,dt}{\int_{t_1}^{t_2} h^2(t)\,dt} \tag{4.67}$$

Similarly,

$$\int_{t_1}^{t_2} \phi_2(t)\phi_0(t)\,dt = B_{20}\int_{t_1}^{t_2} \phi_0^2(t)\,dt + B_{21}\int_{t_1}^{t_2} \phi_1(t)\phi_0(t)\,dt + \int_{t_1}^{t_2} h''(t)\phi_0(t)\,dt = 0$$

gives

$$B_{20} = -\frac{\int_{t_1}^{t_2} h(t)h''(t)\,dt}{\int_{t_1}^{t_2} h^2(t)\,dt} \tag{4.68}$$

since $\phi_1(t)$ and $\phi_0(t)$ are orthogonal. And by setting

$$\int_{t_1}^{t_2} \phi_2(t)\phi_1(t)\,dt = 0$$

we obtain

$$B_{21} = -\frac{\int_{t_1}^{t_2} h''(t)\phi_1(t)\,dt}{\int_{t_1}^{t_2} \phi_1^2(t)\,dt} = -\frac{\int_{t_1}^{t_2} h''(t)[B_{10}h(t) + h'(t)]\,dt}{\int_{t_1}^{t_2} [B_{10}h(t) + h'(t)]^2\,dt} \tag{4.69}$$

By mathematical induction, we may write

$$B_{kj} = -\frac{\int_{t_1}^{t_2} h^{(k)}(t)\phi_j(t)\,dt}{\int_{t_1}^{t_2} \phi_j^2(t)\,dt}, \qquad j = 0, 1, 2, \ldots, k-1 \tag{4.70}$$

Now, let us return to the general expressions of the set of orthogonal functions formed in (4.65). Each function, say

$$\phi_i(t) = h^{(i)}(t) + B_{i,i-1}(t)\phi_{i-1}(t) + \ldots + B_{i2}\phi_2(t) + B_{i1}\phi_1(t) + B_{i0}\phi_0(t) \tag{4.71}$$

may be regarded as an alternative way of writing

$$\epsilon_i(t) = \phi_i(t) = h^{(i)}(t) + a_{i-1}h^{(i-1)}(t) + \ldots + a_2h''(t) + a_1h'(t) + a_0h(t) \tag{4.72}$$

since each is a linear combination of all derivatives of $h(t)$ through the ith. But the expression in (4.72) is the same as that in (4.61), which may be interpreted as the deviation between the giving $h^{(i)}(t)$ and the computed $\hat{h}^{(i)}(t)$ if the latter were to be computed from (4.56).

If the same reasoning used in minimizing E_2 of (4.62) is applied to $\phi_i(t)$ here, we are to set

$$\frac{\partial}{\partial B_{ik}} \int_{t_1}^{t_2} \phi_i^2(t)\,dt = 0, \qquad k = 0, 1, \ldots, i-1 \tag{4.73}$$

But

$$\begin{aligned}\frac{\partial}{\partial B_{ik}} \int_{t_1}^{t_2} \phi_i^2(t)\,dt &= \int_{t_1}^{t_2} 2\phi_i(t)\frac{\partial \phi_i(t)}{\partial B_{ik}}\,dt \\ &= 2\int_{t_1}^{t_2} \phi_i(t)\phi_k(t)\,dt = 0, \qquad k = 0, 1, \ldots, i-1\end{aligned} \tag{4.74}$$

by virtue of (4.66). Hence, if $\phi_0(t), \phi_1(t), \ldots, \phi_i(t)$ in (4.65) are made orthogonal to one another, the error integral of the deviation between ith derivative of the given function $h(t)$ and that of an approximating function $h^*(t)$,

$$E_2 = \int_{t_1}^{t_2} [h^{(i)}(t) - h^{*(i)}(t)]^2 \, dt$$

is automatically minimized with respect to the coefficients B_{ik}, $k = 0, 1, \ldots, i - 1$. Hence, if we generate a set of orthogonal functions through $\phi_i(t)$ as defined in (4.65), rewrite each $\phi_k(t)$ in terms of $h^{(j)}(t)$, $j = 0, 1, \ldots, k - 1$, as in (4.72), and replace each $h^{(j)}(t)$ by s^j, a characteristic equation similar to (4.55) will be obtained for each k. A set of i such equations is available. The roots of each of these characteristic equations are the approximate pole locations for that particular order.

In computing the coefficient B's of (4.65) for a given function $h(t)$ over the interval $t_1 < t < t_2$, it is convenient to evaluate first the individual integrals defined by

$$K_{mn} = \int_{t_1}^{t_2} h^{(m)}(t) h^{(n)}(t) \, dt, \qquad 0 \leq m, n \leq i \tag{4.75}$$

In fact, to obtain $\phi_0(t), \phi_1(t), \ldots, \phi_i(t)$, only those K_{mn} for which $m = n$ need be evaluated through integration.

Since, by integration by parts, we may write

$$\begin{aligned} K_{mn} &= \int_{t_1}^{t_2} h^{(n)}(t) d[h^{(m-1)}(t)] \\ &= h^{(m-1)}(t_2) h^{(n)}(t_2) - h^{(m-1)}(t_1) h^{(n)}(t_1) - K_{m-1,n+1} \end{aligned} \tag{4.76}$$

Also, it is obvious that

$$K_{mn} = K_{nm} \tag{4.77}$$

By letting $n = m - 1$ in (4.76) we obtain

$$K_{m,m-1} = \tfrac{1}{2}\{[h^{(m-1)}(t_2)]^2 - [h^{(m-1)}(t_1)]^2\} \tag{4.78}$$

Furthermore, (4.76) may be written alternatively as

$$K_{mn} + K_{m-1,n+1} = h^{(m-1)}(t_2) h^{(n)}(t_2) - h^{(m-1)}(t_1) h^{(n)}(t_1) \tag{4.79}$$

Hence any two K's whose subscripts have the same sum may be calculated from each other without performing any integration.

Once the necessary K's are known, B's may be calculated successively as follows. Equations (4.67), (4.68), and (4.69) become

$$B_{10} = -\frac{K_{20}}{K_{00}}$$

$$B_{20} = -\frac{K_{20}}{K_{00}}$$

$$B_{21} = -\frac{B_{10}K_{20} + K_{21}}{B_{10}^2K_{00} + 2B_{10}K_{10} + K_{11}}$$

For higher-subscripted B's it would be easier to have $\phi_j(t)$ written in the form of (4.72) first. Then

$$B_{kj} = -\frac{a_0K_{k0} + a_1K_{k1} + \ldots + a_{j-1}K_{k,j-1} + K_{kj}}{a_0^2K_{00} + 2a_0a_1K_{10} + 2a_0a_2K_{20} + \ldots + a_1^2K_{11} + 2a_1a_2K_{21} + \ldots + K_{jj}} \tag{4.80}$$

Example. Suppose

$$h(t) = \frac{1}{1+t}$$

and

$$t_1 = 0, \qquad t_2 = \infty$$

Since

$$K_{kk} = \int_0^\infty \left[\frac{k!}{(1+t)^{k+1}}\right]^2 dt = \frac{(k!)^2}{2k+1}$$

and

$$h^{(k)}(t_2) = 0, \qquad k \geq 0$$

we may write successively

$$K_{00} = 1$$
$$K_{01} + K_{10} = -h^2(0) = -1$$
$$K_{01} = -\tfrac{1}{2}$$
$$K_{11} = \tfrac{1}{3}$$
$$K_{02} + K_{11} = -h(0)h'(0) = 1$$
$$K_{02} = \tfrac{2}{3}$$
$$K_{12} + K_{21} = -[h'(0)]^2 = -1$$
$$K_{12} = -\tfrac{1}{2}$$
$$K_{22} = \tfrac{4}{5}$$
$$K_{03} + K_{12} = -h(0)h''(0) = -2$$
$$K_{03} = -\tfrac{3}{2}$$
$$K_{13} + K_{22} = -h'(0)h''(0) = 2$$

$$K_{13} = \tfrac{6}{5}$$

$$K_{23} + K_{32} = -[h''(0)]^2 = -4$$

$$K_{23} = -2$$

$$K_{33} = \tfrac{36}{7}$$

$$K_{04} + K_{13} = -h(0)h'''(0) = 6$$

$$K_{04} = \tfrac{24}{5}$$

$$K_{14} + K_{23} = -h'(0)h'''(0) = -6$$

$$K_{14} = -4$$

$$K_{24} + K_{33} = -h''(0)h'''(0) = 12$$

$$K_{24} = \tfrac{48}{7}$$

$$K_{34} + K_{43} = -[h'''(0)]^2 = -36$$

$$K_{34} = -18$$

$$K_{44} = 64$$

etc.

$$I_0 = K_{00} = 1$$

$$B_{10} = \frac{K_{10}}{-K_{00}} = \frac{1}{2}$$

$$\phi_1 = \tfrac{1}{2}h(t) + h'(t)$$

$$I_1 = \tfrac{1}{4}K_{00} + K_{10} + K_{11} = \tfrac{1}{12}$$

$$B_{20} = \frac{K_{20}}{-I_0} = -\frac{2}{3}$$

$$B_{21} = \frac{\tfrac{1}{2}K_{20} + K_{21}}{-I_1} = 2$$

$$\phi_2 = \tfrac{1}{3}h(t) + 2h'(t) + h''(t)$$

$$\mathrm{I}_2 = \tfrac{1}{9}K_{00} + 4K_{11} + K_{22} + \tfrac{4}{3}K_{01} + 4K_{12} + \tfrac{2}{3}K_{02} = \tfrac{1}{45}$$

$$B_{30} = \frac{K_{30}}{-I_0} = \frac{3}{2}$$

$$B_{31} = \frac{\tfrac{1}{2}K_{30} + K_{31}}{-I_1} = -\frac{27}{5}$$

$$B_{32} = \frac{\tfrac{1}{3}K_{30} + 2K_{31} + K_{32}}{-I_2} = \frac{9}{2}$$

$$\phi_3 = \tfrac{3}{10}h(t) + \tfrac{18}{5}h'(t) + \tfrac{9}{2}h''(t) + h'''(t)$$

$$I_3 = \tfrac{9}{100}K_{00} + \tfrac{324}{25}K_{11} + \tfrac{81}{4}K_{22} + K_{33} + \tfrac{54}{25}K_{10}$$
$$+ \tfrac{27}{10}K_{20} + \tfrac{3}{5}K_{30} + \tfrac{162}{5}K_{12} + \tfrac{36}{5}K_{13} + 9K_{23} = \tfrac{9}{700}$$

$$B_{40} = \frac{K_{40}}{-I_0} = -\frac{24}{5}$$

$$B_{41} = \frac{\frac{1}{2}K_{40} + K_{41}}{-I_1} = \frac{96}{5}$$

$$B_{42} = \frac{\frac{1}{3}K_{40} + 2K_{41} + K_{42}}{-I_2} = -\frac{144}{7}$$

$$B_{43} = \frac{\frac{3}{10}K_{40} + \frac{18}{5}K_{41} + \frac{9}{2}K_{42} + K_{43}}{-I_3} = 8$$

$$\phi_4 = \frac{12}{35}h(t) + \frac{48}{7}h'(t) + \frac{108}{7}h''(t) + 8h'''(t) + h^{(4)}(t)$$

etc.

Hence the approximate pole locations are given by the following equations for different values of N:

$$s + \tfrac{1}{2} = 0$$

$$s^2 + 2s + \tfrac{1}{3} = 0$$

$$s^3 + \tfrac{9}{2}s^2 + \tfrac{18}{5}s + \tfrac{3}{10} = 0$$

$$s^4 + 8s^3 + \tfrac{108}{7}s^2 + \tfrac{48}{7}s + \tfrac{12}{35} = 0$$

Or the poles are located at

$N = 1, \quad s = -0.5$

$N = 2, \quad s = -0.1835, -1.8165$

$N = 3, \quad s = -0.0942, -0.9115, -3.4943$

$N = 4, \quad s = -0.0570, -0.5503, -2.0363, -5.3563$

4.6 DISCUSSION

In this chapter we have been concerned with the task of determining a set of poles for a realizable $H^*(s)$ when an $h(t)$ or $\hat{h}(t)$ is given. This is equivalent to finding a set of s_ν for the expression

$$h_N^*(t) = \sum_{\nu=1}^{N} A_\nu \epsilon^{-s_\nu t} \tag{4.81}$$

In (4.81) all poles are assumed to be simple. We did this to simplify the discussion. There is no reason why multiple poles cannot be assumed at the outset in each procedure and the same development followed.

A question may have entered the reader's mind in going through this chapter. That is, how can we be sure that the poles found by any of the procedures outlined in this chapter will all lie on the left half of the s plane? There is no theoretical affirmative answer to this question. However, for all practical purposes, this should not be a source of worry. As long as $h(t)$ or $\hat{h}(t)$ is absolutely integrable, chances are that $h^*(t)$ will also be absolutely integrable, if the approximation is good. Hence $H^*(s)$ cannot have right-halfplane or j-axis poles. This assurance is not as certain when the interval of approximation is only finite. In this case right-halfplane poles can indeed be present when one of the procedures in this chapter is used. When this happens, either these unacceptable poles must be discarded or a new set of criteria applied.

The location of pole positions in a time-domain approximation problem is, of course, only half of the task. The other half will involve the location of the zeros of $H^*(s)$ or the evaluation of coefficient A's in $h^*(t)$. This second half will be dealt with in Chapter 5.

It should be emphasized at this juncture that we are most commonly interested in finding *a* (not *the*) set of poles for $h^*(t)$. This is because in almost all engineering problems we do not know in advance that the given $h(t)$ is exactly representable by an $h^*(t)$. Nor do we know the number of poles required. An example of the situation in which advance information on the number of poles is available would be in a classroom quiz, for which the instructor has generated an $h(t)$ based on a chosen network and the student is required to find *the* set of poles of the network function. In such a situation the problem is relatively simple. For instance, the method in Section 4.2 should give *the* answer straightforwardly.

In all the procedures outlined in this chapter, we must decide in advance the number of poles and their orders. Once this has been done, each of the procedures will render a set of poles, optimum according to certain mathematical criteria. Unfortunately, there is no sure scheme by which we can be assured of a good choice in advance. We must rely on such nebulous things as judgment, experience, good luck, and so on. However, these procedures at least enable us to locate some sets of poles with some rationale and usually provide us with a reasonably good starting point from which further improvement or refinement may hopefully be based.

When the given $h(t)$ is not or is known not to be in the form of (4.81), there is no such thing as *the* set, or even *the best* set, of poles for $h^*(t)$. In principle, for any $h(t)$ there is no limit on the number of $h^*(t)$ we can generate. It all depends on what is considered an acceptable approximation. Generally, we may expect that the approximation will be improved by increasing the number of poles. We can defend this statement by saying that an $h^*_{N+1}(t)$ can certainly be made to do no worse than an $h^*_N(t)$, as we can always leave N terms in $h^*_{N+1}(t)$ to be identical to the N terms of $h^*_N(t)$. The extra term in $h^*_{N+1}(t)$ can, at worst, be identically zero. However, this defense is only argumentative at best. For one thing, the application of some of the procedures in this chapter when N is increased renders an entirely new set of poles. Then there is no assurance that this new set, although larger in number, will do any better than the old set. Even when the previous set of poles and their residues are fully retained, it is conceivable that any finite additional term worsens the approximation.

Despite these arguments, increasing the number of poles usually does improve the approximation, except for unusual cases. When a simple $h^*(t)$ fails to meet the accuracy requirements, increasing N is a natural recourse for the next try.

Even the locations of a set of poles for $h^*(t)$ are largely arbitrary. For example, we may require that all poles be negative and real. There is no theoretical reason why we cannot approximate *any* $h(t)$ with $h^*(t)$ that has only negative real s_ν to within any desired accuracy. [This is somewhat analogous to saying that we can approximate any magnitude characteristic $|H(j\omega)|$ by that of an RC network to any desired accuracy.] However, for a certain $h(t)$ it may require a large number of poles to achieve what a pair of complex poles can accomplish. Thus, to approximate an $h(t)$, certain pole locations will enable us to attain the desired approximation much more readily than others. These are the poles we are hoping to locate with methods in this chapter.

In summary, as in most engineering undertakings, which are a matter of compromising between good quality and low cost, we are usually seeking a simple $h^*(t)$ with an acceptable approximation. Procedures in this chapter enable us to locate poles which can subsequently be used to find an approximating function. If the approximation is found to be unacceptable, the whole process may have to be repeated for a new set of parameters until an acceptable approximation is found.

PROBLEMS

4.1. For the impulse response $h(t) = t\epsilon^{-t}$, first obtain a power series for $H(s)$ by replacing ϵ^{-st} with its power series and then integrate term by term. Then use the series obtained this way to find the approximate poles by finding the zero of the denominator Q_{02} of the (0, 2) Padé approximant of the series.

4.2. Obtain the power series for an approximate $H^*(s)$ for $h(t) = t\epsilon^{-t}$ by replacing $h(t)$ by an $\hat{h}(t)$ made up of broken straight lines. Compare this series with the one obtained in Problem 4.1.

4.3. Obtain a power series for $\hat{H}^*(s)$ such that $\hat{h}^*(t)$ approximates $h(t) = t\epsilon^{-t^2}$. From this series estimate the poles of $\hat{H}(s)$ if the latter is assumed to have (a) two poles and (b) three poles.

4.4. For the $h(t)$ of Fig. P4.1, find the poles of an $H^*(s)$ such that (4.33) is satisfied for $N = 3$ and (a) $\Delta t = 0.2$ and (b) $\Delta t = 0.3$.

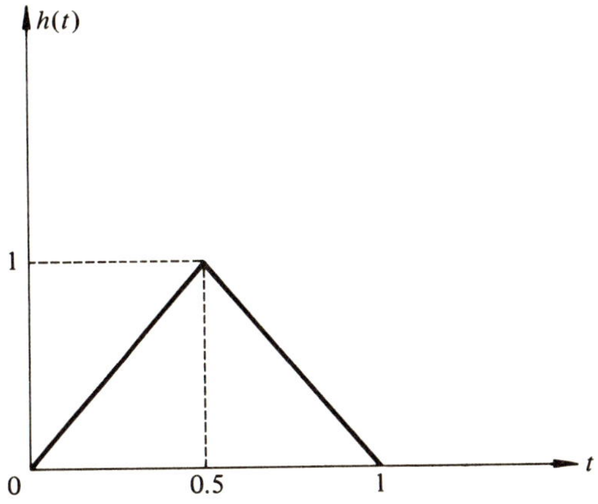

Fig. P4.1

4.5. For the $h(t)$ of Fig. P4.1, find a set of three approximate poles such that the difference equation (4.46) is approximately satisfied in the least-squares sense for $\Delta t = 0.1$ and $M = 10$.

4.6. Repeat Problem 4.5 for a set of four poles.

4.7. Find the orthogonal functions ϕ_0, ϕ_1, and ϕ_2 as defined by (4.65) and (4.66) for the impulse response of Fig. 4.3 for $t_1 = 0$ and $t_2 = \infty$. From these functions determine the set of poles for $H^*(s)$ for $N = 2$.

4.8. Repeat Problem 4.7 for the $h(t)$ of Problem 4.1 and $N = 2, 3, 4$.

5

Determination of an Approximate Network Function When the Poles Are Known

In this chapter we shall discuss the balance of a time-domain synthesis problem when the poles of the desired network function are somehow known. These poles are known either because of a priori knowledge of the network or after one of the procedures described in Chapter 4 has been applied. Hence we are dealing with the determination of the numerator of an $H^*(s)$, whose denominator is already known, such that its inverse Laplace transform approximates a given impulse response $h(t)$. Alternatively, one may say that we are trying to determine the residues in these poles to give the desired approximation.

In several of the methods presented here, they are really continuations of the same problems described in Chapter 4. For example, the point-matching method in Section 5.1 may be viewed as a continuation of the method described in Section 4.3. However, neither of these methods depends on the other. They are really two separate techniques; the similarity between them is incidental. Another way to state this is that the use of one method in one part of the problem does not require the use of the similar method in the other. For example, once the poles have been found by the point-matching method, we may follow it with the method of initial derivatives to find the numerator of $H^*(s)$. Similarly, the method of point matching described in Section 5.1 may be used to find the residues, regardless of how the poles of $H^*(s)$ were determined.

5.1 POINT-MATCHING METHOD

When the set of network function poles are given, we may assume that

$$h^*(t) = \sum_{\nu=1}^{N} A_\nu \epsilon^{s_\nu t} \tag{5.1}$$

in which s_ν are known and may include real and complex-conjugate values. For a real s_ν, the corresponding unknown A_ν is also real. For a pair of conjugate poles, s_ν and $s_{\nu+1} = \bar{s}_\nu$, the corresponding coefficients will also be conjugate (A_ν and $A_{\nu+1} = \bar{A}_\nu$). Here the two unknowns are the real and imaginary parts of A_ν. In either case (5.1) is a suitable representation of the impulse response.

Then we can choose a series of N known points on the impulse response. Call these values $h(t_1), h(t_2), \ldots, h(t_N)$. Substituting these values of h and t into (5.1) yields a set of N simultaneous equations

$$h(t_i) = \sum_{\nu=1}^{N} A_\nu \epsilon^{s_\nu t_i} \tag{5.2}$$

where $i = 1, 2, \ldots, N$. These N equations may then be solved for the A's.

If the N points are chosen from the group of $2N$ equally spaced points Δt apart starting with the origin, then a set of $2N$ equations identical to (4.30) may be formed:

$$\begin{aligned}
h(0) &= \sum_{\nu=1}^{N} A_\nu \\
h(\Delta t) &= \sum_{\nu=1}^{N} A_\nu w_\nu \\
h(2\,\Delta t) &= \sum_{\nu=1}^{N} A_\nu w_\nu^2 \\
&\vdots \\
h([2N-1]\,\Delta t) &= \sum_{\nu=1}^{N} A_\nu w_\nu^{(2N-1)}
\end{aligned} \tag{5.3}$$

where

$$w_\nu = \epsilon^{s_\nu \Delta t}$$

There are only N unknown coefficients in (5.1), but there are $2N$ equations in (5.3). In general, only N of the $2N$ equations can be used. The remaining N equations generally will not be satisfied by the solution from those N equations.

At this point it might be well to point out that there are actually $2N$ unknowns in the expression (5.1) if the pole locations have not been determined. In that case (4.30) or (5.3) may be regarded as a set of $2N$ equations from which A_ν and s_ν may be solved. The fact that s_ν may be found without finding A_ν as well is, of course, incidental. The information contained in (4.33) must be consistent with (4.30). The use of any N equations from (4.30) in addition to (4.33) completes the solution of all $2N$ unknowns in (5.1).

If, however, s_ν have been found by methods other than that corresponding to (4.33), this set of s_ν cannot be expected to be consistent with (4.30) and, therefore, not all $2N$ equations will be satisfied by this set of s_ν.

Example. Using the poles found in the example of Section 4.3 (p. 87) for

$$h(t) = \frac{1}{1+t} \tag{5.4}$$

find an $h^*(t)$ by the point-matching method.

Solution. Here

$$\Delta t = 0.8$$

$$w_1 = 0.89448, \qquad w_2 = 0.52628, \qquad w_3 = 0.13924$$

Using the first three equations of (5.3), we have

$$\begin{aligned} 1.00000A_1 + 1.00000A_2 + 1.00000A_3 &= 1.00000 \\ 0.89448A_1 + 0.52628A_2 + 0.13924A_3 &= 0.55556 \\ 0.80009A_1 + 0.27697A_2 + 0.01939A_3 &= 0.38462 \end{aligned}$$

Solving gives

$$\begin{aligned} A_1 &= 0.31704 \\ A_2 &= 0.45700 \\ A_3 &= 0.22597 \end{aligned}$$

Hence

$$h^*(t) = 0.31704\epsilon^{-0.13940t} + 0.45700\epsilon^{-0.80239t} + 0.22600\epsilon^{-2.46445t} \tag{5.5}$$

Since these w's and A's are obtained from the same set of six equations, (5.3), $h^*(t)$ for $t = 2.4$, 3.2, and 4.0 should agree exactly with the values of $h(t)$ at these instants. The above $h^*(t)$ and $h(t)$ are so close together for almost all values of t that a graphical display will not show their deviations.

5.2 LEAST-SQUARES METHOD

In principle, when the poles of a network function are known, it is very easy to apply the least-mean-square-error criterion to determine the coefficients of the various terms in the approximate impulse response. Let the latter be

$$\begin{aligned} h^*(t) = &\sum_{\nu=1}^{M} [2A_\nu \epsilon^{-\alpha_\nu t} \cos \beta_\nu t + 2B_\nu \epsilon^{-\alpha_\nu t} \sin \beta_\nu t] \\ &+ \sum_{\mu=1}^{N} C_\mu \epsilon^{-\sigma_\mu t} \end{aligned} \tag{5.6}$$

Here each value of ν corresponds to a pair of complex-conjugate poles and each value of μ corresponds to a real pole.

The squared error over a specific range $t_1 < t < t_2$ may be written as

$$E_2 = \int_{t_1}^{t_2} [h(t) - h^*(t)]^2\, dt \tag{5.7}$$

Setting $\partial E_2/\partial A_\nu = 0$, $\partial E_2/\partial B_\nu = 0$, and $\partial E_2/\partial C_\mu = 0$, we obtain a set of $2M + N$ equations:

$$\begin{aligned}
&\int_{t_1}^{t_2} [h(t) - h^*(t)]\epsilon^{-\alpha_\nu t} \cos \beta_\nu t\, dt = 0 \\
&\int_{t_1}^{t_2} [h(t) - h^*(t)]\epsilon^{-\alpha_\nu t} \sin \beta_\nu t\, dt = 0 \\
&\int_{t_1}^{t_2} [h(t) - h^*(t)]\epsilon^{-\sigma_\mu t}\, dt = 0, \qquad \nu = 1, 2, \ldots, M, \quad \mu = 1, 2, , \ldots, N
\end{aligned} \tag{5.8}$$

Since α_ν, β_ν, and σ_μ are constants, (5.6) is a set of $2M + N$ linear equations with A_ν, B_ν, and C_μ as unknowns. The coefficients of these equations can readily be evaluated by integration. The constant terms are given by integrals of the type

$$\begin{aligned}
&\int_{t_1}^{t_2} h(t)\epsilon^{-\alpha_\nu t} \sin \beta_\nu t\, dt \\
&\int_{t_1}^{t_2} h(t)\epsilon^{-\alpha_\nu t} \cos \beta_\nu t\, dt \\
&\int_{t_1}^{t_2} h(t)\epsilon^{-\sigma_\mu t}\, dt
\end{aligned} \tag{5.9}$$

These integrals can be evaluated either by analytical means, when possible, or by numerical methods.

Example. Approximate the $h(t)$ of (5.4) using the same s_ν of (5.5) in the least-squares sense over the range $0 < t < \infty$.

Solution. Since only real poles are present and

$$h^*(t) = C_1\epsilon^{-0.1394t} + C_2\epsilon^{-0.8024t} + C_3\epsilon^{-2.4645t}$$

only the last equation in (5.8) is applicable. We have

$$\begin{aligned}
C_1 \int_0^\infty \epsilon^{-2\alpha_1 t}\, dt + C_2 \int_0^\infty \epsilon^{-(\alpha_1+\alpha_2)t}\, dt + C_3 \int_0^\infty \epsilon^{-(\alpha_1+\alpha_3)t}\, dt &= \int_0^\infty h(t)\epsilon^{-\alpha_1 t}\, dt \\
C_1 \int_0^\infty \epsilon^{-(\alpha_1+\alpha_2)t}\, dt + C_2 \int_0^\infty \epsilon^{-2\alpha_2 t}\, dt + C_3 \int_0^\infty \epsilon^{-(\alpha_2+\alpha_3)t}\, dt &= \int_0^\infty h(t)\epsilon^{-\alpha_2 t}\, dt \\
C_1 \int_0^\infty \epsilon^{-(\alpha_1+\alpha_3)t}\, dt + C_2 \int_0^\infty \epsilon^{-(\alpha_2+\alpha_3)t}\, dt + C_3 \int_0^\infty \epsilon^{-2\alpha_3 t}\, dt &= \int_0^\infty h(t)\epsilon^{-\alpha_3 t}\, dt
\end{aligned}$$

For the $h(t)$ of (5.4) we have†

$$\int_0^\infty h(t)\epsilon^{-\alpha_v t}\,dt = \int_0^\infty \frac{\epsilon^{-\alpha_v t}}{1+t}\,dt = \epsilon^{\alpha_v} Ei(\alpha_v)$$

The definite integrals involving exponential functions are easily computable. The simultaneous equations to be solved are

$$\begin{aligned}
3.6075C_1 + 1.0612C_2 + 0.3846C_3 &= 1.7746 \\
1.0612C_1 + 0.6202C_2 + 0.3058C_3 &= 0.6879 \\
0.3846C_1 + 0.3058C_2 + 0.2029C_3 &= 0.3070
\end{aligned}$$

Solution of these equations give

$$\begin{aligned}
C_1 &= 0.390 \\
C_2 &= 0.326 \\
C_3 &= 0.340
\end{aligned}$$

Hence

$$h^*(t) = 0.390\epsilon^{-0.1394t} + 0.326\epsilon^{-0.8024t} + 0.340\epsilon^{-2.4645t}$$

5.3 DERIVATIVE-MATCHING METHOD

Similar to the method described in Section 5.2, in which $h^*(t)$ is made to match a given $h(t)$ at a series of points along the t axis, we may devise another scheme by which the first few derivatives of $h^*(t)$ are made to match those of a given $h(t)$ at a point $t = t_1$. This scheme will be possible whenever the Taylor's expansion of $h(t)$ is known about $t = t_1$. The selection of the point t_1 is a matter of choice and judgment. In many instances the origin or infinity may prove to be convenient to use. For the purpose of illustrating the scheme, let us choose $t_1 = 0$. The same procedure applies equally well for other values of t_1.

†Here, Ei denotes the exponential integral

$$Ei(z) = \int_z^\infty \frac{\epsilon^{-x}}{x}\,dx$$

The identity

$$\int_0^\infty \frac{\epsilon^{-at}}{b+t}\,dt = \epsilon^{ab} Ei(ab)$$

has also been used. For more details, see M. Abramowitz and I. A. Stegun, *Handbook of Mathematical Functions,* Dover Publications, Inc., New York, 1964, pp. 228–251; or E. Jahnke and F. Emde, *Tables of Functions,* 4th ed., Dover Publications, Inc., New York, 1945, pp. 1–9.

Assume the approximating network function to be

$$H^*(s) = \frac{b_{N-1}s^{N-1} + b_{N-2}s^{N-2} + \cdots + b_1 s + b_0}{s^N + a_{N-1}s^{N-1} + \cdots + a_1 s + a_0} \tag{5.10}$$

in which a's are known numbers. The inverse Laplace transform of $H^*(s)$ is

$$h^*(t) = \sum_{\nu=1}^{N} A_\nu \epsilon^{s_\nu t} \tag{5.11}$$

in which s_ν are known. If (5.11) is expressed in terms of its Taylor's expansion about $t = 0$, it has the form

$$h^*(t) = \sum_{k=0}^{\infty} \frac{h^{*(k)}(0)}{k!} t^k \tag{5.12}$$

Now if the Laplace transform of (5.12) is taken, we have

$$\mathscr{L}[h^*(t)] = \sum_{k=0}^{\infty} h^{*(k)}(0) \frac{1}{s^{k+1}} = H^*(s) \tag{5.13}$$

Equating (5.10) and (5.13) yields

$$\begin{aligned} b_{N-1}s^{N-1} &+ b_{N-2}s^{N-2} + \cdots + b_1 s + b_0 \\ &= (s^N + a_{N-1}s^{N-1} + a_{N-2}s^{N-2} + \cdots + a_1 s + a_0) \\ &\times \left[h^*(0)\frac{1}{s} + h^{*\prime}(0)\frac{1}{s^2} + h^{*\prime\prime}(0)\frac{1}{s^3} + \cdots + h^{*(N-1)}(0)\frac{1}{s^N} + \cdots\right] \end{aligned} \tag{5.14}$$

Multiply the right-hand side of (5.14) and equate coefficients to obtain

$$\begin{aligned} b_{N-1} &= h^*(0) \\ b_{N-2} &= a_{N-1}h^*(0) + h^{*\prime}(0) \\ b_{N-3} &= a_{N-2}h^*(0) + a_{N-1}h^{*\prime}(0) + h^{*\prime\prime}(0) \\ &\vdots \\ b_1 &= a_2 h^*(0) + a_3 h^{*\prime}(0) + \cdots + a_{N-1}h^{*(N-3)}(0) + h^{*(N-2)}(0) \\ b_0 &= a_1 h^*(0) + a_2 h^{*\prime}(0) + \cdots + a_{N-1}h^{*(N-2)}(0) + h^{*(N-1)}(0) \end{aligned} \tag{5.15}$$

This step is seen to be equivalent to making $H^*(s)$ the $(N - 1, N)$ Padé approximant in the variable $1/s$ of the series in (5.13). Hence, if we let

$$h^{*(k)}(0) = h^{(k)}(0)$$

for $k = 0, 1, 2, \ldots, N - 1$, (5.15) enables us to evaluate all necessary coefficients in the numerator of $H^*(s)$.

When $h(t)$ is not given in analytical form, an approximate $\hat{h}(t)$ must first be found such that $\hat{h}(t) \cong h(t)$, and then the initial derivatives of $\hat{h}(t)$ can be used in place of those of $h(t)$.

Example. Assuming three real poles at -0.1361, -0.8062, and -2.4639, find the $H^*(s)$ for

$$h(t) = \frac{1}{1+t} \tag{5.16}$$

by matching $h(t)$ and $h^*(t)$ in the Taylor's sense at the origin.

Solution. From Section 4.3

$$\begin{aligned} H^*(s) &= \frac{b_2 s^2 + b_1 s + b_0}{(s+0.1361)(s+0.8062)(s+2.4639)} \\ &= \frac{b_2 s^2 + b_1 s + b_0}{s^3 + 3.4062s^2 + 2.4315s + 0.2703} \end{aligned}$$

Since

$$\begin{aligned} h(0) &= 1 \\ h'(0) &= -1 \\ h''(0) &= 2 \end{aligned}$$

we have, from (5.15),

$$\begin{aligned} b_2 &= 1 \\ b_1 &= 2.4062 \\ b_0 &= 1.0253 \end{aligned}$$

Thus

$$\begin{aligned} H^*(s) &= \frac{s^2 + 2.4062s + 1.0253}{(s+0.1361)(s+0.8062)(s+2.4639)} \\ &= \frac{0.4592}{s+0.1361} + \frac{0.2382}{s+0.8062} + \frac{0.3025}{s+2.4639} \end{aligned}$$

and

$$h^*(t) = 0.4592\epsilon^{-0.1361t} + 0.2382\epsilon^{-0.8062t} + 0.3025\epsilon^{-2.4639t} \tag{5.17}$$

This $h^*(t)$ is plotted in Fig. 5.1.

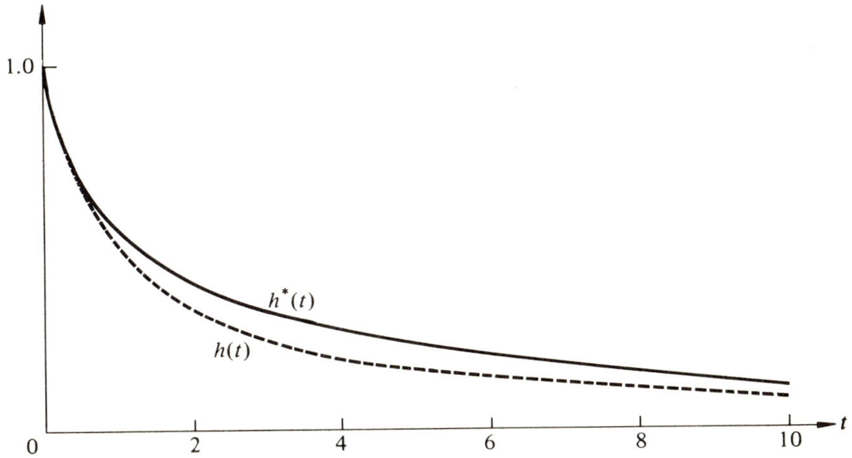

Fig. 5.1 The $h(t)$ of (5.16) and $h^*(t)$ of (5.17).

5.4 ORTHOGONAL-FUNCTION METHOD

This method is based on the assumption that if a set of orthogonal functions can be formed such that they not only have the form of (5.1) but also have the known poles in the exponents of its terms, then the least-squares approximation of these functions can be obtained by the method described in Section 2.4.

For a set of real poles $s = -\alpha_1, -\alpha_2, \ldots, -\alpha_N$, we wish to obtain a set of functions

$$\begin{aligned} \phi_1(t) &= A_{11}\epsilon^{-\alpha_1 t} \\ \phi_2(t) &= A_{21}\epsilon^{-\alpha_1 t} + A_{22}\epsilon^{-\alpha_2 t} \\ &\vdots \\ \phi_N(t) &= A_{N1}\epsilon^{-\alpha_1 t} + A_{N2}\epsilon^{-\alpha_2 t} + \cdots + A_{NN}\epsilon^{-\alpha_N t} \end{aligned} \tag{5.18}$$

such that they are orthogonal. This can be done successively by forcing these functions to satisfy (2.31). Kautz [KA3, KA4, HU1] has shown that this step is equivalent to forming these functions in the frequency domain such that their Laplace transforms are

$$\begin{aligned} \Phi_1(s) &= \sqrt{2\alpha_1}\frac{1}{s+\alpha_1} \\ \Phi_2(s) &= \sqrt{2\alpha_2}\frac{s-\alpha_1}{(s+\alpha_1)(s+\alpha_2)} \\ &\vdots \\ \Phi_N(s) &= \sqrt{2\alpha_N}\frac{(s-\alpha_1)(s-\alpha_2)\cdots(s-\alpha_{N-1})}{(s+\alpha_1)(s+\alpha_2)\cdots(s+\alpha_{N-1})(s+\alpha_N)} \end{aligned} \tag{5.19}$$

The $\phi_i(t)$ corresponding to these $\Phi_i(s)$ are not only orthogonal but also orthonormal.

Similarly, if the set of poles are complex conjugate, $s_{1,2} = -\alpha_1 \pm j\beta_1$, $s_{3,4} = -\alpha_2 \pm j\beta_2, \ldots, s_{2M-1,2M} = -\alpha_M \pm j\beta_M$, then we form

$$
\begin{aligned}
\phi_1(t) &= A_{11}\epsilon^{-\alpha_1 t}\cos(\beta_1 t + \theta_{11}) \\
\phi_2(t) &= A_{21}\epsilon^{-\alpha_1 t}\sin(\beta_1 t + \theta_{11}) \\
\phi_3(t) &= A_{31}\epsilon^{-\alpha_1 t}\cos(\beta_1 t + \theta_{31}) + A_{32}\epsilon^{-\alpha_2 t}\cos(\beta_2 t + \theta_{32}) \\
\phi_4(t) &= A_{41}\epsilon^{-\alpha_1 t}\sin(\beta_1 t + \theta_{31}) + A_{42}\epsilon^{-\alpha_2 t}\sin(\beta_2 t + \theta_{32})
\end{aligned}
\tag{5.20}
$$

etc. Forcing these functions to satisfy (2.31) and (2.32) is equivalent to forming the following functions in the frequency domain:

$$
\begin{aligned}
\Phi_1(s) &= \sqrt{2\alpha_1}\,\frac{s + |s_1|}{(s+\alpha_1)^2 + \beta_1^2} \\
\Phi_2(s) &= \sqrt{2\alpha_1}\,\frac{s - |s_1|}{(s+\alpha_1)^2 + \beta_1^2} \\
\Phi_3(s) &= \sqrt{2\alpha_2}\,\frac{[(s-\alpha_1)^2 + \beta_1^2](s + |s_3|)}{[(s+\alpha_1)^2 + \beta_1^2][(s+\alpha_2)^2 + \beta_2^2]} \\
\Phi_4(s) &= \sqrt{2\alpha_2}\,\frac{[(s-\alpha_1)^2 + \beta_1^2](s - |s_3|)}{[(s+\alpha_1)^2 + \beta_1^2][(s+\alpha_2)^2 + \beta_2^2]} \\
&\;\;\vdots \\
\Phi_{2M-1}(s) &= \sqrt{2\alpha_M}\,\frac{[(s-\alpha_1)^2 + \beta_1^2]\cdots[(s-\alpha_{M-1})^2 + \beta_{M-1}^2](s + |s_{2M-1}|)}{[(s+\alpha_1)^2 + \beta_1^2]\cdots[(s+\alpha_{M-1})^2 + \beta_{M-1}^2][(s+\alpha_M)^2 + \beta_M^2]} \\
\Phi_{2M}(s) &= \sqrt{2\alpha_M}\,\frac{[(s-\alpha_1)^2 + \beta_1^2]\cdots[(s-\alpha_{M-1})^2 + \beta_{M-1}^2](s - |s_{2M-1}|)}{[(s+\alpha_1)^2 + \beta_1^2]\cdots[(s+\alpha_{M-1})^2 + \beta_{M-1}^2][(s+\alpha_M)^2 + \beta_M^2]}
\end{aligned}
\tag{5.21}
$$

When a set of poles includes both real and complex poles, the pattern of forming the Φ's remains the same. For example, if there are five poles at

$$-3 \pm j4,\ -4 \pm j3,\ -5$$

we may have either

$$
\begin{aligned}
\Phi_1(s) &= \sqrt{10}\,\frac{1}{s+5} \\
\Phi_2(s) &= \sqrt{6}\,\frac{(s-5)(s+5)}{(s+5)[(s+3)^2 + 4^2]} \\
\Phi_3(s) &= \sqrt{6}\,\frac{(s-5)(s-5)}{(s+5)[(s+3)^2 + 4^2]} \\
\Phi_4(s) &= \sqrt{8}\,\frac{(s-5)[(s-3)^2 + 4^2](s+5)}{(s+5)[(s+3)^2 + 4^2][(s+4)^2 + 3^2]} \\
\Phi_5(s) &= \sqrt{8}\,\frac{(s-5)[(s-3)^2 + 4^2](s-5)}{(s+5)[(s+3)^2 + 4^2][(s+4)^2 + 3^2]}
\end{aligned}
\tag{5.22}
$$

or

$$
\begin{aligned}
\Phi_1(s) &= \sqrt{6}\,\frac{s+5}{(s+3)^2+4^2}\\
\Phi_2(s) &= \sqrt{6}\,\frac{s-5}{(s+3)^2+4^2}\\
\Phi_3(s) &= \sqrt{8}\,\frac{[(s-3)^2+4^2](s+5)}{[(s+3)^2+4^2][(s+4)^2+3^2]}\\
\Phi_4(s) &= \sqrt{8}\,\frac{[(s-3)^2+4^2](s-5)}{[(s+3)^2+4^2][(s+4)^2+3^2]}\\
\Phi_5(s) &= \sqrt{10}\,\frac{[(s-3)^2+4^2][(s-4)^2+3^2]}{[(s+3)^2+4][(s+4)^2+3^2](s+5)}
\end{aligned}
\tag{5.23}
$$

The actual set of time orthogonal functions is then obtained by finding the inverse Laplace transforms of Φ's. This is usually easiest done by first expanding the Φ's into their partial-fraction expressions. When this is done, a typical $\Phi_k(s)$ may be written as

$$
\Phi_k(s) = \sum_{\mu=1}^{N} \frac{C_{k\mu}}{s+\alpha_\mu} + \sum_{\nu=1}^{M}\left(\frac{A_{k\nu}+j\beta_{k\nu}}{s+\alpha_\nu+j\beta_\nu} + \frac{A_{k\nu}-j\beta_{k\nu}}{s+\alpha_\nu-j\beta_\nu}\right)
\tag{5.24}
$$

The corresponding time function is

$$
\begin{aligned}
\Phi_k(t) &= \sum_{\mu=1}^{N} C_{k\mu}\epsilon^{-\alpha_\mu t} + \sum_{\nu=1}^{M} \operatorname{Re}\,[2(A_{k\nu}+jB_{k\nu})\epsilon^{-(\alpha_\nu+j\beta_\nu)t}]\\
&= \sum_{\mu=1}^{N} C_{k\mu}\epsilon^{-\alpha_\mu t} + \sum_{\nu=1}^{M} (2A_{k\nu}\epsilon^{-\alpha_\nu t}\cos\beta_\nu t + 2B_{k\nu}\epsilon^{-\alpha_\nu t}\sin\beta_\nu t)
\end{aligned}
\tag{5.25}
$$

For example, to the set of Φ's of (5.23), we have the following frequency-domain function:

$$
\begin{aligned}
\Phi_1(s) &= \frac{(\sqrt{6}/4)(2+j)}{s+3+j4} + \frac{(\sqrt{6}/4)(2-j)}{s+3-j4}\\
\Phi_2(s) &= \frac{(\sqrt{6}/2)(1-j2)}{s+3+j4} + \frac{(\sqrt{6}/2)(1+j2)}{s+3-j4}\\
\Phi_3(s) &= \frac{(3\sqrt{8}/2)(2+j)}{s+3+j4} + \frac{(3\sqrt{8}/2)(2-j)}{s+3-j4}\\
&\quad + \frac{(7\sqrt{8}/6)(3+j)}{s+4+j3} + \frac{(7\sqrt{8}/6)(3-j)}{s+4-j3}\\
\Phi_4(s) &= \frac{(3\sqrt{8}/2)(1+j2)}{s+3+j4} + \frac{(3\sqrt{8}/2)(1+j2)}{s+3-j4}\\
&\quad + \frac{(7\sqrt{8}/2)(3+j)}{s+4+j3} + \frac{(7\sqrt{8}/2)(3-j)}{s+4-j3}
\end{aligned}
\tag{5.26}
$$

$$\Phi_5(s) = \frac{(21\sqrt{10}/2)(2+j)}{s+3+j4} + \frac{(21\sqrt{10}/2)(2-j)}{s+3-j4} + \frac{(49\sqrt{10}/3)(-3-j)}{s+4+j3} + \frac{(49\sqrt{10}/3)(-3+j)}{s+4-j3} + \frac{72\sqrt{10}}{s+5}$$

When the inverse Laplace transforms of these functions are performed, we obtain the following orthogonal functions in the time domain:

$$\begin{aligned}
\phi_1(t) &= \frac{\sqrt{30}}{2}\epsilon^{-3t}\cos(4t - 26.57^\circ) \\
\phi_2(t) &= \sqrt{30}\,\epsilon^{-3t}\sin(4t - 26.57^\circ) \\
\phi_3(t) &= 6\sqrt{10}\,\epsilon^{-3t}\cos(4t - 26.57^\circ) \\
&\quad + \frac{28\sqrt{5}}{3}\epsilon^{-4t}\cos(3t - 18.43^\circ) \\
\phi_4(t) &= 6\sqrt{10}\,\epsilon^{-3t}\sin(4t - 26.57^\circ) \\
&\quad + 28\sqrt{5}\,\epsilon^{-4t}\sin(3t - 18.43^\circ) \\
\phi_5(t) &= 105\sqrt{2}\,\epsilon^{-3t}\cos(4t - 26.57^\circ) \\
&\quad - \frac{490}{3}\epsilon^{-4t}\cos(3t - 18.43) + 72\sqrt{10}\,\epsilon^{-5t}
\end{aligned} \tag{5.27}$$

That the $\phi_\nu(t)$ formed this way are orthogonal can also be seen by a theorem on Laplace transform commonly known as the complex convolution theorem, which states that if $f_1(t)$ and $f_2(t)$ are Laplace transformable having the Laplace transforms $F_1(s)$ and $F_2(s)$, respectively, then

$$\int_0^\infty f_1(t)f_2(t)\epsilon^{-st}\,dt = \frac{1}{2\pi j}\int_{c-j\infty}^{c+j\infty} F_1(s-p)F_2(p)\,dp \tag{5.28}$$

$$\max(\sigma_1, \sigma_2, \sigma_1 + \sigma_2) < \sigma, \qquad \sigma_2 < c < \sigma - \sigma_1$$

in which c is a real constant, $\sigma = \operatorname{Re}[s]$, and σ_1 and σ_2 are the abscissas of absolute convergence of $f_1(t)$ and $f_2(t)$, respectively.

For our purposes, we may let $c = 0$ and $\sigma = 0$. Further, if we let $s = 0$ in (5.28), we have

$$\int_0^\infty f_1(t)f_2(t)\,dt = \frac{1}{2\pi j}\int_{s=-j\infty}^{s=j\infty} F_1(-s)F_2(s)\,ds \tag{5.29}$$

If we let $f_1(t) = \phi_k(t)$ and $f_2(t) = \phi_j(t)$, from (5.19) we have

$$\begin{aligned}
\int_0^\infty \phi_k(t)\phi_j(t)\,dt &= \frac{1}{2\pi j}\int_{-j\infty}^{j\infty} \Phi_k(-s)\Phi_j(s)\,ds \\
&= \frac{\sqrt{\alpha_k\alpha_j}}{2\pi j}\int_{-j\infty}^{j\infty} \frac{(-s-\alpha_1)(-s-\alpha_2)\cdots(-s-\alpha_{k-1})(s-\alpha_1)(s-\alpha_2)\cdots(s-\alpha_{j-1})}{(-s+\alpha_1)(-s+\alpha_2)\cdots(-s+\alpha_k)(s+\alpha_1)(s+\alpha_2)\cdots(s+\alpha_j)}\,ds
\end{aligned} \tag{5.30}$$

If $k \neq j$, we may assume that $j > k$; then the integrand in the last integral of (5.30) has no right-halfplane poles. Consequently,

$$\int_0^\infty \phi_k(t)\phi_j(t)\,dt = 0, \qquad k \neq j \tag{5.31}$$

If $k = j$, then

$$\begin{aligned}\int_0^\infty \phi_k^2(t)\,dt &= \frac{1}{2\pi j}\int_{-j\infty}^{j\infty} \Phi_k(s)\Phi_k(-s)\,ds \\ &= \frac{\alpha_k}{\pi j}\int_{-j\infty}^{j\infty} \frac{1}{(s+\alpha_k)(-s+\alpha_k)}\,ds \\ &= \frac{\alpha_k}{\pi_j}\oint \frac{-1/2\alpha_k}{s-\alpha_k}\,ds = 1\end{aligned} \tag{5.32}$$

where the contour integral may be any path around the pole at $s = \alpha_k$. Thus the set of $\phi_\nu(t)$ which are the inverse Laplace transforms of the $\Phi_\nu(s)$ of (5.19) is not only orthogonal but also orthonormal. This proof can easily be extended to include complex poles by using expressions from (5.21) as well as from (5.19).

Inasmuch as the $\phi_\nu(t)$ formed in this manner are orthonormal, the approximation of an $h(t)$ by an $h^*(t)$ in the least-squares sense is easily formulated by virtue of Section 2.4. We simply let

$$h^*(t) = \sum_{k=1}^{N+2M} c_k\phi_k(t) \tag{5.33}$$

where

$$c_k = \int_0^\infty h(t)\phi_k(t)\,dt \tag{5.34}$$

are real and may be evaluated by any method of integration or numerical technique.

If $h(t)$ is given in analytical form and if it is Laplace transformable, c_k may be evaluated through the Laplace transform of $h(t)$. Substituting (5.25) into (5.34) yields

$$\begin{aligned}c_k &= \sum_{\mu=1}^{N} C_{k\mu}\int_0^\infty h(t)\epsilon^{-\alpha_\mu t}\,dt \\ &\quad + \sum_{\nu=1}^{M} \mathrm{Re}\left[2(A_{k\nu} + jB_{k\nu})\int_0^\infty h(t)\epsilon^{-(\alpha_\nu + j\beta_\nu)t}\,dt\right] \\ &= \sum_{\mu=1}^{N} C_{k\nu}H(\alpha_\mu) + \sum_{\nu=1}^{M} \mathrm{Re}\left[2(A_{k\nu} + jB_{k\nu})H(\alpha_\nu + j\beta_\nu)\right]\end{aligned} \tag{5.35}$$

Example 1. Using the poles in the example of the previous section for

$$h(t) = \frac{1}{1+t} \tag{5.36}$$

find an $h^*(t)$ by the orthogonal-function method.

Solution. Since

$$\alpha_1 = 0.1361$$
$$\alpha_2 = 0.8062$$
$$\alpha_3 = 2.4639$$

we form

$$\Phi_1(s) = \frac{0.5217}{s + 0.1361}$$

$$\Phi_2(s) = \frac{1.2698(s - 0.1361)}{(s + 0.1361)(s + 0.8062)}$$
$$= -\frac{0.5158}{s + 0.1361} + \frac{1.7856}{s + 0.8062}$$

$$\Phi_3(s) = \frac{2.2198(s - 0.1361)(s - 0.8062)}{(s + 0.1361)(s + 0.8062)(s + 2.4639)}$$
$$= \frac{0.3649}{s + 0.1361} - \frac{3.0362}{s + 0.8062} + \frac{4.8911}{s + 2.4639}$$

or

$$\phi_1(t) = 0.5217\epsilon^{-0.1361t}$$
$$\phi_2(t) = -0.5158\epsilon^{-0.1361t} + 1.7856\epsilon^{-0.8062t}$$
$$\phi_3(t) = 0.3649\epsilon^{-0.1361t} - 3.0362\epsilon^{-0.8062t} + 4.8911\epsilon^{-2.4639t}$$

From (5.35) we have

$$c_1 = 0.5217 \int_0^\infty \frac{1}{1 + t}\epsilon^{-0.1361t}\,dt$$
$$= 0.5217\epsilon^{0.1361}Ei(0.1361) = 0.9258$$

$$c_2 = -0.5158 \int_0^\infty \frac{1}{1 + t}\epsilon^{-0.1361t}\,dt$$
$$+1.7856 \int_0^\infty \frac{1}{1 + t}\epsilon^{-0.8062t}\,dt$$
$$= -0.5158\epsilon^{0.1361}Ei(0.1361) + 1.7856\epsilon^{0.8062}Ei(0.8062)$$
$$= 0.3130$$

$$c_3 = 0.3649\epsilon^{0.1361}Ei(0.1361) - 3.0362\epsilon^{0.8062}Ei(0.8062)$$
$$+ 4.8911\epsilon^{2.4639}Ei(2.4639)$$
$$= 0.0606$$

Finally,

$$
\begin{aligned}
h^*(t) &= c_1\phi_1(t) + c_2\phi_2(t) + c_3\phi_3(t) \\
&= 0.3437\epsilon^{-0.1361t} + 0.3749\epsilon^{-0.8062t} + 0.2964\epsilon^{-2.4639t}
\end{aligned} \tag{5.37}
$$

This $h^*(t)$ agrees with the desired $h(t)$ so well that a graphical presentation of both functions will not show the deviation.

Example 2. Given the poles†

$$
\begin{aligned}
s_{1,2} &= -1 \mp j3.0777 \\
s_{3,4} &= -2 \mp j2.5440 \\
s_{5,6} &= -3 \mp j1.2133
\end{aligned}
$$

construct the orthogonal functions and find the approximate $h^*(t)$ for the $h(t)$ of Fig. 5.2.

Solution. With the set of six poles given, the following frequency-domain functions are formed according to (5.21):

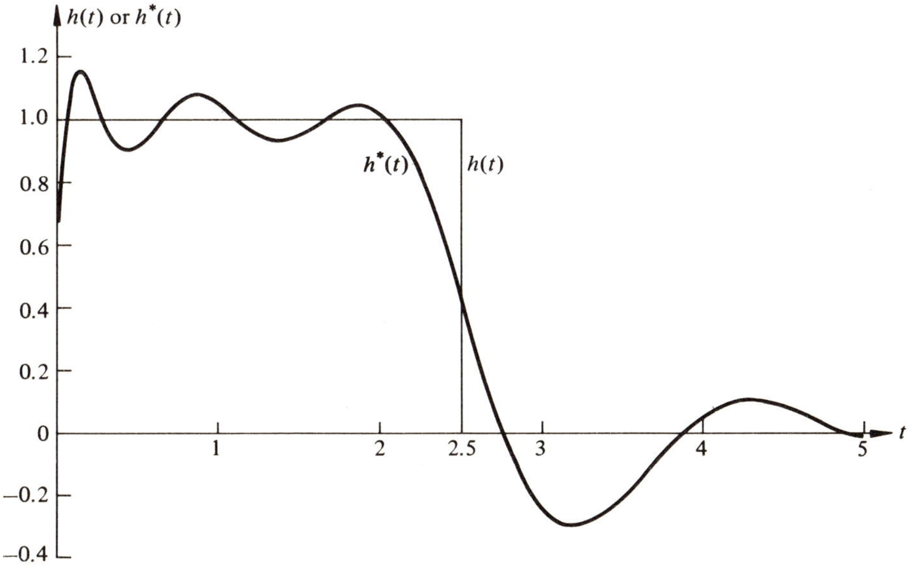

Fig. 5.2 The $h(t)$ and its approximant (5.38), of Example 2.

†These poles are located on a semicircle of radius $1 + \sqrt{5}$ and equally spaced in the vertical direction [KA4].

$$\Phi_1(s) = \frac{\sqrt{2}\,(s + 3.2361)}{(s + 1)^2 + (3.0777)^2} = \frac{0.7071 \pm j0.5137}{s + 1 \pm j3.0777}$$

$$\Phi_2(s) = \frac{\sqrt{2}\,(s - 3.2361)}{(s + 1)^2 + (3.0777)^2} = \frac{-0.7071 \pm j0.9732}{(s + 1)^2 + (3.0777)^2}$$

$$\Phi_3(s) = \frac{2[(s - 1)^2 + (3.0777)^2](s + 3.2361)}{[(s + 1)^2 + (3.0777)^2][(s + 2)^2 + (2.5440)^2]}$$

$$= \frac{-2 \mp j1.4532}{s + 1 \pm j3.0777} + \frac{3 \pm j0.4576}{s + 2 \pm j2.5440}$$

$$\Phi_4(s) = \frac{2[(s - 1)^2 + (3.0777)^2](s - 3.2361)}{[(s + 1)^2 + (3.0777)^2][(s + 2)^2 + (2.5440)^2]}$$

$$= \frac{2 \mp j2.7528}{s + 1 \pm j3.0777} + \frac{-3 \pm j6.1745}{s + 2 \pm j2.5440}$$

$$\Phi_5(s) = \frac{\sqrt{6}\,[(s - 1)^2 + (3.0777)^2][(s - 2)^2 + (2.5440)^2](s + 3.2361)}{[(s + 1)^2 + (3.0777)^2][(s + 2)^2 + (2.5440)^2][(s + 3)^2 + (1.2133)^2]}$$

$$= \frac{3.6742 \pm j2.6695}{s + 1 \pm j3.0777} + \frac{-14.6969 \mp j7.1408}{s + 2 \pm j2.5440} + \frac{12.2474 \pm j2.3829}{s + 3 \pm j1.2133}$$

$$\Phi_6(s) = \frac{\sqrt{6}\,[(s - 1)^2 + (3.0777)^2][(s - 2)^2 + (2.5440)^2](s - 3.2361)}{[(s + 1)^2 + (3.0777)^2][(s + 2)^2 + (2.5440)^2][(s + 3)^2 + (1.2133)^2]}$$

$$= \frac{-3.6742 \pm j5.0572}{s + 1 \pm j3.07777} + \frac{14.6969 \mp j30.2488}{s + 2 \pm j2.5440} + \frac{-12.2475 \pm j62.9841}{s + 3 \pm j1.2133}$$

The inverse Laplace transforms of these functions are

$$\phi_1(t) = 1.7481\epsilon^{-t} \cos(3.0777t - 36^\circ)$$

$$\phi_2(t) = -2.4060\epsilon^{-t} \sin(3.0777t - 36^\circ)$$

$$\phi_3(t) = -4.9443\epsilon^{-t} \cos(3.0777t - 36^\circ) + 6.6707\epsilon^{-2t} \cos(2.5440t - 25.91^\circ)$$

$$\phi_4(t) = 6.8052\epsilon^{-t} \sin(3.0777t - 36^\circ) - 13.7295\epsilon^{-2t} \sin(2.5440t - 25.91^\circ)$$

$$\phi_5(t) = 9.0832\epsilon^{-t} \cos(3.0777t - 36^\circ) - 32.6797\epsilon^{-2t} \cos(2.5440t - 25.91^\circ) + 24.9542\epsilon^{-3t} \cos(1.2133t - 11^\circ)$$

$$\phi_6(t) = -12.5020\epsilon^{-t} \sin(3.0777t - 36^\circ) + 67.2604\epsilon^{-2t} \sin(2.5440t - 25.91^\circ) - 128.2570\epsilon^{-3t} \sin(1.2133t - 11^\circ)$$

The evaluation of the c's of (5.33) is quite simple because the given $h(t)$ is Laplace transformable. Since

$$H(s) = \frac{1 - \epsilon^{-2.5s}}{s}$$

we obtain

$$H(1 + j3.0777) = 0.1181 - j0.2823$$
$$H(2 + j2.5440) = 0.0950 - j0.2413$$
$$H(3 + j1.2133) = 0.0956 - j0.1159$$

Then, by (5.35), we obtain

$$\begin{aligned} c_1 &= 0.4570 \\ c_2 &= -0.3826 \\ c_3 &= 0.5494 \\ c_4 &= -0.7576 \\ c_5 &= 0.9240 \\ c_6 &= -0.5465 \end{aligned}$$

Thus

$$\begin{aligned} h^*(t) &= 0.4570\phi_1(t) - 0.3826\phi_2(t) + 0.5494\phi_3(t) \\ &\quad - 0.7576\phi_4(t) + 0.9240\phi_5(t) - 0.5465\phi_6(t) \\ &= 6.4751\epsilon^{-t}\cos(3.0777t - 36^\circ) \\ &\quad + 2.5974\epsilon^{-t}\sin(3.0777t - 36^\circ) \\ &\quad - 26.5300\epsilon^{-2t}\cos(2.5440t - 25.91^\circ) \\ &\quad - 26.3580\epsilon^{-2t}\sin(2.5440t - 25.91^\circ) \\ &\quad + 23.0569\epsilon^{-3t}\cos(1.2133t - 11^\circ) \\ &\quad + 70.0960\epsilon^{-3t}\sin(1.2133t - 11^\circ) \end{aligned} \tag{5.38}$$

This $h^*(t)$ is plotted in Fig. 5.2 together with the given $h(t)$.

5.5 CONCLUSIONS

The methods given in this chapter enable us to find the numerator of $H^*(s)$ for a given set of poles such that the corresponding $h^*(t)$ approximates a given $h(t)$. Any of these methods combined with any of the methods of pole estimation given in Chapter 4 constitute a complete procedure for finding $H^*(s)$ for a given $h(t)$.

We have divided the finding of an $H^*(s)$ into two steps mainly for convenience in presentation. In a partical problem it is seldom that the two halves can be isolated. If an initial try on obtaining an $H^*(s)$ does not render a satisfactory answer, both the possibilities of modifying the pole locations and the numerator should be considered.

PROBLEMS

5.1. Using the poles obtained in Problem 4.4, find the complete $h^*(t)$ to approximate the $h(t)$ given in Fig. P4.1 by the point-matching method.

5.2. Assume the poles of $H^*(s)$ to be located at -5 and $-3 \pm j4$. Find the approximate $h^*(t)$ such that it matches the $h(t)$ given in Fig. P.4.1 at $t=0.2$, 0.4, and 0.8. Plot this $h^*(t)$.

5.3. Using the poles obtained in the example of Section 4.1 [or zeros of (4.6)], find an $h^*(t)$ to approximate the $h(t)$ of Fig. 4.1 in the least-squared-error sense in the range $0 < t < \infty$.

5.4. Using the method of matching initial derivatives, find the approximate $h^*(t)$ with poles at -1 and $-2 \pm j1$ for $h(t) = t\epsilon^{-t}$.

5.5. Repeat Problem 5.4 for $h(t) = t\epsilon^{-t^2}$ with poles at -2.5 and $-2 \pm j2$.

5.6. Repeat Problem 5.4 by forming the set of orthogonal functions based on the assumed poles there. Then approximate the $h(t)$ by $h^*(t)$ in the least-squared-error sense.

6

Direct Determination of an Approximate Network Function

In Chapters 4 and 5 we have divided the problem of finding a rational $H^*(s)$ so that its corresponding $h^*(t)$ approaches a given $h(t)$ into two separate steps —determination of the denominator of $H^*(s)$ and then determination of the numerator. In this chapter we shall discuss some procedures by which an approximate $h^*(t)$ may be found without taking these two separate steps.

6.1 APPROXIMATION IN THE FREQUENCY DOMAIN

In some time-domain synthesis problems the form of the problem strongly suggests that the approximation be carried out in the frequency domain. This is frequently the case when the given $h(t)$ or an approximate $\hat{h}(t)$ is Laplace transformable. In that case $H(s)$ or $\hat{H}(s)$ can readily be determined. Thus it becomes extremely tempting to apply an approximation technique to obtain a realizable $H^*(s)$ in the frequency domain. This approach does indeed prove to be very simple and effective. However, two cautionary remarks should be mentioned in this connection.

First, there is no simple, direct correlation between the accuracy of approximation in the frequency and time domains. In other words, if $H^*(s)$ is a good approximation to $H(s)$, there is no assurance that the corresponding $h^*(t)$ will be a good approximation of $h(t)$. Hence, if the approximation is carried out in the frequency domain, the corresponding approximation in the time domain should be checked to see if the accuracy is satisfactory.

Second, despite the large amount of work done in the frequency domain, little is available in making the frequency-domain approximation between an $\hat{H}(s)$ in analytical form and a realizable $H^*(s)$. This is partly because in most frequency-domain work we are more concerned with finding an $H^*(s)$ to give certain characteristics, usually either magnitude or phase, along the real-frequency axis. Hence the outlook for performing the approximation in the frequency domain is not as promising as it might appear at first.

However, although no advance assurance can be expected, problems do exist in which a good time-domain approximation is obtained via the frequency domain. For instance, the Padé approximants can be used in the approximation in the frequency domain.

Example 1. Suppose it is desired to find an $H^*(s)$ for the $h(t)$ of Fig. 6.1. Since $h(t)$ is already a composite of singularity functions, the step of finding an $\hat{h}(t)$ may be eliminated. We have

$$h(t) = u_{-2}(t) - 2u_{-2}(t - 0.5) + u_{-2}(t - 1) \tag{6.1}$$

and

$$\begin{aligned} H(s) &= \frac{1 - 2\epsilon^{-0.5s} + \epsilon^{-s}}{s^2} \\ &= 0.25 - 0.125s + 0.03646s^2 - 0.007813s^3 \\ &\quad + 0.001345s^4 - 0.0001953s^5 \\ &\quad + 0.00002461s^6 \ldots \end{aligned} \tag{6.2}$$

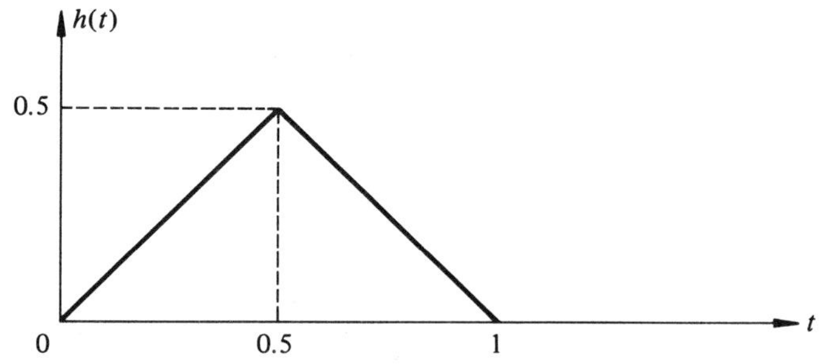

Fig. 6.1 The desired impulse response of (6.1).

As a first trial we shall find the (1, 3) Padé approximant of (6.2). Let

$$H^*_{13}(s) = \frac{a_0 + a_1 s}{1 + b_1 s + b_2 s^2 + b_3 s^3} \tag{6.3}$$

From (2.52) we have

$$\begin{aligned} a_0 &= 0.25 \\ a_1 &= -0.125 + 0.25b_1 \end{aligned}$$

and, from (2.53),

$$\begin{aligned} -0.125b_1 + 0.25b_2 \qquad\qquad\qquad &= -0.03646 \\ 0.03646b_1 - 0.125b_2 + 0.25b_3 \qquad &= 0.007813 \\ -0.007813b_1 + 0.03646b_2 - 0.125b_3 &= -0.001345 \end{aligned}$$

Solution of the last three equations yields

$$\begin{aligned} b_1 &= 0.4750 \\ b_2 &= 0.09166 \\ b_3 &= 0.007812 \end{aligned}$$

whence

$$a_1 = -0.00625$$

and

$$\begin{aligned} H_{13}^*(s) &= \frac{0.25 - 0.00625s}{1 + 0.4750s + 0.09166s^2 + 0.007812s^3} \\ &= \frac{-0.8001(s - 40)}{(s + 4.5555)[(s + 3.5890)^2 + (3.9009)^2]} \end{aligned} \tag{6.4}$$

which corresponds to the time-domain function

$$\begin{aligned} h_{13}^*(t) = 2.2071\epsilon^{-4.5555t} \\ + 2.2334\epsilon^{-3.5890t} \sin(3.9009t - 81.2°) \end{aligned} \tag{6.5}$$

This time function and the desired $h(t)$ are plotted in Fig. 6.2.

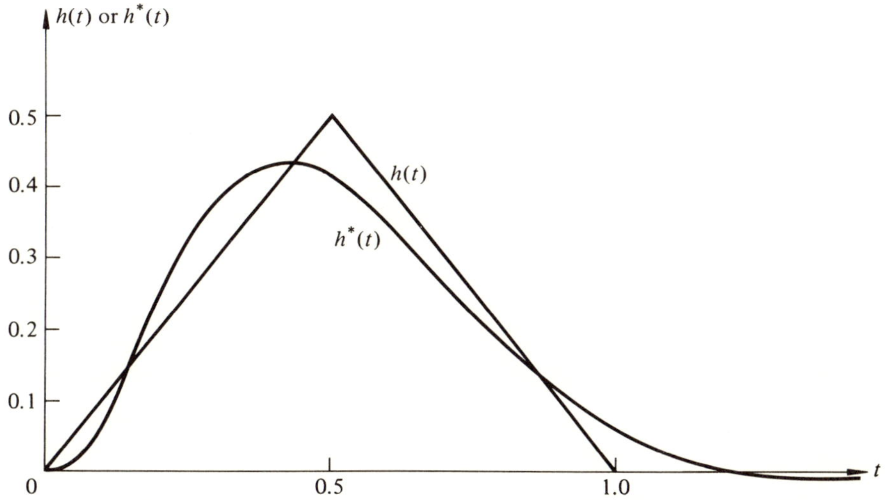

Fig. 6.2 Functions $h(t)$ of (6.1) and $h_{13}^*(t)$ of (6.5).

The maximum deviation between $h(t)$ and $h^*(t)$ occurs at $t = 0.5$, at which time $h(t) - h^*(t) = 0.085$. If this error is excessive, we may try next increasing the order of the Padé approximant, because we know, in this case, that all the inaccuracy is caused by the frequency-domain approximation. Let us find the (2, 4) approximant of (6.2). Here we let

$$H_{24}^*(s) = \frac{a_0 + a_1 s + a_2 s^2}{1 + b_1 s + b_2 s^2 + b_3 s^3 + b_4 s^4} \tag{6.6}$$

for which we make

$$\begin{aligned} a_0 &= 0.25 \\ a_1 &= -0.125 + 0.25b_1 \\ a_2 &= 0.03646 - 0.125b_1 + 0.25b_2 \end{aligned}$$

and

$$\begin{aligned} 0.03646b_1 - 0.125b_2 + 0.25b_3 &= 0.007813 \\ -0.007813b_1 + 0.03646b_2 - 0.125b_3 + 0.25b_4 &= -0.001345 \\ 0.001345b_1 - 0.007813b_2 + 0.03646b_3 - 0.125b_4 &= 0.0001953 \\ -0.0001953b_1 + 0.001345b_2 - 0.007813b_3 + 0.03646b_4 &= -0.00002461 \end{aligned}$$

Solution of the above equations yields

$$H_{24}^*(s) = \frac{0.25 - 0.007539s + 0.001231s^2}{1 + 0.4698s + 0.09401s^2 + 0.009737s^3 + 0.0004591s^4} \tag{6.7}$$

which corresponds to the time-domain function

$$\begin{aligned} h_{24}^*(t) = {} & 12.0286\epsilon^{-6.0370t} \sin(1.8698t + 2.54^\circ) \\ & - 3.3180\epsilon^{-4.5674t} \sin(5.8028t + 9.25^\circ) \end{aligned} \tag{6.8}$$

This time function and the $h(t)$ of (6.1) are plotted in Fig. 6.3. The maximum deviation has been reduced to 0.0425, which occurs at $t = 0.37$.

The reader may have recognized at this point that this method is actually identical to the procedure outlined in Section 4.1 for estimating the poles of $H^*(s)$. In Section 4.1 only poles of $H^*(s)$ need be found, because we were only interested in estimating the locations of those poles. Here we simply carry the frequency-domain approximation slightly further, so that the entire $H^*(s)$ is found.

Sometimes it is more convenient to approximate one of the derivatives of $h(t)$ by the corresponding derivative of $h^*(t)$ instead of dealing with the function itself. The following example should illustrate this alternative approach to the problem.

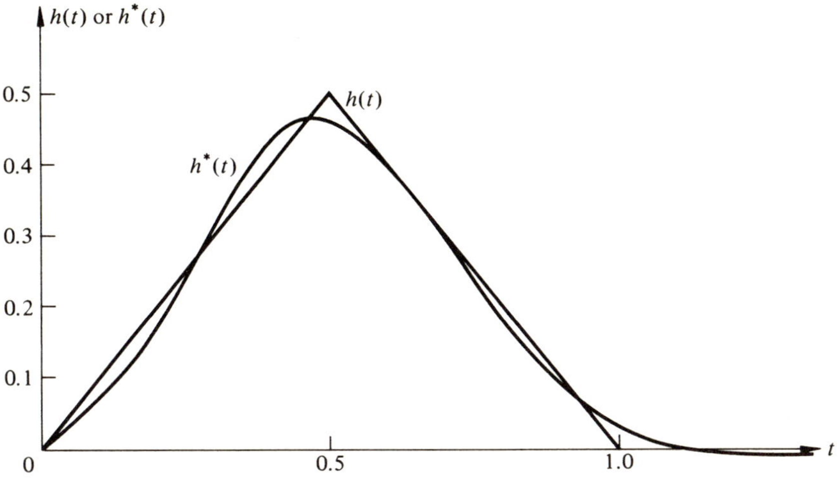

Fig. 6.3 Functions $h(t)$ of (6.1) ond $h^*_{24}(t)$ of (6.8).

Example 2. The previous example bypasses the time-domain part of approximating a given curve by another, because $h(t)$ is readily transformable. Suppose now we wish to find an $H^*(s)$ for

$$h(t) = t\epsilon^{-t^2} \tag{6.9}$$

This function is plotted in Fig. 6.4 as a solid curve.

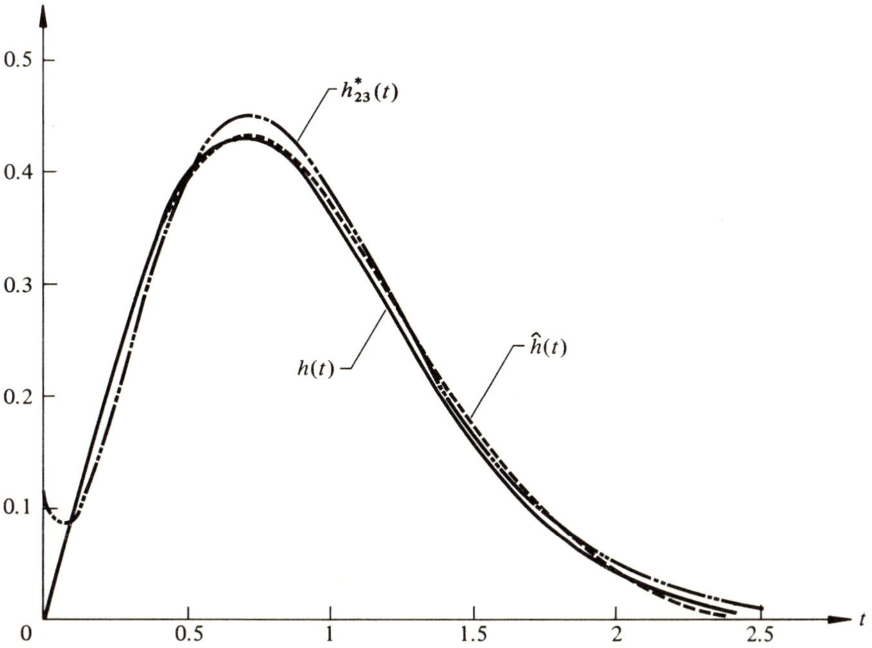

Fig. 6.4 The $h(t)$ of (6.9), $\hat{h}(t)$ of (6.12), and $h^*_{23}(t)$ of (6.16).

Because of the shape of this curve, it might be expected that a series of parabolas might be used to form the segments of $\hat{h}(t)$. To accomplish this is equivalent to approximate the derivative of $h(t)$ and $h'(t)$ by a series of straight lines. Since the given $h(t)$ is in analytic form, it is easy to find $h'(t)$ in this case. If, instead, $h(t)$ is given in graphical or numerical form, $h'(t)$ could still be obtained either graphically or numerically. Here

$$h'(t) = (1 - 2t^2)\epsilon^{-t^2} \tag{6.10}$$

This $h'(t)$ is plotted as the solid curve of Fig. 6.5. Then the straight-line approximation $\hat{h}'(t)$ is drawn in by inspection. [The negative peak is located at the point of inflection, $h''(t) = 0$, at $t = \sqrt{1.5}$.] From Fig. 6.5 we obtain

$$\begin{aligned}\hat{h}'(t) = u_{-1}(t) &- 0.5833u_{-2}(t) - 1.1980u_{-2}(t - 0.24)\\ &+ 1.2989u_{-2}(t - 0.88) + 0.8324u_{-2}(t - 1.2247)\\ &- 0.35u_{-2}(t - 2.5)\end{aligned} \tag{6.11}$$

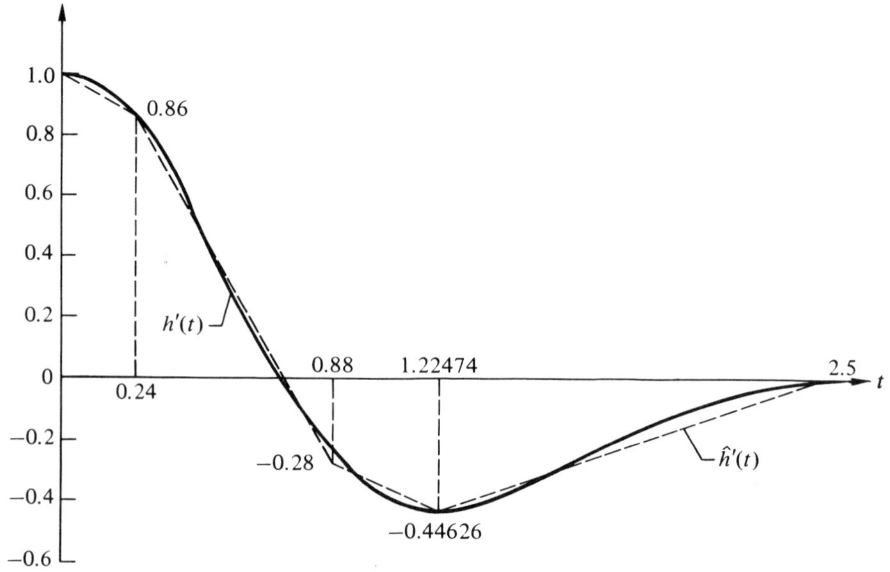

Fig. 6.5 The $h'(t)$ of (6.10) and $\hat{h}'(t)$ of (6.11).

which corresponds to

$$\begin{aligned}\hat{h}(t) = u_{-2}(t) &- 0.5833u_{-3}(t) - 1.1980u_{-3}(t - 0.24)\\ &- 1.2989u_{-3}(t - 0.88) + 0.8324u_{-3}(t - 1.2247)\\ &- 0.35u_{-3}(t - 2.5)\end{aligned} \tag{6.12}$$

Explicitly (6.12) gives

$$
\begin{aligned}
\hat{h}(t) &= -0.2917t^2 + t, && 0 < t < 0.24 \\
\hat{h}(t) &= -0.8907t^2 + 1.2875t - 0.0345, && 0.24 < t < 0.88 \\
\hat{h}(t) &= -0.2412t^2 + 0.1445t + 0.4685, && 0.88 < t < 1.2247 \\
\hat{h}(t) &= 0.175t^2 - 0.8749t + 1.0927, && 1.2247 < t < 2.5 \\
\hat{h}(t) &= 0, && 2.5 < t
\end{aligned}
\tag{6.13}
$$

and this $h(t)$ is plotted as the dashed curve in Fig. 6.4.

Transforming $\hat{h}(t)$ of (6.12) into the frequency domain yields

$$
\begin{aligned}
\hat{H}(s) = \frac{1}{s^2} - \frac{0.5833}{s^3} - \frac{1.1980}{s^3}\epsilon^{-0.24s} - \frac{1.2989}{s^3}\epsilon^{-0.88s} \\
+ \frac{0.8324}{s^3}\epsilon^{-1.2247s} - \frac{0.35}{s^3}\epsilon^{-2.5s}
\end{aligned}
\tag{6.14}
$$

Applying the series expansion of ϵ^{-as} and gathering like terms, we obtain

$$
\begin{aligned}
\hat{H}(s) = 0.511709 - 0.459248s + 0.259965s^2 - 0.113920s^3 + 0.041590s^4 \\
- 0.013127s^5 + \dots
\end{aligned}
\tag{6.15}
$$

A Padé approximant of the appropriate degrees may then be found for the series of (6.15) and used as an $H^*(s)$. For instance, if the approximant (2, 3) is used, we have

$$
H^*_{23}(s) = \frac{12.6599 - 0.08744s + 0.11236s^2}{24.7403 + 22.0331s + 7.42490s^2 + 0.97802s^3}
$$

which gives a corresponding impulse response,

$$
\begin{aligned}
h^*_{23}(t) = 4.0752\epsilon^{-2.74t} - 3.9603\epsilon^{-2.4259t}\cos(1.8302t) \\
+ 0.32631\epsilon^{-2.4259t}\sin(1.8302t)
\end{aligned}
\tag{6.16}
$$

This $h^*_{23}(t)$ is plotted in Fig. 6.4.

Although this example is designed to show the simplicity of the procedure by using singularity functions in the step of finding an $\hat{h}(t)$ for a given $h(t)$, the function (6.9) is actually Laplace transformable. Thus the approximation procedure can be compared with the exact answer. For (6.9)

$$
\begin{aligned}
H(s) = \mathscr{L}[h(t)] &= -\frac{d}{ds}\left[\frac{\sqrt{\pi}}{2}\epsilon^{s^2/4}\operatorname{erfc}\left(\frac{s}{2}\right)\right] \\
&= 0.5 - 0.4431s + 0.25s^2 - 0.1108s^3 + 0.0417s^4 - \dots
\end{aligned}
$$

which is seen to be comparable to the series in (6.15).

6.2 APPROXIMATION OF THE REAL PART OF $H(s)$ BY IMPULSE FUNCTIONS ALONG A CONTOUR IN THE s PLANE

It is well known that the impulse response $h(t)$ and the frequency response of a network are related by the Laplace transform pair

$$H(s) = \int_0^\infty h(t)\epsilon^{-(\sigma + j\omega)t}\,dt, \qquad \sigma > \sigma_0 \tag{6.17}$$

$$h(t) = \frac{1}{2\pi j}\int_{c-j\infty}^{c+j\infty} H(s)\epsilon^{st}\,ds, \qquad c < \sigma_0, \quad t > 0 \tag{6.18}$$

where σ_0 is the abscissa of convergence.

Now suppose we fix our contour to be a vertical line $\sigma = -\alpha$ to the right of $s = \sigma_0$ (Fig. 6.6). On this contour the network function written in rectangular form reads

$$\begin{aligned} H(-\alpha + j\omega) &= R(\omega) + jX(\omega) \\ &= \int_0^\infty h(t)\epsilon^{\alpha t}\cos\omega t\,dt + j\int_0^\infty h(t)\epsilon^{\alpha t}\sin\omega t\,dt, \end{aligned} \tag{6.19}$$

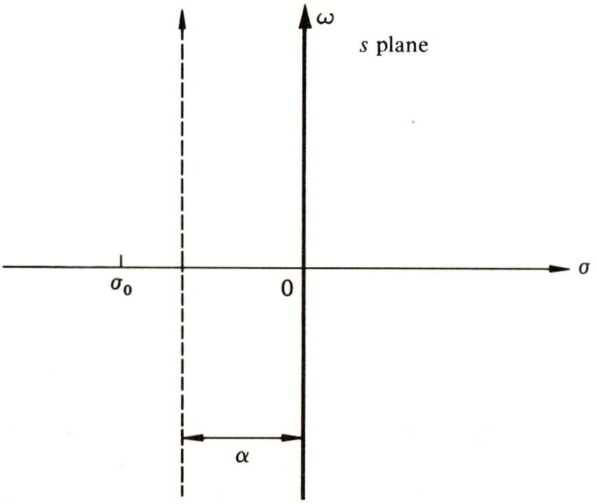

Fig. 6.6 The vertical contour $\sigma = -\alpha$ in the s plane.

Hence the real part of $H(s)$ along this contour may be expressed as

$$R(\omega) = \int_0^\infty h(t)\epsilon^{\alpha t}\cos\omega t\,dt \tag{6.20}$$

On this same vertical contour, (6.18) may be written as

$$h(t) = \frac{1}{2\pi}\epsilon^{-\alpha t}\int_{-\infty}^{\infty}[R(\omega) + jX(\omega)](\cos\omega t + j\sin\omega t)\,d\omega \tag{6.21}$$

This integral is the inverse Fourier transform of the function $[R(\omega) + jX(\omega)]$ and is frequently written as

$$h(t) = \epsilon^{-\alpha t}\int_{0}^{\infty}\frac{R(\omega)\cos\omega t - X(\omega)\sin\omega t}{\pi}\,d\omega \tag{6.22}$$

by halving the interval of integration and dropping those parts of the integrand which are odd functions of ω. It can be shown [GU1, p. 664] that the two terms of the integrand each contributes one half to the total integral. Hence we may write

$$h(t) = \frac{2}{\pi}\int_{0}^{\infty}\epsilon^{-\alpha t}R(\omega)\cos\omega t\,d\omega \tag{6.23}$$

Equations (6.20) and (6.23) may be regarded as another transform pair which relates the real part of $H(s)$ along the contour of Fig. 6.6 and a time-domain function $h(t)$, and vice versa.

Now suppose we have a given impulse response $h(t)$. The behavior of the real part of the corresponding $H(s)$ along the contour of Fig. 6.6, $R(\omega)$, can be computed through (6.20). If another realizable network function $H^*(s)$ can be found such that its real part along the same contour $R^*(\omega)$ approximates $R(\omega)$, then there is all the likelihood that the impulse response $h^*(t)$ will also approximate $h(t)$.

To arrive at a specific procedure for effecting this approximation scheme, suppose we first look at the contribution of a pole with a real residue k_ν situated a distance a_ν from and to the left of the contour as shown in Fig. 6.7. Such a pole will contribute the following term to $H(s)$:

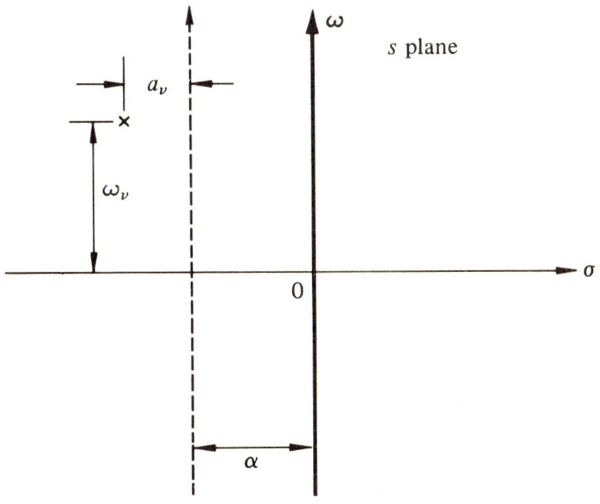

Fig. 6.7 Relative position of a pole to the contour $\sigma = -\alpha$.

$$H_\nu(s) = \frac{k_\nu}{s + a_\nu + \alpha - j\omega_\nu} \tag{6.24}$$

On the contour, this term has the value

$$H_\nu(-\alpha + j\omega) = \frac{k_\nu}{a_\nu + j(\omega - \omega_\nu)} \tag{6.25}$$

The real part of (6.25) is

$$R_\nu(\omega) = \frac{a_\nu k_\nu}{a_\nu^2 + (\omega - \omega_\nu)^2} \tag{6.26}$$

The variation of $R_\nu(\omega)$ depends on the value of a_ν. Figure 6.8 shows a typical variation of $R_\nu(\omega)$ for a moderate value of a_ν. When a_ν is large, the curve tends to be very flat and broad. When a_ν is small, $R_\nu(\omega)$ becomes a very tall and narrow pulse. Regardless of the value of a_ν, however, the area under $R_\nu(\omega)$ remains constant, that is,

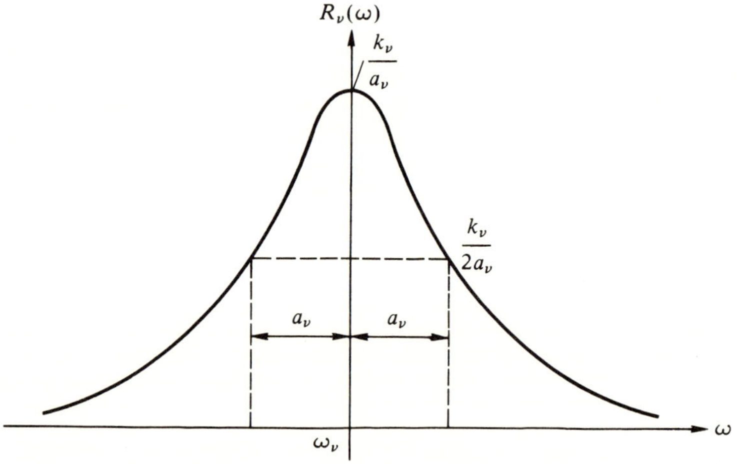

Fig. 6.8 Contribution of a pole to $R(\omega)$ along the contour $\sigma = -\alpha$.

$$\int_{-\infty}^{\infty} R_\nu(\omega)\, d\omega = \int \frac{a_\nu k_\nu\, d\omega}{a_\nu^2 + (\omega - \omega_\nu)^2} = k_\nu \pi \tag{6.27}$$

In the limit, as a_ν approaches zero, $R_\nu(\omega)$ approaches an impulse function or

$$\lim_{a_\nu \to 0} R_\nu(\omega) = \pi k_\nu\, \delta(\omega - \omega_\nu) \tag{6.28}$$

and

$$\lim_{a_\nu \to 0} H_\nu(s) = \frac{k_\nu}{s - j\omega_\nu} \tag{6.29}$$

If the poles of an $H(s)$ are all fairly close to the contour and are well separated from one another, the variation of its corresponding $R(\omega)$ will consist of a series of positive and negative pulses, depending on the signs of k_ν. Conversely, if a given $R(\omega)$ is a series of distinct pulses, it should be fairly easy to find a set of poles together with their residues to approximate this $R(\omega)$, because the contribution of each pole is quite independent of the others. We simply consider the given $R(\omega)$ as being made up of several components, one corresponding to each pulse, and then each pulse is approximated by a curve of the type shown in Fig. 6.8 and given by a term of (6.26) or (6.24).

If a given $R(\omega)$ is not readily divisible into several isolated pulses approximately of the shape of Fig. 6.8, the following scheme will be helpful.

The principle of this scheme is very simple. We integrate $R(\omega)$ twice, or

$$R^{(-2)}(\omega) = \int_{-\infty}^{\omega} \int_{-\infty}^{\omega} R(\omega')\, d\omega'\, d\omega' \tag{6.30}$$

Then the curve of $R^{(-2)}(\omega)$ is approximated by a series of line segments (with even symmetry, of course). The broken-line variation is then taken as the variation of $R^{*(-2)}(\omega)$. Thus $R^*(\omega)$ is obtained as a series of impulse functions by differentiating $R^{*(-2)}(\omega)$ twice. This scheme is illustrated in Fig. 6.9, in which the given $R(\omega)$ is shown in part (c). The $R^{(-2)}(\omega)$ is computed and shown in (a). Then $R^{(-2)}(\omega)$ is approximated by the broken-line variation in (a). The weights and locations of the impulses representing $R^*(\omega)$ are then readily obtained from (a) and are shown in (c). Once $R^*(\omega)$ is obtained in this form, each impulse is then represented by a pole on the contour $\sigma = -\alpha$.

Although this scheme is simple in principle, it suffers from one practical difficulty. This difficulty arises from the necessity of integrating $R(\omega)$. When $h(t)$ is given in analytic form, the corresponding $R(\omega)$ is usually too complicated for integration in closed form. If $R(\omega)$ is obtained numerically, inaccuracy will be magnified after each integration. This difficulty can be somewhat reduced by the following artifice, which circumvents each integration of $R(\omega)$ by replacing it with an equivalent step in the time domain.

For a given $h(t)$, suppose we form another time function,

$$q(t) = -\frac{h(t)}{t^2} \tag{6.31}$$

and call its Laplace transform $Q(s)$. We choose a contour $\sigma = -\alpha$ to the right of all poles of $Q(s)$. Then we have, on this contour,

$$Q(-\alpha + j\omega) = Q_r(\omega) + jQ_i(\omega) \tag{6.32}$$

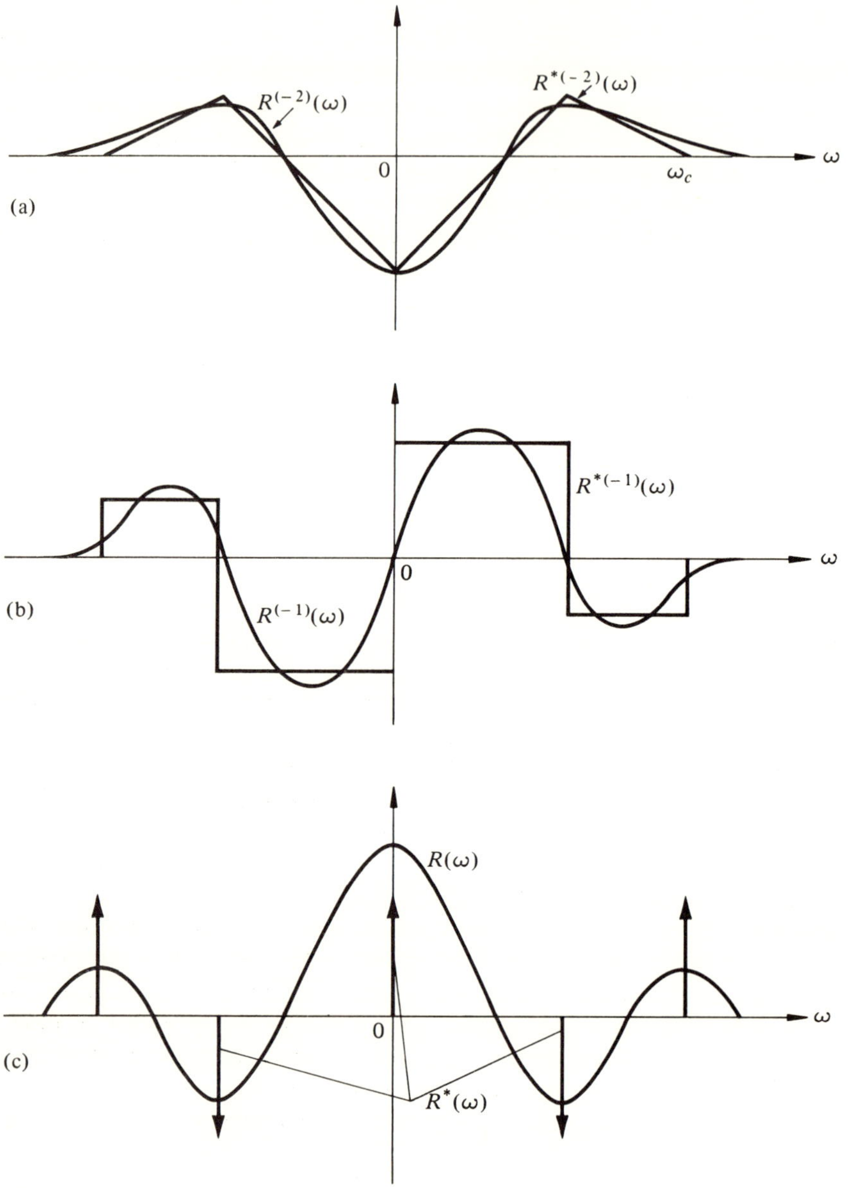

Fig. 6.9 Approximation of an $R(\omega)$ by a series of impulse functions through the straight-line approximation of the second integral.

and

$$Q_r(\omega) = \int_0^\infty q(t)\epsilon^{\alpha t} \cos \omega t \, dt \tag{6.33}$$

where $Q_r(\omega)$ and $Q_i(\omega)$ are the real and imaginary parts of $Q(-\alpha + j\omega)$, respectively. Differentiating $Q_r(\omega)$ twice yields

$$\frac{d^2 Q_r}{d\omega^2} = \int_0^\infty h(t)\epsilon^{\alpha t} \cos \omega t \, dt = R(\omega) \tag{6.34}$$

Hence

$$Q_r(\omega) = R^{(-2)}(\omega) \tag{6.35}$$

The inverse transform of (6.33) is

$$\begin{aligned} q(t) &= \frac{1}{\pi} \int_{-\infty}^{\infty} Q_r(\omega)\epsilon^{-\alpha t} \cos \omega t \, d\omega \\ &= \frac{\epsilon^{-\alpha t}}{\pi t} \int_{\omega=-\infty}^{\omega=\infty} Q_r(\omega) \, d(\sin \omega t) \\ &= \frac{\epsilon^{-\alpha t}}{\pi t} \left[Q_r(\omega) \sin \omega t \Big|_{\omega=-\infty}^{\omega=\infty} - \int_{\omega=-\infty}^{\omega=\infty} Q_r'(\omega) \sin \omega t \, d\omega \right] \\ &= \frac{Q_r(\omega)\epsilon^{-\alpha t} \sin \omega t}{\pi t} \Big|_{\omega=-\infty}^{\omega=\infty} + \frac{\epsilon^{-\alpha t}}{\pi t^2} \int_{\omega=-\infty}^{\omega=\infty} Q_r'(\omega) \, d(\cos \omega t) \\ &= \frac{Q_r(\omega)\epsilon^{-\alpha t} \sin \omega t}{\pi t} \Big|_{\omega=-\infty}^{\omega=\infty} + \frac{Q_r'(\omega)\epsilon^{-\alpha t} \cos \omega t}{\pi t^2} \Big|_{\omega=-\infty}^{\omega=\infty} \\ &\quad - \frac{\epsilon^{-\alpha t}}{\pi t^2} \int_{-\infty}^{\infty} Q_r''(\omega) \cos \omega t \, d\omega \end{aligned} \tag{6.36}$$

From (6.36) it is seen that for this inverse transform to give back

$$q(t) = -\frac{h(t)}{t^2}$$

it is necessary that

$$\lim_{\omega \to \pm\infty} Q_r(\omega) = 0 \tag{6.37}$$

and

$$\lim_{\omega \to \pm\infty} Q_r'(\omega) = 0 \tag{6.38}$$

But

$$\lim_{t\to 0}\left[-\frac{h(t)}{t^2}\right] = \lim_{s\to\infty} sQ(s) \tag{6.39}$$

Hence (6.37) and (6.38) are satisfied if

$$-\frac{h(t)}{t^2}$$

is finite at $t = 0$.

Therefore,

$$R^{(-2)}(\omega) = \int_0^\infty \left[-\frac{h(t)}{t^2}\right]\epsilon^{\alpha t} \cos \omega t \, dt \tag{6.40}$$

$$-\frac{h(t)}{t^2} = \frac{1}{\pi}\int_{-\infty}^{\infty} R^{(-2)}(\omega)\epsilon^{-\alpha t} \cos \omega t \, d\omega \tag{6.41}$$

form a transform pair only if $h(t)$ approaches zero at least as far as t^2 for $t \to 0$. And when $h(t)$ does behave this way at the vicinity of the origin, $R^{(-2)}(\omega)$ may be obtained by using (6.40) instead of integrating $R(\omega)$ twice.

When an $h(t)$ does not satisfy this requirement, then $h(t)$ must first be modified. This can easily be accomplished by subtracting from the given $h(t)$ a time function, say $b(t)$, such that the difference $h(t) - b(t) = h_1(t)$ satisfies this requirement. Then the process may be applied to $h_1(t)$. If the chosen $b(t)$ is Laplace transformable, the function may be restored by adding $B(s)$ to $H_1(s)$. One useful $b(t)$ is

$$b(t) = (A + Bt)\epsilon^{-\beta t} \tag{6.42}$$

Once $R^{(-2)}(\omega)$ has been found and a series of half-lines have been determined by inspection, the vertical positions (ω_ν) and the strength (k_ν) of the series of impulse functions that approximate $R(\omega)$ along the contour can be determined. Their real-axis displacements (a_ν) remain to be determined. The following schemes might be employed:

1. From (6.26) it is seen that

$$R_\nu(\omega_\nu) = \frac{k_\nu}{a_\nu} \tag{6.43}$$

   Assuming little or no interference among the contributions of all poles, each a_ν may be calculated according to (6.43), in which each $R_\nu(\omega_\nu)$ is the maximum value of each peak of $R(\omega)$, and each k_ν is the area of each impulse divided by π by virtue of (6.27).
2. Simply let $a_\nu = 0$ for all poles. This places all poles squarely on the contour $\sigma = -\alpha$. Experience has shown that the effect of slight variations in the value of α is not serious in the time domain.

1. Approximate Computation of $R^{(-2)}(\omega)$

When the desired $h(t)$ is given in the form of a graph or tabulated data or when the given analytic expression of $h(t)$, when substituted into (6.40), leads to an integral so complicated that closed-form expression of $R^{(-2)}(\omega)$ is not practical, the following approximate method will be useful.

We compute a new function

$$h_a(t) = -\frac{h(t)}{t^2} \epsilon^{\alpha t} \tag{6.44}$$

From (6.40) we have

$$R^{(-2)}(\omega) = \int_0^\infty h_a(t) \cos \omega t \, dt \tag{6.45}$$

Now suppose that we approximate the computed $h_a(t)$ by another function $\hat{h}_a$, which is a series of line segments. In theory $\hat{h}_a(t)$ can be made as close to $h_a(t)$ as necessary. In practice, since we are going to approximate $R^{(-2)}(\omega)$ by $R^{*(-2)}(\omega)$ anyhow, it would seem that we should be able to tolerate a large error in computing $R^{(-2)}(\omega)$. Generally, $\hat{h}_a(t)$ may be expressed as

$$\hat{h}_a(t) = \sum_{k=1}^{p} a_k u_{-1}(t - t_k) + \sum_{j=1}^{q} b_j u_{-2}(t - t_j) \tag{6.46}$$

where each t_k corresponds to an instant at which $\hat{h}_a(t)$ has a discontinuity and each t_j corresponds to an instant at which $\hat{h}_a(t)$ changes its slope. Then an approximate second integral of $R(\omega)$ may be found according to [GU1, p. 674]

$$\begin{aligned} \hat{R}^{(-2)}(\omega) &= \int_0^\infty \hat{h}_a(t) \cos \omega t \, dt \\ &= \frac{1}{\omega} \sum_{k=1}^{p} a_k \sin \omega t_k + \frac{1}{\omega^2} \sum_{j=1}^{q} b_j \cos \omega t_j \end{aligned} \tag{6.47}$$

This $\hat{R}^{(-2)}(\omega)$ can be computed easily.

2. Selection of the Contour

For a given $h(t)$ the choice of the vertical contour (or the value of α) has a great bearing on the final outcome of the approximation. Qualitatively, we may say that in most situations, a moderate value of α should be used. This can be justified as follows. If α is made zero, then $h^*(t)$ will have an undamped oscillatory wave shape, which usually will not be suitable for approximation. On the other hand, if α is large, terms in $h^*(t)$ will be highly damped (either oscillatory or nonoscillatory) and $h^*(t)$ will be a spike with very short duration. Hence the value of α should be somewhere between these two extremes.

Within these extremes, ideally, the value of α should be chosen such that the variation of $R(\omega)$ (1) shows clearly defined peaks and yet (2) does not have too many overall valleys and peaks. In some situations it may be advisable to make sample calculations of $R(\omega)$ for several values of α before proceeding with the approximation procedure.

3. Estimate of Time-Domain Errors

The deviation between the given $h(t)$ and the approximate $h^*(t)$ computed by the method outlined in this section,

$$\epsilon(t) = h(t) - h^*(t) \tag{6.48}$$

may be expressed through (6.41) as

$$\epsilon(t) = -\frac{2t^2}{\pi}\epsilon^{-\alpha t}\int_0^{\infty}[R^{(-2)}(\omega) - R^{*(-2)}(\omega)]\cos\omega t\, d\omega \tag{6.49}$$

Let ω_c denote the value of ω above which $R^{*(-2)}(\omega) = 0$ [see Fig. 6.9(a)]. Then

$$\epsilon(t) = \epsilon_A(t) + \epsilon_T(t) \tag{6.50}$$

where

$$\epsilon_A(t) = -\frac{2t^2}{\pi}\epsilon^{-\alpha t}\int_0^{\omega_c}[R^{(-2)}(\omega) - R^{*(-2)}(\omega)]\cos\omega t\, d\omega \tag{6.51}$$

$$\epsilon_T(t) = -\frac{2t^2}{\pi}\epsilon^{-\alpha t}\int_{\omega_c}^{\infty}R^{(-2)}(\omega)\cos\omega t\, d\omega \tag{6.52}$$

The time function $\epsilon_A(t)$ is the error due to approximation in the frequency band $|\omega| < |\omega_c|$. This error is usually small. The upper bound of this integral may be written as

$$|\epsilon_A(t)| \leq \frac{2t^2\epsilon^{-\alpha t}}{\pi}\int_0^{\omega_c}|R^{(-2)}(\omega) - R^{*(-2)}(\omega)|\, d\omega \tag{6.53}$$

Since the maximum value of $(2t^2/\pi)\epsilon^{-\alpha t}$ is $0.3446/\alpha^2$, we may further say that

$$|\epsilon_A(t)| \leq \frac{0.3446}{\alpha^2}\int_0^{\omega_c}|R^{(-2)}(\omega) - R^{*(-2)}(\omega)|\, d\omega \tag{6.54}$$

The time function $\epsilon_T(t)$ is the error due to truncating the frequency band above $\omega = \omega_c$. This error may be estimated by replacing $R^{(-2)}(\omega)$ by $\hat{R}^{(-2)}(\omega)$ of (6.47). Thus

$$\epsilon_T \cong \frac{2t^2\epsilon^{-\alpha t}}{\pi}\left[\sum_{k=1}^{p} a_k \int_{\omega_c}^{\infty} \frac{\sin t_k\omega \cos t\omega}{\omega}\, d\omega \right.$$
$$\left. + \sum_{j=1}^{q} b_j \int_{\omega_c}^{\infty} \frac{\cos t_k\omega \cos t\omega}{\omega}\, d\omega\right] \tag{6.55}$$

Both integrals in (6.55) can be expressed in closed form in terms of the sine integral.

Example. Suppose the desired $h(t)$ is the triangular pulse given in Fig. 6.10(a). Since

$$\lim_{t\to 0}\left[\frac{h(t)}{t^2}\right] = \lim_{t\to 0}\left(\frac{1}{t}\right) \tag{6.56}$$

is not bounded, (6.33) cannot be used on this $h(t)$ as it is. We choose, arbitrarily,

$$b(t) = t\epsilon^{-2.5t} \tag{6.57}$$

for which

$$B(s) = \frac{1}{(s+2.5)^2} \tag{6.58}$$

Substituting

$$h_1(t) = h(t) - t\epsilon^{-2.5t} \tag{6.59}$$

in place of $h(t)$ in (6.40) and using a numerical scheme of integration, $R_1^{(-2)}(\omega)$ is obtained and shown in Fig. 6.10(b) for $\alpha = 2$. The ramp-function approximation $R_1^{*(-2)}(\omega)$ is then inserted by inspection on the same graph. It is seen that

$$R_1^{*(-2)}(\omega) = \frac{4}{1.5}u_{-2}(\omega + 4.3) - \left(\frac{4}{1.5} + \frac{13}{2.8}\right)u_{-2}(\omega + 2.8)$$
$$+ \frac{26}{2.8}u_{-2}(\omega) - \left(\frac{4}{1.5} + \frac{13}{2.8}\right)u_{-2}(\omega - 2.8)$$
$$+ \frac{4}{1.5}u_{-2}(\omega - 4.3) \tag{6.60}$$

or

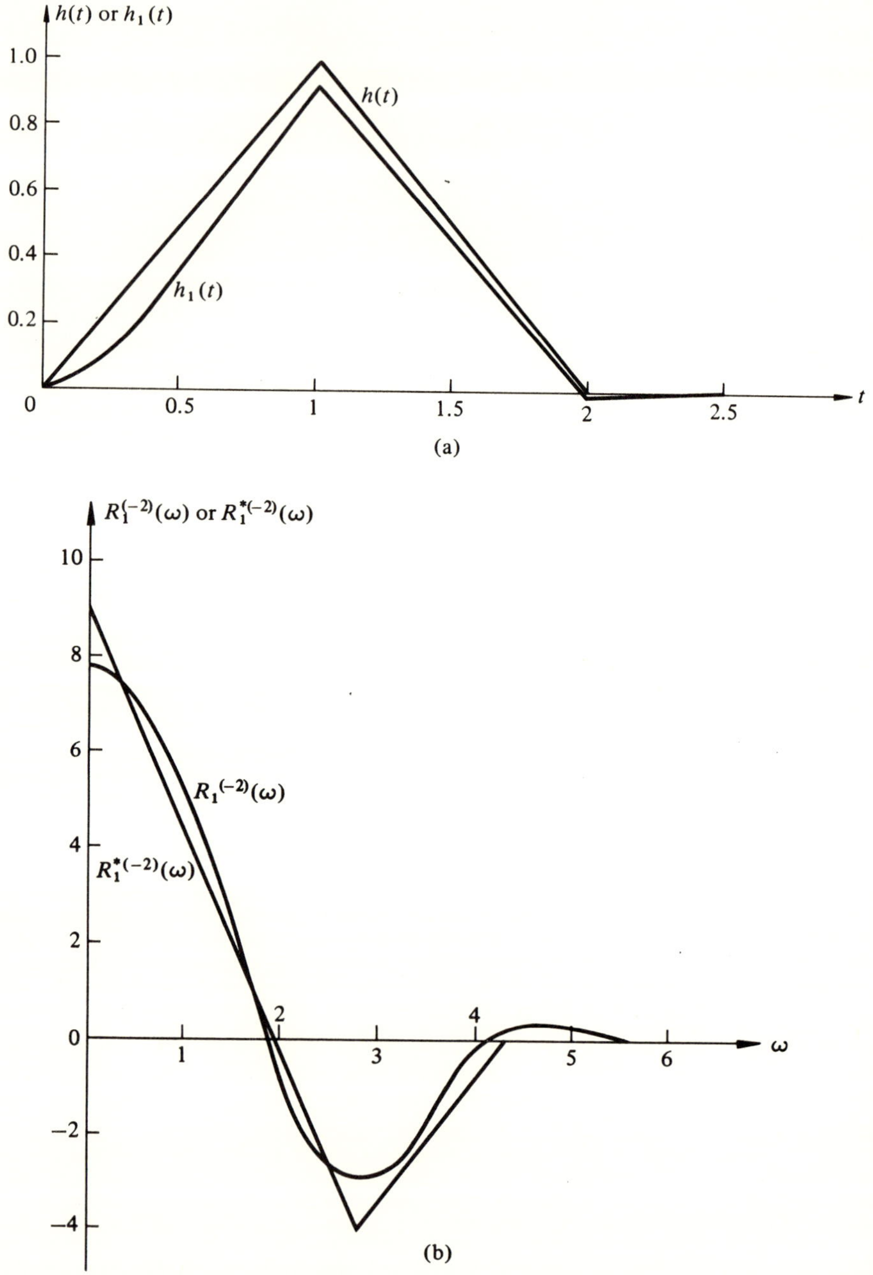

Fig. 6.10 Various functions used in the example in Section 6.2.

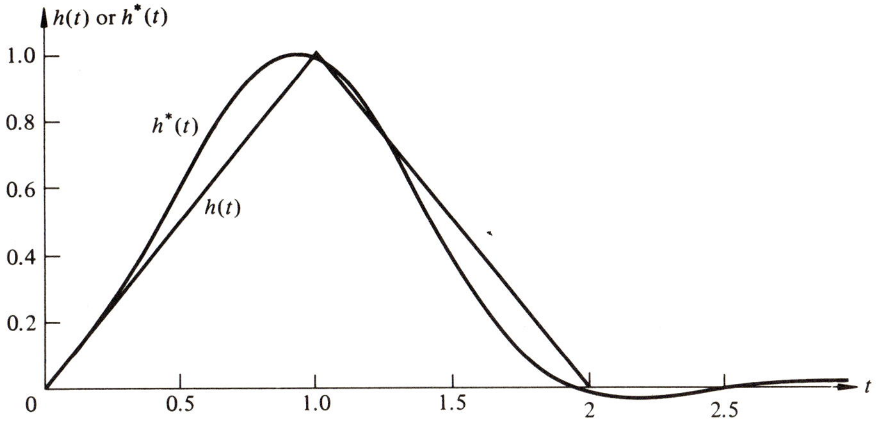

(c)

Fig. 6.10 (continued)

$$R_1^*(\omega) = 2.6667u_0(\omega + 4.3) - 7.3095u_0(\omega + 2.8) + 9.2857u_0(\omega) - 7.3095u_0(\omega - 2.8) + 2.6667u_0(\omega - 4.3) \tag{6.61}$$

Recalling that $\alpha = 2$ and setting $a_v = 0$, we thus have

$$H_1^*(s) = \frac{0.8488}{s + 2 \pm j4.3} - \frac{2.3267}{s + 2 \pm j2.8} + \frac{2.9557}{s + 2} \tag{6.62}$$

and

$$H^*(s) = \frac{1.6976s + 3.3952}{s^2 + 4s + 22.49} - \frac{4.6534s + 9.3068}{s^2 + 4s + 11.84} + \frac{2.9557}{s + 2} + \frac{1}{(s + 2.5)^2} \tag{6.63}$$

Hence

$$h^*(t) = 1.6976\epsilon^{-2t} \cos 4.3t - 4.6534\epsilon^{-2t} \cos 2.8t + 2.9557\epsilon^{-2t} + t\epsilon^{-2.5t} \tag{6.64}$$

This $h^*(t)$ is shown in Fig. 6.10(c) together with the desired $h(t)$.

6.3 FOURIER-SERIES METHOD

The Fourier-series method of approximation is well known to engineers. It is an orthogonal-function approximation and, therefore, has all the advantages of the use of orthogonal functions. In addition, the series is readily transformable into the frequency domain; for

$$f(t) = \sum_{k=-\infty}^{\infty} a_k \epsilon^{jk\omega_0 t} \tag{6.65}$$

$$F(s) = \sum_{k=-\infty}^{\infty} \frac{a_k}{s - jk\omega_0} \tag{6.66}$$

The disadvantage of the Fourier-series method lies in the fact that $f(t)$ is a periodic function. Thus the Fourier series cannot always be used to approximate an impulse response $h(t)$ directly. This and the succeeding two sections deal with methods of overcoming this shortcoming of the Fourier-series method.

For example, suppose we have a given impulse response $h(t)$ which may be considered sufficiently close to zero after $t = T$ to be negligible. Then we may assume that $h(t) = 0$ for $t < 0$ and $t > T$, as shown in Fig. 6.11. A periodic function $h_p(t)$ with period T can be created by repeating the original $h(t)$ every T seconds along the entire t axis. The $h_p(t)$ so formed for the $h(t)$ of Fig. 6.11 is illustrated in Fig. 6.12. The Fourier series of $h_p(t)$ can now be found by various analytical, graphical, or computational techniques and may be expressed as

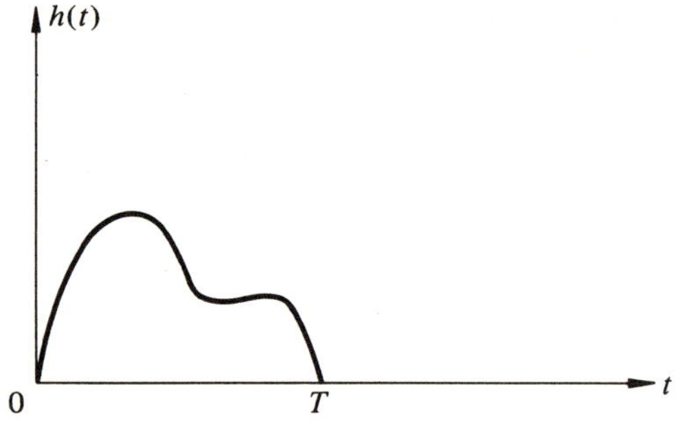

Fig. 6.11 An $h(t)$ that is zero for $0 > t > T$.

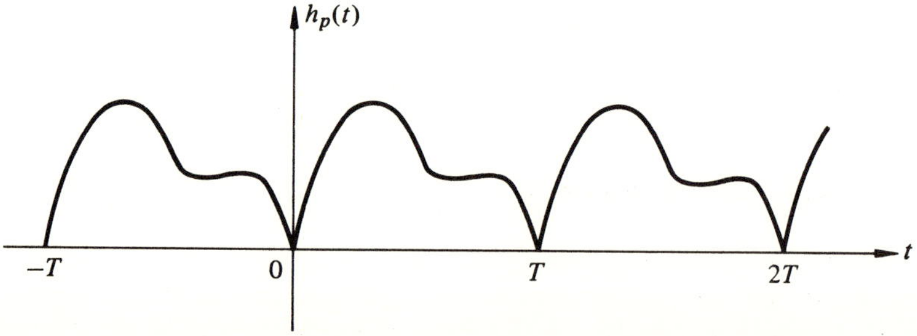

Fig. 6.12 A periodic $h_p(t)$ formed from the $h(t)$ of Fig. 6.11.

$$h_p(t) = \sum_{k=-\infty}^{\infty} a_k \epsilon^{jk\omega_0 t}, \qquad \omega_0 = \frac{2\pi}{T} \tag{6.67}$$

When this Fourier series is truncated to include only terms inside $k = \pm N$, an approximate periodic function $\hat{h}_p(t)$ is obtained, where

$$\hat{h}_p(t) = \sum_{k=-N}^{N} a_k \epsilon^{jk\omega_0 t} \tag{6.68}$$

If a method can now be found such that only that part of $\hat{h}_p(t)$ inside the range $0 < t < T$ is retained, a function approximating $h(t)$ would have been found.

To accomplish this, we may first multiply $\hat{h}_p(t)$ by the unit step $u(t)$. This cuts off that part of $\hat{h}_p(t)$ in the range $t < 0$. Actually, this step may be omitted, because we are going to take the Laplace transform† of these time functions.

To eliminate that part of $\hat{h}_p(t)$ for $t > T$, we may use the double-impulse weighting function $w(t)$ shown in Fig. 6.13 and given by

$$w(t) = u_0(t) - u_0(t - T) \tag{6.69}$$

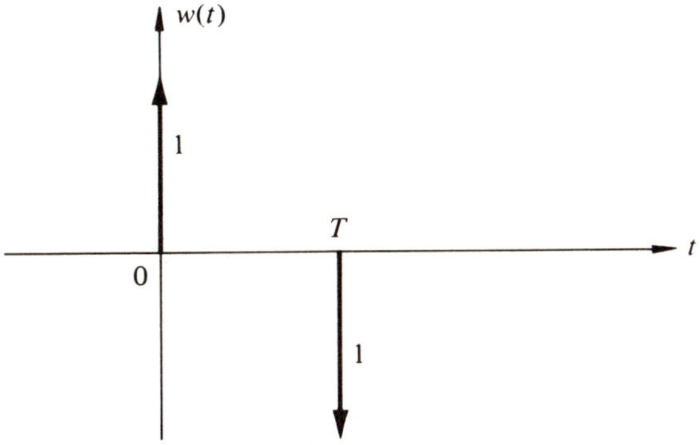

Fig. 6.13 The weighting function $w(t)$ of (6.69).

Since when a function $f(t)$ is convolved with a unit impulse located at t_0, the result is $f(t)$ delayed by t_0, or $f(t - t_0)$, the effect of convolving a function with $w(t)$ is to add to the function its own negative delayed by T. This process may be represented schematically as in Fig. 6.14. It is easy to see that when $w(t)$ is convolved into $h_p(t)$, it gives $h(t)$, or

$$h(t) = w(t) \circledast h_p(t) \tag{6.70}$$

Now since $\hat{h}_p(t)$ approximates $h_p(t)$, if the convolution with $w(t)$ is performed on $\hat{h}_p(t)u(t)$, the result would be an approximate function of $h(t)$. Or

†The transform in question is the unilateral Laplace transform as defined in (6.17).

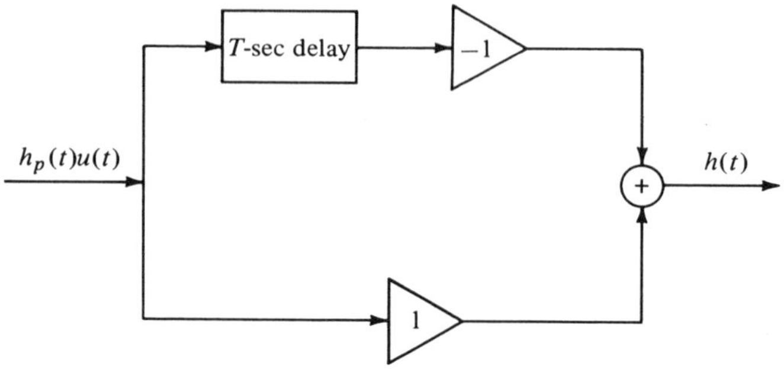

Fig. 6.14 A circuit interpretation of Equation (6.70).

$$\hat{h}(t) = w(t) \circledast [\hat{h}_p(t)u(t)] \tag{6.71}$$

When this $\hat{h}(t)$ is transformed into the frequency domain it becomes

$$\hat{H}(s) = (1 - \epsilon^{-Ts}) \sum_{k=-N}^{N} \frac{a_k}{s - jk\omega_0} \tag{6.72}$$

This function is not rational and normally cannot be realized without other frequency-domain approximations, which may be unacceptable as far as the time-domain accuracy is concerned. To make use of this Fourier-series technique and its associated properties, some additional mathematical artifices are necessary. Two such schemes are discussed in the following two sections.

6.4 FOURIER-SERIES APPROXIMATION METHOD I

To overcome the difficulties arisen from the Laplace transform of the direct Fourier-series approximation of an impulse response function, the following scheme may be employed [GU1, p. 707; ST4].

Assuming that $h(t)$ may be considered zero for $t > T$, we may create two periodic functions, $h_{p1}(t)$ and $h_{p2}(t)$, such that

$$h_{p1}(t) = h_{p2}(t) = \tfrac{1}{2}h(t), \qquad 0 < t < T \tag{6.73}$$

and $h_{p1}(t)$ repeats every T with alternate sign while $h_{p2}(t)$ repeats every T with the same sign. This arrangement is illustrated in Fig. 6.15. These functions may each be regarded as periodic functions of period $2T$. The Fourier series of these functions can then be found. Because of the symmetry possessed by these functions, $h_{p1}(t)$ will have only odd harmonics and $h_{p2}(t)$ will have only even harmonics. Thus we may write

$$h_{p1}(t) = \sum_{k=-\infty}^{\infty} a_{2k+1}\epsilon^{j(2k+1)\omega_0 t} \tag{6.74}$$

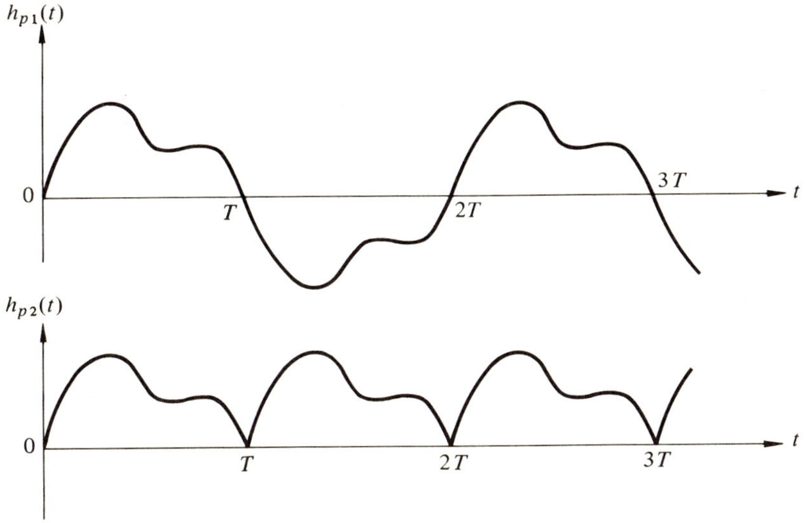

Fig. 6.15 Formation of $h_{p1}(t)$ and $h_{p2}(t)$ for the $h(t)$ of Fig. 6.11.

$$h_{p2}(t) = \sum_{k=-\infty}^{\infty} a_{2k} \epsilon^{j(2k)\omega_0 t} \tag{6.75}$$

$$\omega_0 = \frac{2\pi}{2T} \tag{6.76}$$

Now

$$h_{p1}(t) + h_{p2}(t) = h_p(t) = \sum_{k=-\infty}^{\infty} a_k \epsilon^{jk\omega_0 t} \tag{6.77}$$

is another periodic function $h_p(t)$ of period $2T$ and

$$h_p(t) = h(t), \qquad 0 < t < T \tag{6.78}$$

$$h_p(t) = 0, \qquad T < t < 2T \tag{6.79}$$

Also

$$h_{p2}(t) - h_{p1}(t) = h_p(t - T) = h_{pa}(t) \tag{6.80}$$

is $h_p(t)$ delayed by T seconds. The arrangements of these two periodic functions are depicted in Fig. 6.16.

We may next obtain an approximate periodic function $\hat{h}_p(t)$ by truncating the series of (6.77). Or if we let

$$\hat{h}_p(t) = \sum_{k=-N}^{N} a_k \epsilon^{jk\omega_0 t} \tag{6.81}$$

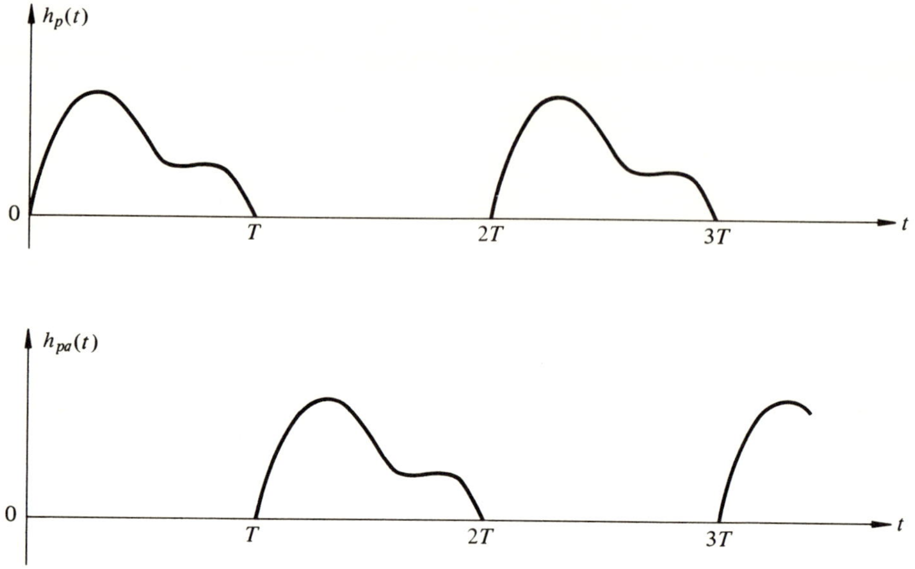

Fig. 6.16 Function $h_p(t)$ as defined by (6.77) and $h_{pa}(t)$ as defined by (6.80) for the $h(t)$ of Fig. 6.11.

then

$$\hat{h}_p(t) = \hat{h}(t) \cong h(t), \qquad 0 < t < T \tag{6.82}$$

$$\hat{h}_p(t) \cong 0, \qquad T < t < 2T \tag{6.83}$$

If the $h_{pa}(t)$ series is truncated we may designate it as

$$\begin{aligned} \hat{h}_{pa}(t) &= \sum_{k=-N}^{N} (-1)^k a_k \epsilon^{jk\omega_0 t} \\ &= \sum_{k=-N}^{N} a_k \epsilon^{jk\omega_0 (t-T)} \end{aligned} \tag{6.84}$$

Suppose now we let

$$\hat{H}_{p1}(s) = \mathscr{L}\left[\sum_{\substack{k=-N \\ k \text{ odd}}}^{N} a_k \epsilon^{jk\omega_0 t}\right] = \mathscr{L}[\hat{h}_{p1}(t)] \tag{6.85}$$

$$\hat{H}_{p2}(s) = \mathscr{L}\left[\sum_{\substack{k=-N \\ k \text{ even}}}^{N} a_k \epsilon^{jk\omega_0 t}\right] = \mathscr{L}[\hat{h}_{p2}(t)] \tag{6.86}$$

Then

$$\hat{H}_p(s) = \mathscr{L}[\hat{h}_p(t)] = \hat{H}_{p2}(s) + \hat{H}_{p1}(s) \tag{6.87}$$

$$\hat{H}_{pa}(s) = \mathscr{L}[\hat{h}_{pa}(t)] = \hat{H}_{p2}(s) - \hat{H}_{p1}(s) \tag{6.88}$$

We also create

$$\hat{H}_{pb}(s) = \hat{H}_p(s)\epsilon^{-sT} \tag{6.89}$$

and thus

$$\hat{h}_{pb}(t) = \mathscr{L}^{-1}[H_{pb}(s)] \tag{6.90}$$

Then $\hat{h}_{pb}(t)$ and $\hat{h}_{pa}(t)$ will be identical for all t except $0 < t < T$. The typical behavior and relative main pulses of $\hat{h}_p(t)$, $\hat{h}_{pa}(t)$, and $\hat{h}_{pb}(t)$ are illustrated in Fig. 6.17.

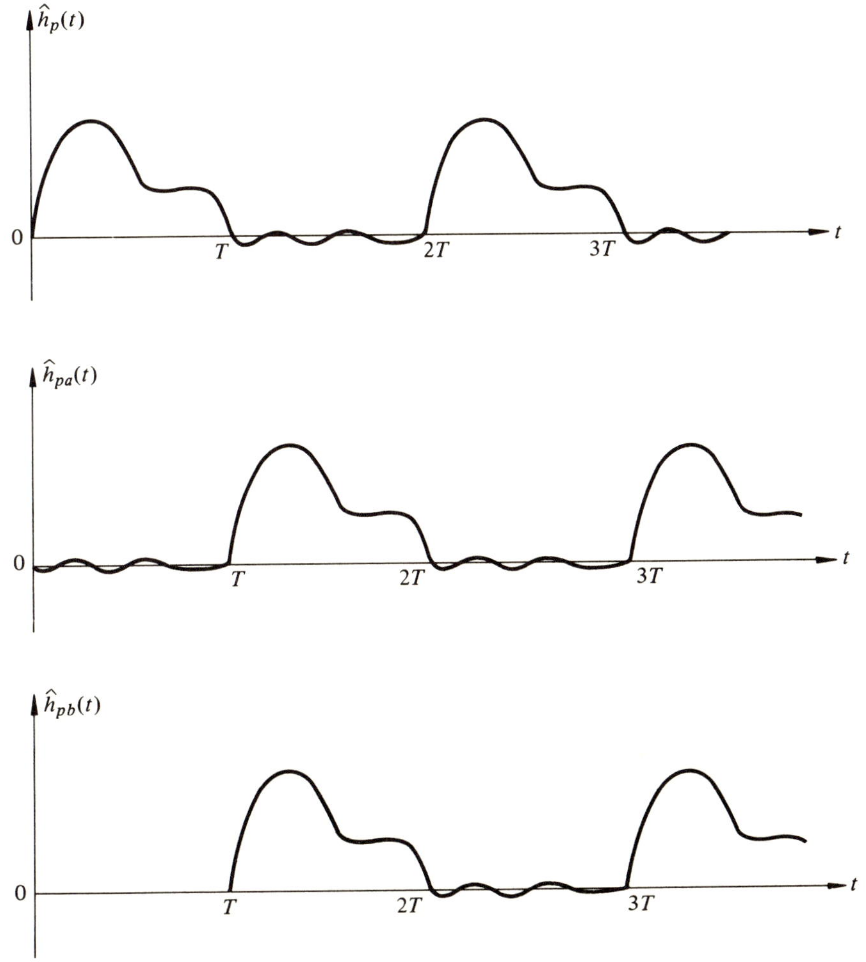

Fig. 6.17 Functions $\hat{h}_p(t)$, $\hat{h}_{pa}(t)$, and $\hat{h}_{pb}(t)$ as defined by (6.81), (6.84), and (6.90).

Let $\epsilon(t)$ represent the difference between $\hat{h}_{pa}(t)$ and $\hat{h}_{pb}(t)$, or

$$\epsilon(t) = \hat{h}_{pa}(t) - \hat{h}_{pb}(t) \tag{6.91}$$

Clearly

$$\begin{aligned} \epsilon(t) &= \hat{h}_{p2}(t) - \hat{h}_{p1}(t), && 0 < t < T \\ &= 0, && \text{elsewhere} \end{aligned} \tag{6.92}$$

Thus it is seen that the error in substituting $\hat{h}_{pb}(t)$ and $\hat{h}_{pa}(t)$ for each other is small if the Fourier series includes a sufficient number of terms. This function is useful in estimating the error of this method in the time domain.

From (6.89) we have

$$\epsilon^{-sT} = \frac{\hat{H}_{pb}(s)}{\hat{H}_p(s)} \tag{6.93}$$

Thus

$$\epsilon^{-sT} \cong \frac{\hat{H}_{pa}(s)}{\hat{H}_p(s)} \tag{6.94}$$

Substituting (6.87) and (6.88) into (6.94) and squaring both sides gives

$$\epsilon^{-2sT} = \left[\frac{\hat{H}_{p2}(s) - \hat{H}_{p1}(s)}{\hat{H}_{p2}(s) + \hat{H}_{p1}(s)}\right]^2 \tag{6.95}$$

and

$$1 - \epsilon^{-2sT} \cong \frac{4\hat{H}_{p2}(s)\hat{H}_{p1}(s)}{[\hat{H}_{p2}(s) + \hat{H}_{p1}(s)]^2} \triangleq \hat{G}(s) \tag{6.96}$$

Thus a rational function that approximates the factor $(1 - \epsilon^{-2sT})$ [analogous to the factor $(1 - \epsilon^{-sT})$ of (6.72)] has been generated for the particular $h(t)$. In other words, the inverse Laplace transform of $\hat{G}(s)$ is an approximation of $[u_0(t) - u_0(t - 2T)]$. Since

$$\begin{aligned} h(t) &= [u_0(t) - u_0(t - 2T)] \circledast h_p(t) \\ &= [u_0(t) - u_0(t - 2T)] \circledast [h_{p1}(t) + h_{p2}(t)] \end{aligned} \tag{6.97}$$

and

$$H(s) = (1 - \epsilon^{-2sT})H_p(s) \tag{6.98}$$

we may presume that

$$H^*(s) = \frac{4\hat{H}_{p1}(s)\hat{H}_{p2}(s)}{\hat{H}_{p1}(s) + \hat{H}_{p2}(s)} \tag{6.99}$$

will give an $h^*(t)$ that is an approximation of $h(t)$.

Example. For the triangular $h(t)$ given in Fig. 6.18(a), $h_{p1}(t)$ and $h_{p2}(t)$ are shown in Fig. 6.18(b) and (c). The Fourier series of these periodic functions are readily found to be

$$h_{p1}(t) = \frac{4}{\pi^2}\left(\sin t - \frac{1}{9}\sin 3t + \frac{1}{25}\sin 5t + \dots\right)$$

$$h_{p2}(t) = \frac{1}{4} - \frac{2}{\pi^2}\left(\cos 2t + \frac{1}{9}\cos 6t + \frac{1}{25}\cos 10t + \dots\right)$$

Truncating after the fourth harmonic, we have

$$\hat{h}_{p1}(t) = \frac{4}{\pi^2}\left(\sin t - \frac{1}{9}\sin 3t\right)$$

$$\hat{h}_{p2}(t) = \frac{1}{4} - \frac{2}{\pi^2}\cos 2t$$

and

$$\hat{H}_{p1}(s) = \frac{4/\pi^2}{s^2+1} - \frac{4/9\pi^2}{s^2+9}$$

$$\hat{H}_{p2}(s) = \frac{\frac{1}{4}}{s} - \frac{(2/\pi^2)s}{s^2+4}$$

Substituting these functions into (6.99) we get

$$\begin{aligned} H^*(s) &\\ &= \frac{1.0808(s^2+13)(s^2+21.1160)}{s^6+5.7053s^5+31.1160s^4+96.9904s^3+220.160s^2+296.677s+190.044} \\ &= -\frac{0.3074s - 5.6194}{(s+1.3691)^2+(0.7312)^2} + \frac{0.3324s - 4.8093}{(s+1.1349)^2+(2.2005)^2} \\ &\quad - \frac{0.02503s - 0.1660}{(s+0.3487)^2+(3.5704)^2} \end{aligned} \tag{6.100}$$

This $H^*(s)$ gives

$$\begin{aligned} h^*(t) = &-8.2667\epsilon^{-1.3691t}\cos(0.7312t + 87.87°) + 2.3803\epsilon^{-1.1349t} \\ &\times \cos(2.2005t + 81.97°) - 0.05497\epsilon^{-0.3487t}\cos(3.5704t + 69.91°) \end{aligned} \tag{6.101}$$

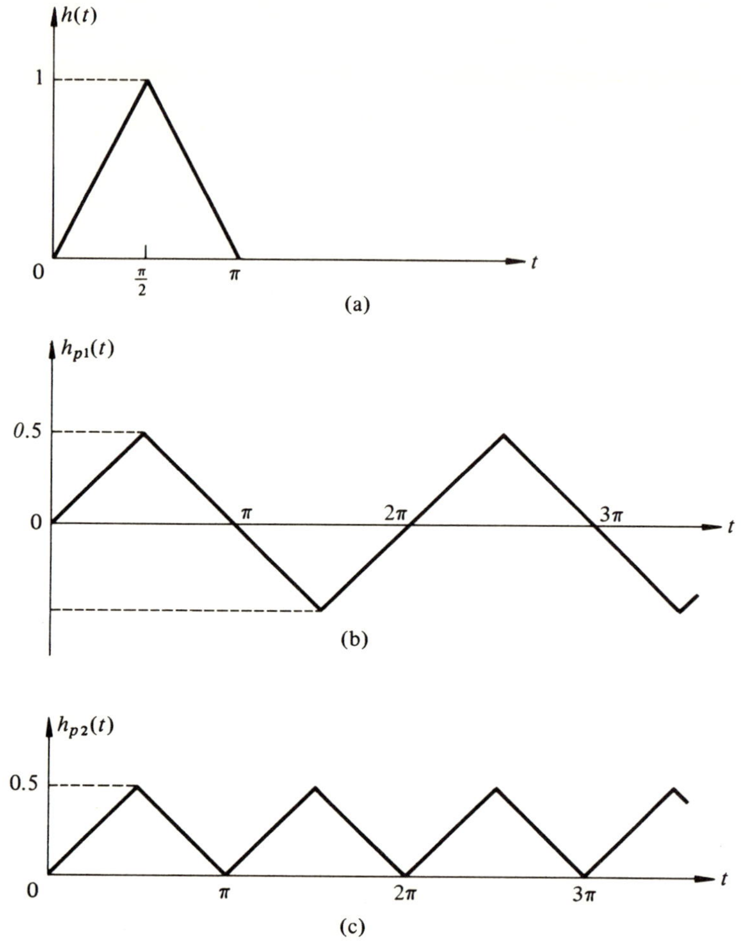

Fig. 6.18 Functions $h(t)$, $h_{p1}(t)$, and $h_{p2}(t)$ for the example of Section 6.4.

The function in (6.101) is plotted in Fig. 6.19 together with the desired $h(t)$.

From this example it is seen that this method is very easy to implement, especially when means of computing the Fourier series of a given function is already available. The expression of (6.99) is, however, not always theoretically realizable in that it can have right-halfplane poles. Although the poles of both $\hat{H}_1(s)$ and $\hat{H}_2(s)$ are all on the j axis, the poles of $H^*(s)$ are the zeros of $[\hat{H}_2(s) + \hat{H}_1(s)]$ which could be anywhere on the s plane. Also the poles of $H^*(s)$ are introduced by the approximation of the delay factor ϵ^{-2sT} (6.95). Referring to Fig. 6.14, it is seen that if right-halfplane poles do exist in $H^*(s)$, they are

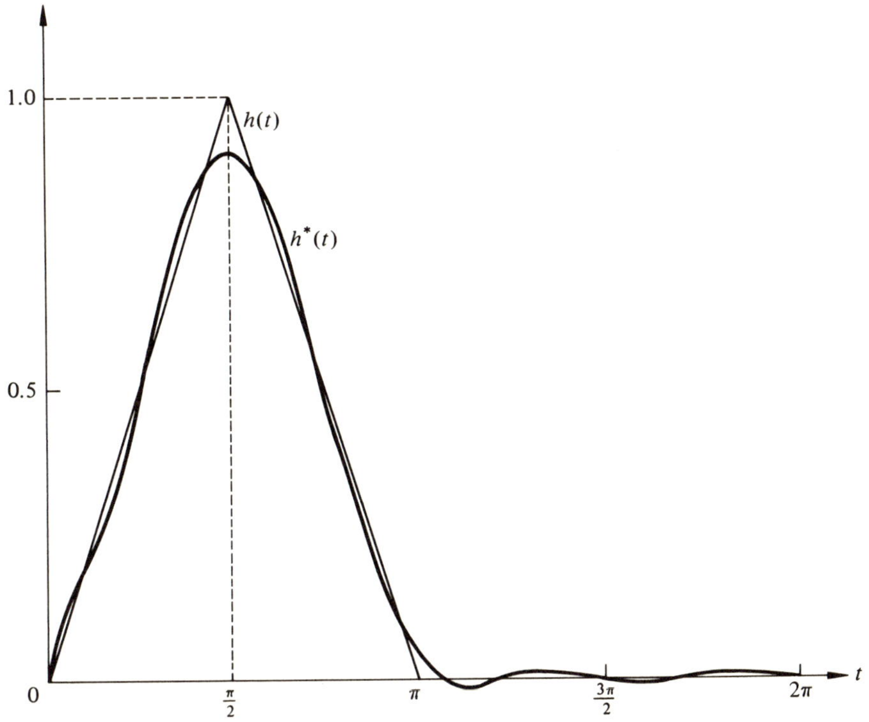

Fig. 6.19 The $h^*(t)$ of (6.101) which approximates the $h(t)$ of Fig. 6.18(a).

introduced into the system by the delay network. In other words, if the approximation for the delay network is realizable, then $H^*(s)$ is also realizable.

When right-halfplane poles do occur in $H^*(s)$, we may either try other combinations of terms of the Fourier expansions or simply omit those right-halfplane poles from $H^*(s)$. The latter scheme was discussed in Section 3.1.

Estimate of Time-Domain Errors

Other than the truncation error inherent with the Fourier-series approximation, the method presented in this section introduces an additional source of error represented by $\epsilon(t)$ of (6.91). From (6.87) we have

$$\hat{H}_p(s) = \hat{H}_{p1}(s) + \hat{H}_{p2}(s) \tag{6.102}$$

Substituting (6.88) and (6.89) into the Laplace transform of (6.91) yields

$$\hat{H}_p(s)\epsilon^{-sT} + E(s) = \hat{H}_{p2}(s) - \hat{H}_{p1}(s) \tag{6.103}$$

where $E(s) = \mathscr{L}[\epsilon(t)]$. Solution of (6.102) and (6.103) yields

$$2\hat{H}_{p2}(s) = \hat{H}_p(s) + \hat{H}_p(s)\epsilon^{-sT} + E(s) \tag{6.104}$$

$$2\hat{H}_{p1}(s) = \hat{H}_p(s) - \hat{H}_p(s)\epsilon^{-sT} - E(s) \tag{6.105}$$

Substitution of (6.104) and (6.105) into (6.99) renders

$$H^*(s) = \hat{H}_p(s) - \hat{H}_p(s)\epsilon^{-2sT} - 2E(s)\epsilon^{-sT} - \frac{E^2(s)}{\hat{H}_p(s)} \tag{6.106}$$

In the time domain, we have

$$h^*(t) = h_p(t) - h_p(t - 2T) - 2\epsilon(t - T) - \mathscr{L}^{-1}\left[\frac{E^2(s)}{\hat{H}_p(s)}\right] \tag{6.107}$$

The combination of the first two terms gives an approximation of $h(t)$ in the interval $0 < t < 2T$ and is identically zero for $t > 2T$. The third term is small if sufficient number of terms are included in the Fourier series. The last term is the triple convolution

$$\epsilon(t) \circledast \epsilon(t) \circledast \mathscr{L}^{-1}\left[\frac{1}{\hat{H}_p(s)}\right] \tag{6.108}$$

The last function in (6.108) cannot be estimated in general terms. However, it can be reasonably presumed that it will usually be of the same order of magnitude as that of $1/\hat{h}_p(t)$. Since $\epsilon(t)$ is usually small and it appears twice in the triple convolution (6.108), the last term in (6.107) may be expected to be of the same order of magnitude as that of $\epsilon(t)$.

6.5 FOURIER-SERIES APPROXIMATION METHOD II

An alternative scheme that will enable us to take advantage of the Fourier-series method of approximation of a periodic function is given here [LI1]. Again, assume that $h(t)$ is negligible after $t = T$. We form a periodic function $h_p(t)$ such that $h(t)$ is repeated every T with alternate signs. This arrangement is illustrated in Fig. 6.20. Then if a function that is obtained by delaying $h_p(t)u(t)$ by T seconds is added to $h_p(t)$ itself, $h(t)$ will result. This operation is illustrated in Fig. 6.21. Mathematically this operation is equivalent to

$$h(t) = h_p(t)u(t) + h_p(t - T)u(t - T) \tag{6.109}$$

This relationship can be effected if we create the weighting function

$$w(t) = u_0(t) + u_0(t - T) \tag{6.110}$$

which is depicted in Fig. 6.22, and then convolve it with $h_p(t)$.

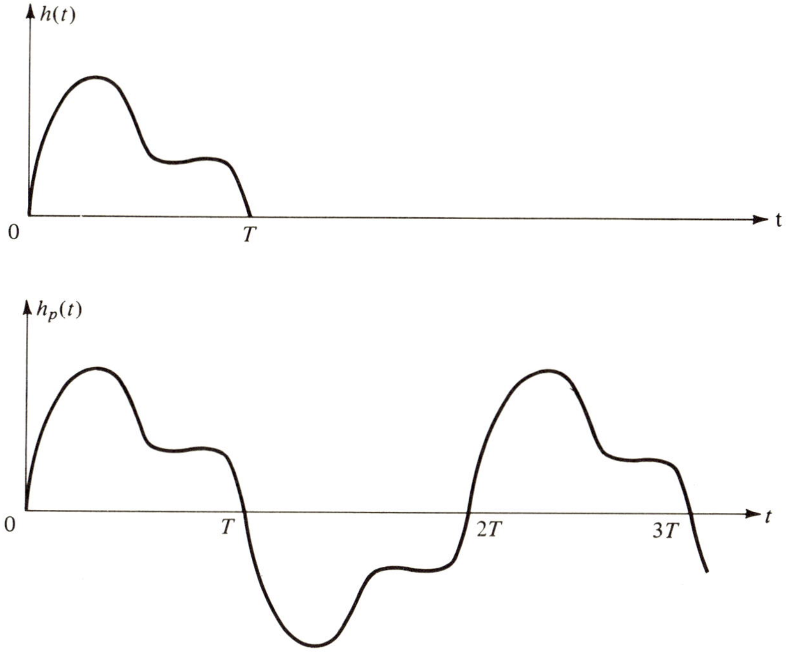

Fig. 6.20 Illustration of the $h_p(t)$ for a given $h(t)$ for Fourier series method II.

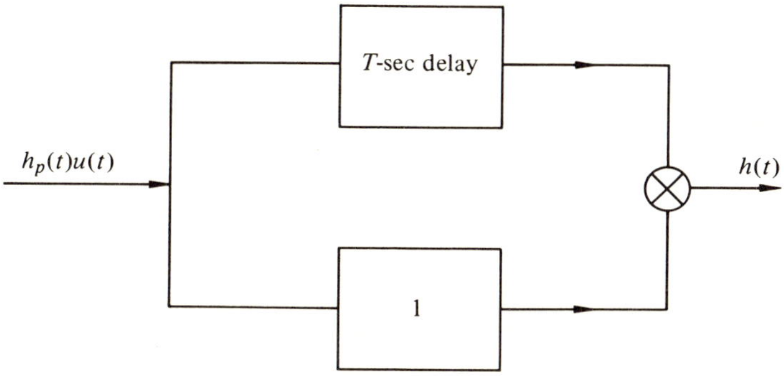

Fig. 6.21 Circuit interpretation of (6.111).

That is,

$$h(t) = h_p(t) \circledast w(t) \tag{6.111}$$

In the frequency domain, we have

$$H(s) = H_p(s)W(s) = H_p(s)(1 + \epsilon^{-sT}) \tag{6.112}$$

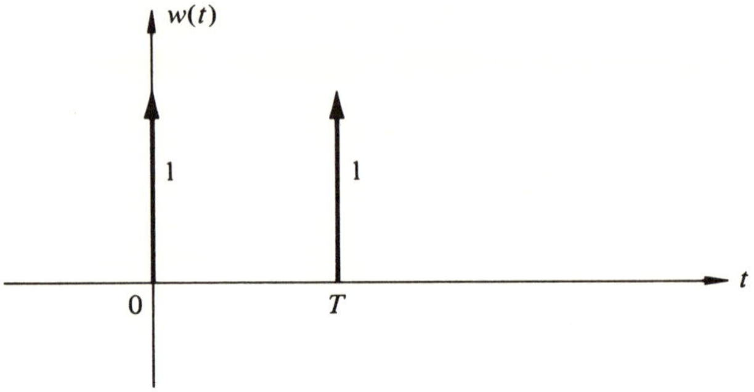

Fig. 6.22 Weighting function $w(t)$ of (6.110).

Now, if we can find a function $\hat{H}(s)$ that approximates $H_p(s)$ and another function $\hat{W}(s)$ that approximates $W(s)$, and if these two functions are such that

$$H^*(s) = \hat{H}_p(s)\hat{W}(s) \tag{6.113}$$

is a realizable network function, then the corresponding $h^*(t)$ is an approximation of $h(t)$.

Since $h_p(t)$ is a periodic function, it can be represented by a Fourier series. Since its second half and its first half of each cycle are negative of each other, its Fourier-series representation will have only odd terms or

$$h_p(t) = \sum_{k=1}^{\infty} [a_k \sin (2k - 1)\omega_0 t + b_k \cos (2k - 1)\omega_0 t] \tag{6.114}$$

where

$$\omega_0 = \frac{2\pi}{2T}$$

If this series is truncated after the mth term, we have an approximating series

$$\hat{h}_p(t) = \sum_{k=1}^{m} [a_k \sin (2k - 1)\omega_0 t + b_k \cos (2k - 1)\omega_0 t] \cong h_p(t) \tag{6.115}$$

Transforming (6.115) into the frequency domain, we get

$$\hat{H}_p(s) = \frac{A(s)}{\prod_{k=1}^{m} [s^2 + (2k - 1)^2\omega_0^2]} = \frac{A(s)}{D(s)} \tag{6.116}$$

The denominator polynomial, $D(s)$, is of degree $2m$ and has only j-axis zeros. The numerator polynomial, $A(s)$, is of degree $2m - 1$ or lower.

To generate a $\hat{W}(s)$ such that $\hat{w}(t)$ approximates $w(t)$ of (6.110), we utilize the Fourier-series approximation of impulse trains as described in Appendix B. Specifically, if we use the relationship

$$\frac{1}{1+\epsilon^{-sT}} = 1 - \epsilon^{-sT} + \epsilon^{-2sT} - \epsilon^{-3sT} + \epsilon^{-4sT} - \dots \tag{6.117}$$

we see that

$$\frac{2}{W(s)} - 1 = \frac{2}{1+\epsilon^{-sT}} - 1 = 1 - 2\epsilon^{-sT} + 2\epsilon^{-2sT} - 2\epsilon^{-3sT} + 2\epsilon^{-4sT} - \dots$$
$$\triangleq I(s) \tag{6.118}$$

has an inverse Laplace transform

$$i(t) = u_0(t) - 2u_0(t-T) + 2u_0(t-2T) - 2u_0(t-3T) + 2u_0(t-4T) - \dots \tag{6.119}$$

This $i(t)$ is depicted in Fig. 6.23. This $i(t)$ is also identical to $f'(x)$ of Fig. B-3 for $t > 0$. Actually, the range over which $i(t)$ and $f'(x)$ are identical may be extended to $t \geq 0$ if we consider the impulse of $f'(x)$ at $x = 0$ as being situated symmetrically about that point. Since

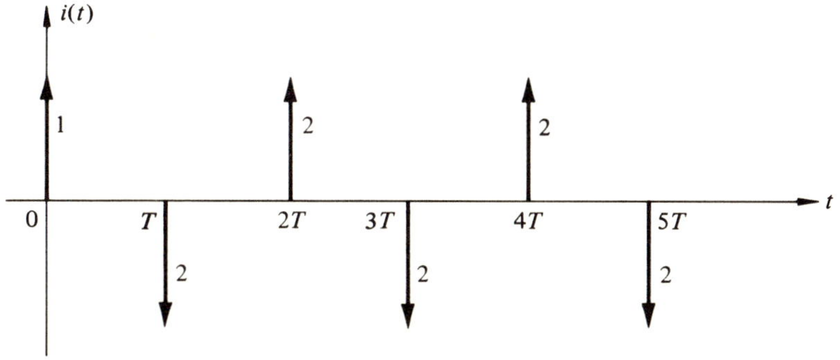

Fig. 6.23 Function $i(t)$ of (6.119).

$$i(t) = \frac{4}{T}\sum_{k=1}^{\infty} \cos(2k-1)\omega_0 t \tag{6.120}$$

we may either truncate this series or introduce the σ factors explained in Appendix B to reduce the ripples before truncation. If these σ factors are used and we truncate the series after the $2n$th term, we have

$$\hat{\imath}_{2n}(t) = \frac{4}{T} \sum_{k=1}^{n} \sigma_{2k-1} \cos (2k - 1)\omega_0 t \tag{6.121}$$

where

$$\sigma_{2k-1} = \frac{\sin [(2k - 1)\pi/2n]}{(2k - 1)\pi/2n} \tag{6.122}$$

Except for a scale difference, $\hat{\imath}_{2n}(t)$ of (6.121) is identical to $\bar{f}'_m(x)$ of (B-19). When $\hat{\imath}_{2n}(t)$ is transformed into the frequency domain, we have

$$\hat{I}_{2n}(s) = \sum_{k=1}^{n} \frac{\sigma_{2k-1} s}{s^2 + (2k - 1)^2\omega_0^2} \tag{6.123}$$

Since the series of $h_p(t)$ and $i(t)$ are made to have the same period, they have the same fundamental frequency, or ω_0. Unless we truncate the two series differently, $\hat{H}_p(s)$ and $\hat{I}_{2n}(s)$ should have the same poles. If we let $n > m$, then $\hat{I}_{2n}(s)$ will have all the poles of $\hat{H}_p(s)$ plus a few more of its own. Then we may write

$$\hat{I}_{2n}(s) = \frac{N(s)}{C(s)D(s)} \tag{6.124}$$

From (6.118) we get

$$W(s) = \frac{2}{1 + I(s)} \tag{6.125}$$

Hence we may make

$$\hat{W}(s) = \frac{2}{1 + \hat{I}_{2n}(s)} = \frac{2C(s)D(s)}{C(s)D(s) + N(s)} \tag{6.126}$$

Substituting (6.116) and (6.126) into (6.113) gives

$$H^*(s) = \frac{2A(s)D(s)}{C(s)D(s) + N(s)} \tag{6.127}$$

Since $\hat{I}_{2n}(s)$ is a reactance function [σ's are positive in (6.121)], the denominator of the $H^*(s)$ of (6.127) is always Hurwitz.

Example. Let $h(t)$ be the triangular pulse of Fig. 6.18(a). Then $h_p(t)$ is shown in Fig. 6.24. The truncated Fourier series of $h_p(t)$ for $m = 3$ is

$$\hat{h}_p(t) = \frac{8}{\pi^2}\left(\sin t - \frac{1}{9} \sin 3t + \frac{1}{25} \sin 5t\right)$$

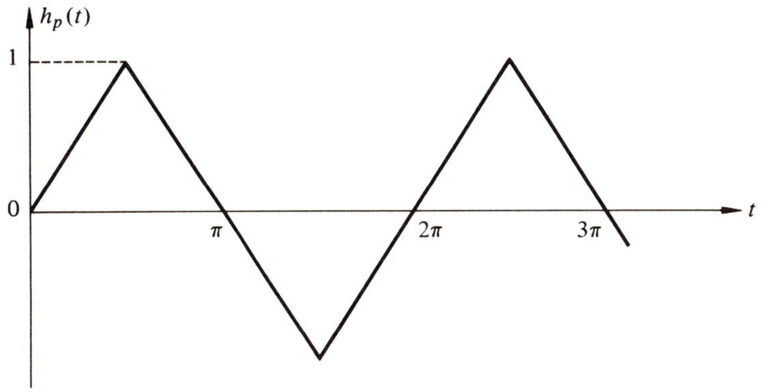

Fig. 6.24 Periodic function $h_p(t)$ for the example in Section 6.5.

for which

$$\hat{H}_p(s) = \frac{8/\pi^2}{s^2+1} - \frac{8/3\pi^2}{s^2+9} + \frac{8/5\pi^2}{s^2+25}$$

For the Fourier-series approximation of the impulse train, we also take three terms, or let $n = 3$ in (6.123). Then

$$\hat{I}_6(s) = \frac{(3/\pi)s}{s^2+1} + \frac{(2/\pi)s}{s^2+9} + \frac{(3/5\pi)s}{s^2+25}$$
$$= \frac{4}{5\pi}\frac{7s^5 + 200s^3 + 913s}{(s^2+1)(s^2+9)(s^2+25)}$$

Substituting this $\hat{I}_6(s)$ into (6.126) and the resulting $\hat{W}(s)$ and the above $\hat{H}_p(s)$ into (6.113) we obtain

$$H^*(s) = \frac{16(13s^4 + 350s^2 + 3277)}{15\pi^2(s^2+1)(s^2+9)(s^2+25) + 12\pi(7s^5 + 200s^3 + 913s)}$$
$$= -\frac{0.02731s - 1.8285}{(s+0.5214)^2 + (0.9037)^2} - \frac{0.02736s + 1.1055}{(s+0.2875)^2 + (2.8805)^2}$$
$$+ \frac{0.05467s + 0.6467}{(s+0.08229)^2 + (4.9658)^2} \tag{6.128}$$

and

$$h^*(t) = -2.03932\epsilon^{-0.5214t}\cos(0.9037t + 89.22°)$$
$$- 0.38204\epsilon^{-0.2875t}\cos(2.8805t - 85.89°)$$
$$+ 0.1404\epsilon^{-0.08229t}\cos(4.9658t - 67.02°) \tag{6.129}$$

The $h^*(t)$ of (6.129) is shown in Fig. 6.25.

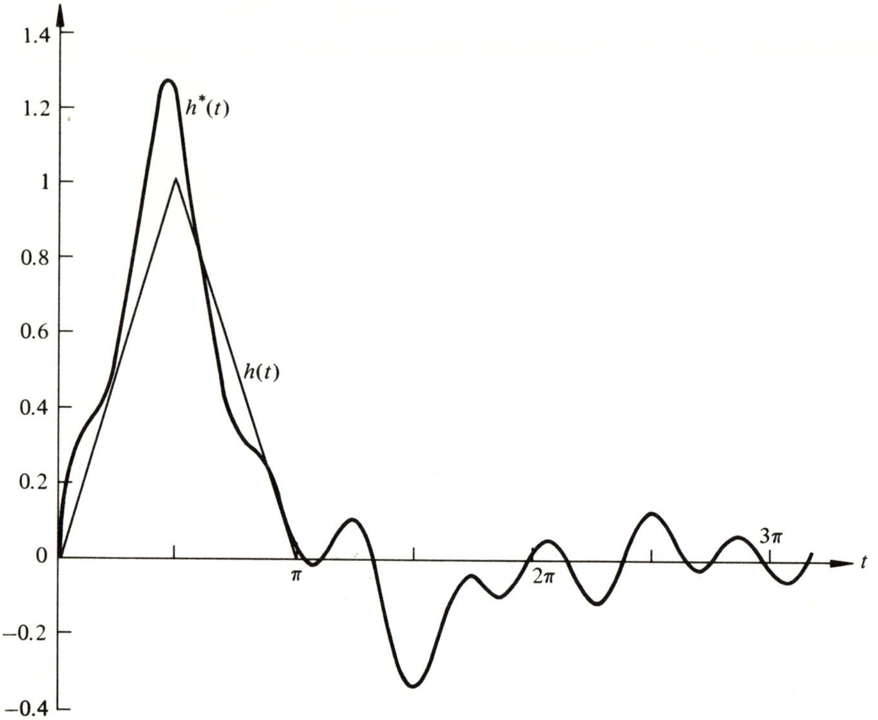

Fig. 6.25 The $h^*(t)$ of (6.129).

Estimate of Time-Domain Errors

There are two sources of error in this scheme. One is the inherent error in the Fourier-series approximation of $h_p(t)$ by $\hat{h}_p(t)$. This error can always be reduced by including more terms in the series. The other error stems from the approximation of $w(t)$ by $\hat{w}(t)$, where

$$\hat{w}(t) = \mathscr{L}^{-1}\left[\frac{2}{1 + \hat{I}_{2n}(s)}\right] \tag{6.130}$$

Since

$$H^*(s) = \frac{2\hat{H}_p(s)}{1 + \hat{I}_{2n}(s)} \tag{6.131}$$

and

$$W(s)[1 + I(s)] = 2 \tag{6.132}$$

we have

$$H^*(s)[1 + \hat{I}_{2n}(s)] = \hat{H}_p(s)W(s)[1 + I(s)] \tag{6.133}$$

In the time domain (6.133) reads

$$\begin{aligned} h^*(t) \circledast [u_0(t) + \hat{\imath}_{2n}(t)] &= [\hat{h}_p(t) \circledast w(t)] \circledast [u_0(t) + i(t)] \\ &= \bar{h}(t) \circledast [u_0(t) + i(t)] \end{aligned} \tag{6.134}$$

where

$$\bar{h}(t) = \hat{h}_p(t) \circledast w(t) \tag{6.135}$$

The time function $\bar{h}(t)$ may be expressed explicitly as

$$\begin{aligned} \bar{h}(t) &= \hat{h}_p(t), \qquad 0 < t < T \\ &= 0, \qquad \text{otherwise} \end{aligned} \tag{6.136}$$

Now let us limit our attention to the time region $0 < t < T - \tau_{2n}$, where τ_{2n} is the first instant at which $\hat{\imath}_{2n}(t)$ crosses zero, as illustrated in Fig. 6.26. In this region we may define a new function

$$\begin{aligned} a(t) &= \hat{\imath}_{2n}(t), \qquad 0 < t < \tau_{2n} \\ &= 0, \qquad \text{elsewhere} \end{aligned} \tag{6.137}$$

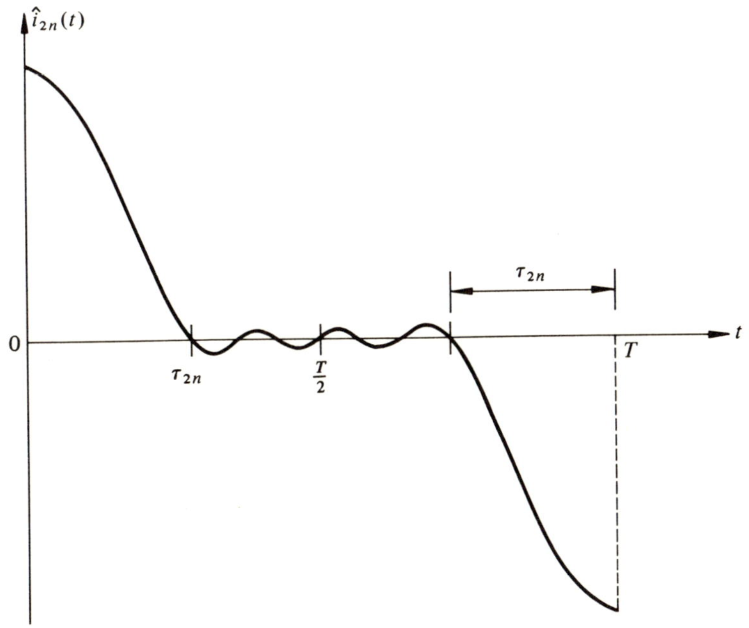

Fig. 6.26 Definition of τ_{2n} in connection with $\hat{\imath}_{2n}(t)$.

We may claim that

$$a(t) \cong i_{2n}(t), \qquad 0 < t < T - \tau_{2n} \tag{6.138}$$

Equation (6.134) may be expressed as

$$h^*(t) + h^*(t) \circledast a(t) \cong 2\hat{h}_p(t), \qquad 0 < t < T - \tau_{2n} \tag{6.139}$$

Since $a(t)$ is an approximation of $i(t)$ near the origin, the area under $a(t)$ is approximately unity and $h^*(t) \circledast a(t)$ will lie somewhere between $h^*(t)$ and $h^*(t) \circledast u_0(t - \tau_{2n}) = h^*(t - \tau_{2n})$. Graphically this statement is illustrated in Fig. 6.27, in which $h^*(t) \circledast a(t)$ must lie in the shaded region. Since $\hat{h}_p(t)$ is the average of $h^*(t)$ and $h^*(t) \circledast a(t)$, it must also lie in this region.

Conversely, if we take $\hat{h}_p(t)$ and shift it to the left by τ_{2n} to obtain $\hat{h}_p(t + \tau_{2n})$, then $h^*(t)$ must lie between these two curves. This is illustrated in Fig. 6.28. Thus the error between $h^*(t)$ and $h(t)$ may be estimated. For a smooth $h(t)$, it can be expected that $h^*(t)$ will lie approximately midway between $\hat{h}_p(t)$ and

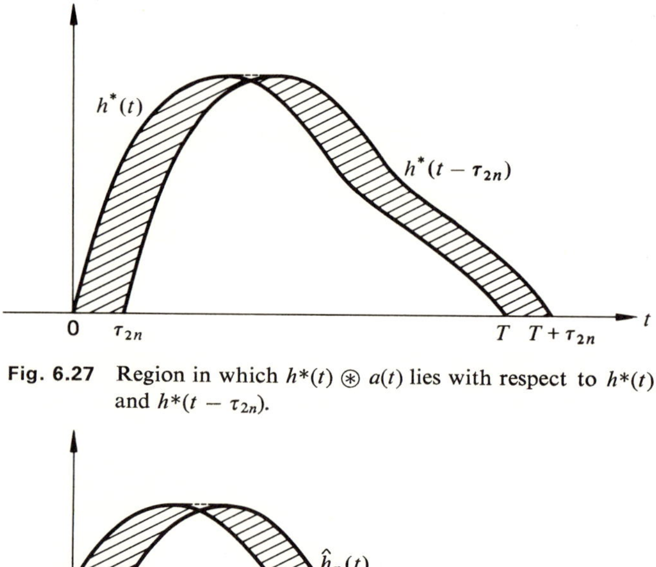

Fig. 6.27 Region in which $h^*(t) \circledast a(t)$ lies with respect to $h^*(t)$ and $h^*(t - \tau_{2n})$.

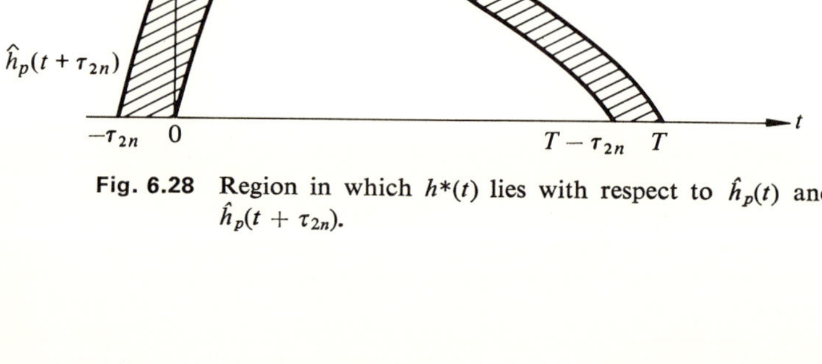

Fig. 6.28 Region in which $h^*(t)$ lies with respect to $\hat{h}_p(t)$ and $\hat{h}_p(t + \tau_{2n})$.

$\hat{h}_p(t + \tau_{2n})$ in the region $0 < t < T - \tau_{2n}$. For large values of n, τ_{2n} becomes very small and this error becomes negligible.

6.6 INDIRECT PADÉ APPROXIMANT METHOD

In Section 6.1 an $h^*(t)$ is obtained by first obtaining its Laplace transform, $H^*(s)$, which is a frequency-domain approximation of $H(s)$. Specifically, $H^*(s)$ is made to be simply a Padé approximant of $H(s)$. This approximation does not give any consideration to the accuracy in the time domain. In this section a function other than $H(s)$ is approximated by another function—its Padé approximant—in the frequency domain. Such an approximation minimizes an upper bound of the error in the time domain [TE1].

The magnitude of the error in the time domain may be expressed as

$$|\epsilon(t)| = |h(t) - h^*(t)| = \frac{1}{2\pi}\left|\int_{-j\infty}^{j\infty} [H(s) - H^*(s)]\epsilon^{ts}\, ds\right| \tag{6.140}$$

To obtain an estimate of the integral in (6.140), the following transformation will be employed:

$$s = \frac{1 - z}{1 + z} \qquad \text{or} \qquad z = \frac{1 - s}{1 + s} \tag{6.141}$$

This transformation maps the right halfplane of one variable onto the interior of the unit circle centered at the origin of the other variable, the j axis onto the unit circle, the left halfplane onto the exterior of the circle, and the point at infinity onto -1.

Assuming that neither $H(s)$ nor $H^*(s)$ has poles in the right halfplane and on the j axis, the contour for the integral in (6.140) can be chosen to be $C_1 + C_2$ as shown in Fig. 6.29, where C_1 is the unit circle and C_2 is a semicircle centered at $z = -1$ with a radius ρ. Let

$$H\left(\frac{1 - z}{1 + z}\right) = H_1(z) \tag{6.142}$$

$$H^*\left(\frac{1 - z}{1 + z}\right) = H_1^*(z) \tag{6.143}$$

Thus (6.140) becomes

$$|\epsilon(t)| = \frac{1}{\pi}\left|\oint_{C_1+C_2} [H_1(z) - H_1^*(z)]\epsilon^{(1-z/1+z)t} \frac{dz}{(1 + z)^2}\right| \tag{6.144}$$

If we further assume that both $H(s)$ and $H^*(s)$ have no less than N zeros at

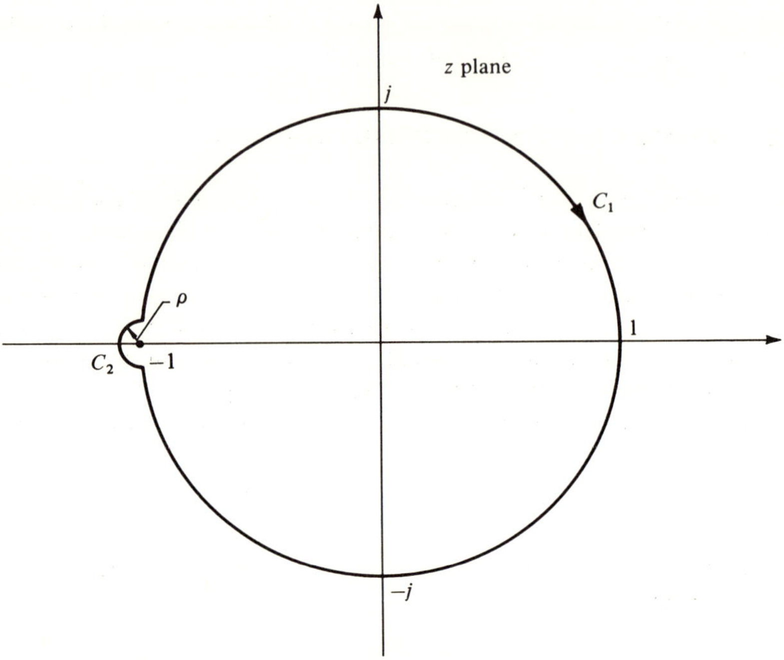

Fig. 6.29 Contour of integration used in (6.144).

$s = \infty$, then both $H_1(z)$ and $H_1^*(z)$ contain a factor $(1 + z)^N$ in their numerators. And we can let

$$H_2(z) = \frac{H_1(z)}{(1 + z)^N} \tag{6.145}$$

$$H_2^*(z) = \frac{H_1^*(z)}{(1 + z)^N} \tag{6.146}$$

Both $H_2(z)$ and $H_2^*(z)$ are finite at $z = -1$. Thus

$$|\epsilon(t)| = \frac{1}{\pi}\left|\oint_{C_1+C_2} [H_2(z) - H_2^*(z)]\epsilon^{(1-z/1+z)t}(1 + z)^{N-2}\, dz\right| \tag{6.147}$$

For $N \geq 2$ the integrand in (6.147) is a bounded function on C_1 and C_2. Furthermore, the contribution along C_2 is zero if we let $\rho \to 0$. On C_1

$$z = \epsilon^{j\theta} \qquad dz = jz\, d\theta \tag{6.148}$$

and

$$|\epsilon^{(1-z/1+z)t}| = |\epsilon^{j\omega t}| = 1 \tag{6.149}$$

and the integral of (6.147) has the upper bound

$$|\epsilon(t)| \leq \frac{1}{\pi}\left|\oint_{C_1} [H_2(z) - H_2^*(z)](1+z)^{N-2}\,dz\right| \tag{6.150}$$

By applying Schwartz's inequality we may write [WI1]

$$|\epsilon(t)|^2 \leq \frac{1}{\pi^2}\int_{-\pi}^{\pi} |H_2(\epsilon^{j\theta}) - H_2^*(\epsilon^{j\theta})|^2\,d\theta \int_{-\pi}^{\pi} |1+\epsilon^{j\theta}|^{2N-4}\,d\theta \tag{6.151}$$

But

$$\begin{aligned}\int_{-\pi}^{\pi} |1+\epsilon^{j\theta}|^{2N-4}\,d\theta &= \int_{-\pi}^{\pi}\left(2\cos\frac{\theta}{2}\right)^{2N-4} d\theta \\ &= 2^{2N-2}\int_{0}^{\pi/2} \cos^{2N-4} x\,dx \\ &= 2^{2N-3}\sqrt{\pi}\,\frac{\Gamma(N-\frac{3}{2})}{\Gamma(N-1)}, \qquad N>1\end{aligned} \tag{6.152}$$

Therefore,

$$|\epsilon(t)|^2 \leq \frac{2^{2N-3}}{3/\pi^2}\frac{\Gamma(N-\frac{3}{2})}{\Gamma(N-1)}\int_{-\pi}^{\pi}\left|H_2(z) - H_2^*(z)\right|^2_{z=\epsilon^{j\theta}} d\theta \tag{6.153}$$

For a given $N(\geq 2)$, the right-hand side of (6.153) is independent of t. It represents an upper bound (not necessarily reached) of $|\epsilon(t)|^2$. Hence, if this quantity can be minimized, the peak error between $h(t)$ and $h^*(t)$ is likely to be reduced. The $|\epsilon(t)|^2$ of (6.153) is also proportional to the total squared error between $H_2(z)$ and $H_2^*(z)$ along the unit circle centered at the origin in the z plane. Thus if $H(s)$ is known and the degrees of the numerator and denominator of $H^*(s)$ have been selected, minimizing $|\epsilon(t)|^2$ with respect to the coefficients in $H^*(s)$ in the least-squares sense is equivalent to making $H_2^*(z)$ a Padé approximant of the proper order of $H_2(z)$, as was shown in Section 2.5. Once $H_2^*(z)$ has been found, $H^*(s)$ may be obtained through (6.143) and (6.146), or

$$H^*(s) = \frac{2}{1+s}H_2^*\left(\frac{1-s}{1+s}\right) \tag{6.154}$$

The above reasoning does not apply for $N<2$ directly because the contribution along C_2 is no longer zero. However, the argument is no less valid qualitatively. Experiences have shown that if the same method is used for $N<2$, a reasonably good approximation is still obtained in the time domain.

As an example, suppose it is desired to find an $H^*(s)$ for

$$h(t) = \frac{J_1(t)}{t} \tag{6.155}$$

where J_1 is the Bessel function of the first kind, first order. This $h(t)$ corresponds to

$$H(s) = \sqrt{1 + s^2} - s = \frac{1}{\sqrt{1 + s^2} + s} \tag{6.156}$$

in the frequency domain. This $H(s)$ has a simple zero at $s = \infty$. From (6.142),

$$H_1(z) = \frac{\sqrt{2}\,\sqrt{1 + z^2} - (1 - z)}{1 + z}$$

Here $N = 1$ and

$$\begin{aligned} H_2(z) &= \frac{H_1(z)}{1 + z} = \frac{\sqrt{2}\,\sqrt{1 + z^2} - (1 - z)}{(1 + z)^2} \\ &= 0.4142 + 0.1716z - 0.0503z^2 - 0.0710z^3 + \dots \end{aligned}$$

For this $H_2(z)$,

$$(0, 1) = \frac{-1}{z - 2.414} = H_2^*(z)$$

From (6.146)

$$H_1^*(z) = -\frac{z + 1}{z - 2.414}$$

By (6.143)

$$H^*(s) = \frac{0.5858}{s + 0.4142} \tag{6.157}$$

Hence

$$h^*(t) = 0.5858\epsilon^{-0.4142t} \tag{6.158}$$

Both $J_1(t)/t$ and $h^*(t)$ are shown in Fig. 6.30.

The approximation can be further improved if a higher-order Padé approximant in z is used. For example, for the same $H_2(z)$

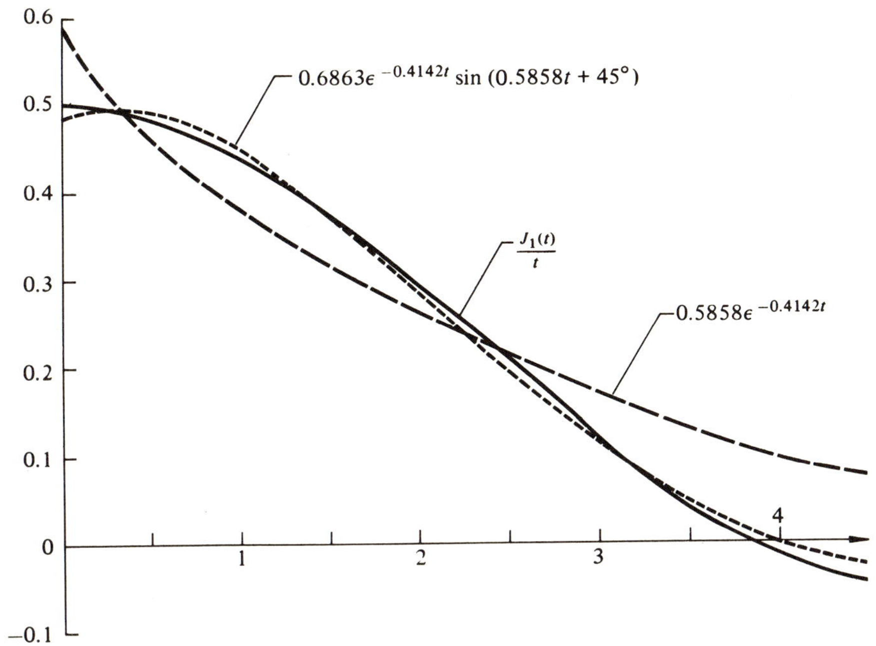

Fig. 6.30 Time functions (6.155), (6.158), and (6.160).

$$(1, 2) = \frac{1.4142}{z^2 - 1.4142z + 3.4142}$$

which gives

$$H^*(s) = \frac{0.4853(s + 1)}{s^2 + 0.8284s + 0.5148} \qquad \textbf{(6.159)}$$

The inverse Laplace transform gives

$$h^*(t) = 0.6863\epsilon^{-0.4142t} \sin (0.5858t + 45°) \qquad \textbf{(6.160)}$$

This $h^*(t)$ is also plotted in Fig. 6.30. The improved approximation of this higher-order $H^*(s)$ is quite evident.

6.7 DISCRETE OPTIMAL CHEBYSHEV APPROXIMATION

When the desired impulse response is given at a number of equally spaced points, the method described in this section gives an $h^*(t)$ that will not only approximate the given $h(t)$ in the equiripple manner at the discrete points but also render the approximation in the minimum-maximum error manner. The significance of these descriptions is explained below.

Suppose $h(t)$ is given at $t = 0, \Delta t, 2\,\Delta t, \ldots, (M-1)\,\Delta t$ and let

$$h_k = h(k\,\Delta t), \qquad k = 0, 1, 2, \ldots, M-1 \tag{6.161}$$

as in (4.37). The values of the approximating function $h^*(t)$ at these points are denoted by

$$h_k^* = h^*(k\,\Delta t), \qquad k = 0, 1, 2, \ldots, M-1 \tag{6.162}$$

Define the errors at these points as

$$\epsilon_k = h_k - h_k^* \tag{6.163}$$

Let $\epsilon_{\max} = \max(|\epsilon_0|, |\epsilon_1|, |\epsilon_2|, \ldots, |\epsilon_{M-1}|)$. The term "equiripple" means that $|\epsilon_k| = \epsilon_{\max}$ for several values of k and $|\epsilon_k| < \epsilon_{\max}$ for all other k. The term "minimum-maximum error" refers to the fact that any other $h^*(t)$ will have a larger $\epsilon_{\max}$. Such an $h^*(t)$ is said to render the *optimum Chebyshev* approximation to $h(t)$ at the discrete points of t. Figure 6.31 is an example of such an approximation. Such an approximation may or may not be equiripple or minimum-maximum error for the continuous variation of t.

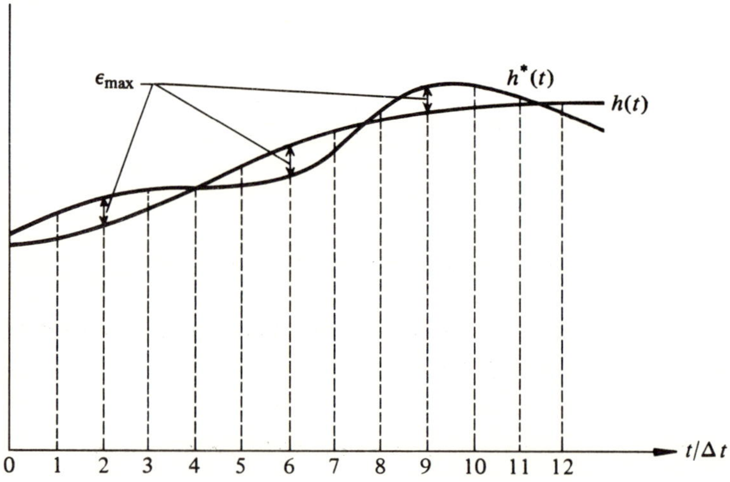

Fig. 6.31 An $h^*(t)$ that is the discrete optimal Chebyshev approximation of $h(t)$.

We assume that

$$h^*(t) = \sum_{\nu=1}^{N} A_\nu \epsilon^{s_\nu t} \tag{6.164}$$

Hence it is desired to determine a set of s_ν and a set of A_ν such that the h_k^* approximate h_k in the optimum Chebyshev manner for $k = 0, 1, 2, \ldots, M - 1$; $M < 2N$. The existence and uniqueness of such a solution have been established [FI1]. Furthermore, it has been shown that such an $h^*(t)$ will give a $2N + 1$ value of k for which $|\epsilon_k| = \epsilon_{\max}$ (and $|\epsilon_k| < \epsilon_{\max}$ for other k's) and the signs of any two such consecutive ϵ's alternate. Based on the assumptions of the uniqueness and existence of such a solution, we shall proceed to show a method for obtaining the solutions.

Let

$$w_\nu = \epsilon^{s_\nu \Delta t}, \qquad \nu = 1, 2, \ldots, N \tag{6.165}$$

Rewrite (6.162) and (6.164) in matrix form:

$$\begin{bmatrix} h_0^* \\ h_1^* \\ h_2^* \\ \vdots \\ h_{M-1}^* \end{bmatrix} = \begin{bmatrix} 1 & 1 & \ldots & 1 \\ w_1 & w_2 & \ldots & w_N \\ w_1^2 & w_2^2 & \ldots & w_N^2 \\ \vdots & \vdots & & \vdots \\ w_1^{M-1} & w_2^{M-1} & \ldots & w_N^{M-1} \end{bmatrix} \begin{bmatrix} A_1 \\ A_2 \\ A_3 \\ \vdots \\ A_N \end{bmatrix} \tag{6.166}$$

There are $2N$ unknowns in (6.166)—N values of A_ν and N values of w_ν. The rectangular matrix involving w's is of the dimension $M \times N$ with a rank N. Its rows are related to each other in a special manner in that rows $N + 1$ through M are obtainable from a linear combination of the N rows immediately preceding them. Furthermore, this relationship is identical among any group of $N + 1$ consecutive rows. For example, the $(N + 1)$th row may be written as

$$[w_1^N \, w_2^N \ldots w_N^N] = -[r_0 \, r_1 \ldots r_{N-1}] \begin{bmatrix} 1 & 1 & \ldots & 1 \\ w_1 & w_2 & \ldots & w_N \\ w_1^2 & w_2^2 & \ldots & w_N^2 \\ \vdots & \vdots & & \vdots \\ w_1^{N-1} & w_2^{N-1} & \ldots & w_N^{N-1} \end{bmatrix} \tag{6.167}$$

where r's are a set of real numbers determinable from a set of N w's. In general

$$[w_1^{N+k}\, w_2^{N+k} \dots w_N^{N+k}] = [w_1^N\, w_2^N \dots w_N^N]\begin{bmatrix} w_1^k & & & & 0 \\ & w_2^k & & & \\ & & \cdot & & \\ & & & \cdot & \\ 0 & & & & \cdot \\ & & & & w_N^k \end{bmatrix}$$

$$= -[r_0\, r_1 \dots r_{N-1}]\begin{bmatrix} 1 & 1 & \dots & 1 \\ w_1 & w_2 & \dots & w_N \\ w_1^2 & w_2^2 & \dots & w_N^2 \\ \vdots & \vdots & & \vdots \\ w_1^N & w_2^N & \dots & w_N^N \end{bmatrix}\begin{bmatrix} w_1^k & & & & 0 \\ & w_2^k & & & \\ & & \cdot & & \\ & & & \cdot & \\ 0 & & & & \cdot \\ & & & & w_N^k \end{bmatrix}$$

$$= -[r_0\, r_1 \dots r_{N-1}]\begin{bmatrix} w_1^k & w_2^k & \dots & w_N^k \\ w_1^{k+1} & w_2^{k+1} & \dots & w_N^{k+1} \\ w_1^{k+2} & w_2^{k+2} & \dots & w_N^{k+2} \\ \vdots & \vdots & & \vdots \\ w_1^{k+N-1} & w_2^{k+N-1} & \dots & w_N^{k+N-1} \end{bmatrix},$$

$$k = 0, 1, 2, \dots, M - N - 1 \qquad \textbf{(6.168)}$$

Equation (6.168) may be rewritten as

$$[r_0\, r_1 \dots r_{N-1}\, 1]\begin{bmatrix} w_1^k & w_2^k & \dots & w_N^k \\ w_1^{k+1} & w_2^{k+1} & \dots & w_N^{k+1} \\ \vdots & \vdots & & \vdots \\ w_1^{k+N} & w_2^{k+N} & \dots & w_N^{k+N} \end{bmatrix} = 0,$$

$$k = 0, 1, 2, \dots, M - N - 1 \qquad \textbf{(6.169)}$$

Equation (6.169) implies that

$$w_\nu^k(r_0 + r_1 w_\nu + r_2 w_\nu^2 + \dots + r_{N-1} w_\nu^{N-1} + w_\nu^N) = 0,$$
$$k = 0, 1, 2, \dots, M - N - 1 \qquad \textbf{(6.170)}$$

Hence w_ν's are the roots of the equation

$$w^N + r_{N-1} w^{N-1} + \dots + r_2 w^2 + r_1 w + r_0 = 0 \qquad \textbf{(6.171)}$$

An alternative representation of (6.171) is

$$(w - w_1)(w - w_2) \dots (w - w_N) = 0 \qquad \textbf{(6.172)}$$

Thus if w_ν are known, r_ν are readily obtained, and vice versa. Since s_ν are either real or in conjugate pairs, all r's are real.

Now if we define the error vector to be

$$\begin{bmatrix} \epsilon_0 \\ \epsilon_1 \\ \epsilon_2 \\ \vdots \\ \epsilon_{M-1} \end{bmatrix} = \begin{bmatrix} h_0 \\ h_1 \\ h_2 \\ \vdots \\ h_{M-1} \end{bmatrix} - \begin{bmatrix} h_0^* \\ h_1^* \\ h_2^* \\ \vdots \\ h_{M-1}^* \end{bmatrix} \tag{6.173}$$

we may write $M - N$ equations

$$\begin{bmatrix} \epsilon_k \\ \epsilon_{k+1} \\ \epsilon_{k+2} \\ \vdots \\ \epsilon_{k+N} \end{bmatrix} = \begin{bmatrix} h_k \\ h_{k+1} \\ h_{k+2} \\ \vdots \\ h_{k+N} \end{bmatrix} - \begin{bmatrix} h_k^* \\ h_{k+1}^* \\ h_{k+2}^* \\ \vdots \\ h_{k+N}^* \end{bmatrix}, \qquad k = 0, 1, 2, \ldots, M - N - 1 \tag{6.174}$$

Premultiplying (6.174) by the row matrix $[r_0\, r_1 \ldots r_{N-1}\, 1]$ we have

$$[r_0\, r_1 \ldots r_{N-1}\, 1] \begin{bmatrix} \epsilon_k \\ \epsilon_{k+1} \\ \epsilon_{k+2} \\ \vdots \\ \epsilon_{k+N} \end{bmatrix} = [r_0\, r_1 \ldots r_{N-1}\, 1] \begin{bmatrix} h_k \\ h_{k+1} \\ h_{k+2} \\ \vdots \\ h_{k+N} \end{bmatrix}, \qquad k = 0, 1, 2, \ldots, M - N - 1 \tag{6.175}$$

by virtue of (6.166) and (6.169).

The relationship in (6.175) is independent of the coefficients A's. For a given set of M known points on $h(t)$, the values of r_ν (which in turn determine w_ν and s_ν) determine a relationship that must be satisfied by the M errors ϵ_ν. Conversely, if a certain conditions are required of the values of ϵ's the values of r's must be such that they satisfy (6.175).

Equation (6.175) is a set of $M - N$ equations. There are $M + N$ unknowns —N values of r's and M values of ϵ's. If $(2N + 1)$ of the ϵ's are assumed to be equal in magnitude and alternate in sign, $2N$ additional equation may be written. Thus all $M + N$ unknowns can be solved. For the problem at hand, if we had advance information on which set of $(2N + 1)$ values of the M values of ϵ's should be made of equal magnitude to lead to an optimal Chebyshev approximation, the problem would be relatively simple. This is, of course, not the case. However, it is fortunate that we know that such a set does exist and is unique. Thus if a systematic search is conducted such that all possible combinations of $2N + 1$ of the M ϵ's are assumed to be of equal magnitude and any combinations that give one or more ϵ's larger than the equal error be discarded, the set that does lead to the optimal approximation will eventually be found. Actually, as soon as an optimal set of ϵ's is found there is no need to proceed further. The corresponding set of r's may be entered into (6.171), from which the values of w's are determined via (6.172). The poles s_ν are then computable through (4.43). In addition, $h_k^* = h_k - \epsilon_k$ may be computed for all values of k. The values of A's may then be found by solving (6.166).

Example. Suppose we have

$$h(t) = \frac{1}{(1 + t)^2}, \qquad \Delta t = 0.5, \quad M = 9 \tag{6.176}$$

For a one-pole $H^*(s)$, $N = 1$, we let $r_0 = r$. The eight equations given by (6.175) are

$$\begin{aligned}
r\epsilon_0 + \epsilon_1 &= rh_0 + h_1 = 1.0000r + 0.4444 \\
r\epsilon_1 + \epsilon_2 &= rh_1 + h_2 = 0.4444r + 0.2500 \\
r\epsilon_2 + \epsilon_3 &= rh_2 + h_3 = 0.2500r + 0.1600 \\
r\epsilon_3 + \epsilon_4 &= rh_3 + h_4 = 0.1600r + 0.1111 \\
r\epsilon_4 + \epsilon_5 &= rh_4 + h_5 = 0.1111r + 0.0816 \\
r\epsilon_5 + \epsilon_6 &= rh_5 + h_6 = 0.0816r + 0.0625 \\
r\epsilon_6 + \epsilon_7 &= rh_6 + h_7 = 0.0625r + 0.0494 \\
r\epsilon_7 + \epsilon_8 &= rh_7 + h_8 = 0.0494r + 0.0400
\end{aligned} \tag{6.177}$$

As a first try, let

$$\epsilon_0 = -\epsilon_3 = \epsilon_6 \tag{6.178}$$

Solution of the 10 equations in (6.177) and (6.178) gives $r = -0.5767$ and $\epsilon_0 = 0.02669$. These values gives $\epsilon_1 = -0.1169$. This trial does not lead to an optimal approximation.

For the assumption that

$$\epsilon_0 = -\epsilon_1 = \epsilon_5 \tag{6.179}$$

we have $r = -0.5156$ and $\epsilon_0 = 0.0468$. Successive applications of equations in (6.177) yield

$$\begin{aligned}
\epsilon_0 &= 0.0468 \\
\epsilon_1 &= -0.0468 \\
\epsilon_2 &= -0.0038 \\
\epsilon_3 &= 0.0290 \\
\epsilon_4 &= 0.0434 \\
\epsilon_5 &= 0.0468 \\
\epsilon_6 &= 0.0445 \\
\epsilon_7 &= 0.0401 \\
\epsilon_8 &= 0.0353
\end{aligned}$$

This is the optimal set. From (4.43) we obtain

$$s_1 = \frac{1}{0.5} \ln (0.5156) = -1.325$$

and, from (6.166),

$$A_1 = h_0^* = h_0 - \epsilon_0 = 0.9532$$

Thus

$$h^*(t) = 0.9532\epsilon^{-1.325t} \tag{6.180}$$

Two points can be noted from this example. First, the number of equations that have to be solved and the number of possible combinations of equations that have to be screened before the optimal Chebyshev set is obtained can be staggeringly high. Furthermore, these equations are not linear. Therefore, if N and M are large, the computational effort can be extremely laborious. Generally, the solution of these equations will require the aid of an electronic computer.

Second, the optimal Chebyshev characteristic is true only at discrete points at which $t = k\,\Delta t$, with k an integer. For instance, in the example above, $h^*(0.4) = 0.5611$, $h(0.4) = 0.5102$, $\epsilon(0.4) = -0.0509$, and $|\epsilon(0.4)| > \epsilon_{\max} = 0.0468$. Hence this approximation is optimal Chebyshev only at the discrete points and should not be presumed to be existing between $h(t)$ and $h^*(t)$ for all intermediate values of t.

6.8 DISCUSSION

Several methods of directly obtaining an $H^*(s)$ for a given $h(t)$ are given in this chapter. As contrasted to the two-step approach outlined in Chapters 4 and 5, methods in this chapter give the poles and zeros of $H^*(s)$ all in one step. The various methods given here are each quite unique in approach and are generally useful for certain particular situations.

The frequency-domain approach of Section 6.1 is applicable to any $h(t)$ or $\hat{h}(t)$ that is Laplace transformable. [The approximate $\hat{h}(t)$ can always be designed such that $\hat{H}(s)$ is easily obtainable.] But no control can be exercised on the time-domain accuracy of $h^*(t)$.

The real-part approximation of $H(s)$ given in Section 6.2 is a hybrid of the frequency-domain approximation and the time-domain approximation. The frequency-domain approximation is done by a semigraphical process. The procedure is applicable when $h(t)$ is given in analytical, graphical, or tabular form. The time-domain accuracy can be estimated. In some steps of the procedure, arbitrary choices of certain parameters are necessary.

The Fourier-series methods of Sections 6.3, 6.4, and 6.5 are easiest in terms of the computational effort. However, these methods usually lead to the same order of accuracy as the Fourier-series approximation of $h(t)$ in the time interval $0 < t < T$. Hence it is most suitable for those impulse responses that do not have any discontinuity or in situations that a large number of poles is acceptable. The given $h(t)$ can be in any of the three forms.

The approximation in the indirect Padé approximant method of Section 6.6 is carried out in a transformed frequency plane wherein some justification can be given to reduce the time-domain error. Its usefulness is limited to situations in which not only $H(s)$ or $\hat{H}(s)$ is obtainable but also either $H(s)$ or $\hat{H}(s)$ has zeros at $s = \infty$.

The discrete optimal Chebyshev approximation method of Section 6.7 assumes that only $h(t)$ at discrete equally spaced points are of interest. To obtain a good approximation for a large number of points, the computational effort is usually large. However, when $h(t)$ is only known at these discrete points, this method bypasses the effort of finding an intermediate $\hat{h}(t)$ either by inspection or by means of analytic substitution.

PROBLEMS

6.1. For the $h(t)$ of Fig. 4.1, find an approximating $h^*(t)$ by obtaining the (2, 3) Padé approximant of the series of (4.5).

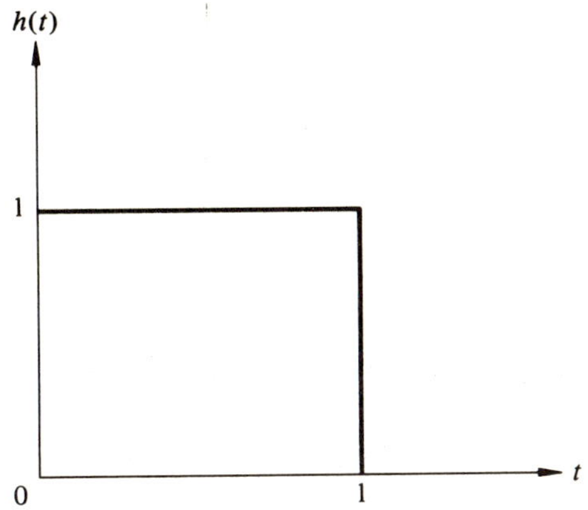

Fig. P6.1

6.2. For the $h(t)$ of Fig. P6.1 and a value of $\alpha = 2$, calculate $R^{(-2)}(\omega)$ for $0 < \omega < 12$. Based on this $R^{(-2)}(\omega)$, find an $H^*(s)$ for the $h(t)$. Compute and plot the corresponding $h^*(t)$.

6.3. For the $h(t)$ of Fig. P6.1, find a six-pole $H^*(s)$ by Fourier-series method I.

6.4. For the $h(t)$ of Fig. P6.1, find a six-pole $H^*(s)$ by Fourier-series method II.

6.5. For $h(t) = J_1(t)$, find an $h^*(t)$ corresponding to a four-pole network function through the indirect Padé approximant. This $h(t)$ has a Laplace transform

$$H(s) = \frac{1}{\sqrt{s^2+1}(\sqrt{s^2+1}+s)}$$

6.6. Obtain a one-pole $H^*(s)$ such that its $h^*(t)$ approximates $h(t) = 1/(1+t)$ at $t = 0, 0.8, 1.6, 2.4, 3.2, 4.0, 4.8$, and 5.6.

6.7. Repeat Problem 6.6 for a two-pole $H^*(s)$.

7

The Constant-Delay Network Functions

A network whose transfer characteristic is described in the frequency domain by

$$H(s) = \epsilon^{-\alpha}\epsilon^{-T_D s} \tag{7.1}$$

will render what is known as distortionless transmission. This network causes the same attenuation (or gain) and delay to all signal components. The response has exactly the same shape as the excitation, but it is scaled by the constant $\epsilon^{-\alpha}$ and delayed by T_D seconds. In the time domain, a distortionless transmission is characterized by the relationship

$$f_2(t) = \epsilon^{-\alpha} f_1(t - T_D) \tag{7.2}$$

Here we are not concerned about the value of α, because it is always possible to compensate this factor by either an amplifier or an attenuator. Hence, we shall assume, for convenience, that $\alpha = 0$. The response of a distortionless-transmission network is an exact replica of the excitation T_D seconds later. Such a network is known as a *constant-delay network.*

A distributed network is generally required to realize either a distortionless transmission or a constant delay. A uniform transmission line whose series impedance per unit length $(R + j\omega L)$ and shunt admittance per unit length $(G + j\omega C)$ satisfy the relationship

$$\frac{R}{L} = \frac{G}{C} \tag{7.3}$$

gives a distortionless transmission when it is terminated in its characteristic impedance. A lossless transmission line when properly terminated is a constant-delay network whose delay time T_D depends on the parameters L and C, and on the length of the line [JO1].

Constant-delay networks find many applications in electronics, controls, communication systems, and the like. A distributed delay network is not always feasible in many situations for practical reasons. Hence lumped networks are frequently designed to approximately achieve a constant delay. These lumped delay networks are nevertheless frequently referred to as constant-delay networks.

The problem of designing an approximate constant-delay lumped network is to find an $H^*(s)$ such that its corresponding $h^*(t)$ approximates the impulse response

$$h(t) = u_0(t - T_D) \tag{7.4}$$

in the time domain. Although several methods discussed in this chapter appear to be methods for finding an $H^*(s)$ such that it approximates

$$H(s) = \epsilon^{-T_D s} \tag{7.5}$$

in the frequency domain, the reader should keep in mind that this is merely an intermediate step. Each frequency-domain technique can be justified only if it renders an acceptable time-domain approximation of the impulse response of (7.4).

In the frequency domain, two properties of the $H(s)$ of (7.5) are important. First, it is an all-pass function, as its magnitude is unity for all real frequencies ($s = j\omega$). Second, the phase-lag function along the j axis, $\beta(\omega)$, where $H(j\omega) = |H(j\omega)|\epsilon^{-j\beta(\omega)}$, is a linear function of ω, or

$$\beta(\omega) = T_D\omega \tag{7.6}$$

Hence a constant-delay network is also known as a linear-phase network.

7.1 ALL-PASS, LOW-PASS, AND BAND-PASS CONSTANT-DELAY NETWORKS

In designing a finite lumped network such that its transfer characteristic approximates the ideal delay of (7.5), several real-world problems arise. Since (7.5) is an all-pass function, it appears natural to use a finite all-pass network for this approximation. Such a network may be represented by its transfer function,

$$H_{\text{ap}}(s) = \frac{M(s) - N(s)}{M(s) + N(s)} \tag{7.7}$$

where $M(s)$ is an even polynomial and $N(s)$ is an odd polynomial. For this network to be stable, $M(s) + N(s)$ must be a Hurwitz polynomial.

Functions of the type of (7.7) are very easy to work with, as they have many desirable analytical features. These functions in two-port networks are somewhat analogous to the reactance functions in one-port networks. One of the properties is that the poles and zeros of such a function are always the negative of each other.

The transfer function $H_{\text{ap}}(s)$ can always be realized by the constant-resistance lattice of Fig. 7.1, in which it is assumed that z_a and z_b are reactive and that

$$z_a z_b = R^2 \tag{7.8}$$

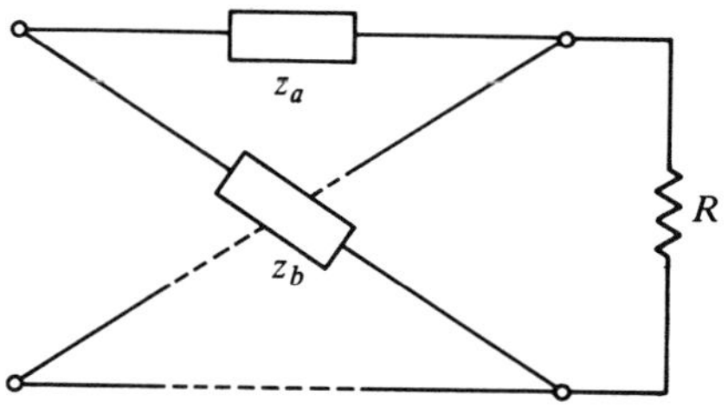

Fig. 7.1 A properly terminated constant-resistance lattice two-port.

The input impedance of such a network when terminated in the resistance R is always R. Since the network is lossless, one can easily see that

$$|H_{\text{ap}}(j\omega)| \equiv 1 \tag{7.9}$$

The value of the resistance R will be normalized to 1 ohm here without any loss of generality.

The basic realization procedures of such networks are also very simple. Writing (7.7) as

$$H_{\text{ap}}(s) = \frac{1 - N(s)/M(s)}{1 + N(s)/M(s)} \tag{7.10}$$

and comparing it with the transfer function of the network of Fig. 7.1,

$$\frac{E_2}{E_1} = \frac{1 - z_a}{1 + z_a} \tag{7.11}$$

we see that it is sufficient to make

$$z_a = \frac{M(s)}{N(s)} \tag{7.12}$$

Thus

$$z_b = \frac{N(s)}{M(s)} \tag{7.13}$$

Because of the reciprocal relationship between z_a and z_b, it is seen that if we interchange the identifications of (7.12) and (7.13), then

$$\frac{E_2}{E_1} = \frac{1 - M(s)/N(s)}{1 + M(s)/N(s)} \tag{7.14}$$

which is the negative of (7.10). This interchange is thus seen to be equivalent to reversing the two terminals at either port of the constant-resistance network. This reversal does not alter the unity-magnitude feature of the network function, while a reversal in sign of E_2/E_1 (or a 180° in phase shift) always accompanies each reversal.

An alternative realization procedure makes use of the properties that a number of constant-resistance twoports can be cascaded without disturbing the constant-resistance property. The overall transfer function is simply the product of all transfer functions. Thus any given $H_{ap}(s)$ can always be factored into components each of which is no more complicated than a biquadratic. Thus any all-pass network may be realized by the arrangement of Fig. 7.2 to within possibly a minus sign.

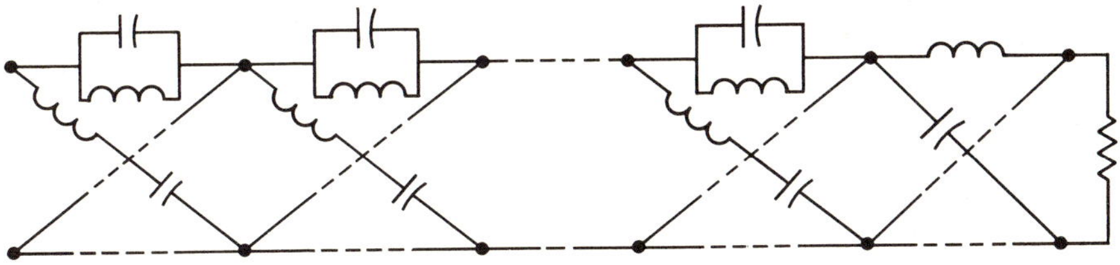

Fig. 7.2 A constant-resistance twoport made up of a cascade of first- and second-order constant-resistance symmetrical lattices.

Although functions of the type of (7.7) are easy to deal with analytically and the network of Fig. 7.1 or 7.2 is easy to realize on paper, these networks suffer many practical shortcomings. First, these networks are balanced and have no common grounded terminal between the input and the output. Unless one uses either transformers or active realizations, these networks cannot be unbalanced. Second, these networks generally require large numbers of elements. Third, the step responses of these networks have a value of either $+1$ or -1 at $t = 0$. This is undesirable for many applications.

The last of these properties can be explained if we realize that $H_{ap}(s)$ has the form

$$H_{\text{ap}}(s) = \pm\frac{1 - b_1 s + b_2 s^2 - b_3 s^3 + \cdots (-1)^m b_m s^m}{1 + b_1 s + b_2 s^2 + b_3 s^3 + \cdots + b_m s^m} \tag{7.15}$$

The Laplace transform of the step response, $g_{\text{ap}}(t)$, of such a network is $1/s$ times the $H_{\text{ap}}(s)$ of (7.15). Hence the final value of $g_{\text{ap}}(t)$ is

$$\lim_{t\to\infty} g_{\text{ap}}(t) = \lim_{s\to 0} [sH_{\text{ap}}(s)] = \pm 1 \tag{7.16}$$

And the initial value of $g_{\text{ap}}(t)$ is

$$\lim_{t\to 0} g_{\text{ap}}(t) = \lim_{s\to\infty} [sH_{\text{ap}}(s)] = \pm(-1)^m \tag{7.17}$$

Thus if the all-pass network is connected such that the final value of its step response is $+1$, then for even m the step response starts with $+1$ at $t = 0$ and returns to $+1$ as $t \to \infty$; for odd m, it starts with a value of -1 at $t = 0$ and approaches $+1$ as $t \to \infty$. These nonzero initial values of the step responses make the all-pass networks undesirable for many pulse applications.

From the bandwidth utilization point of view, an all-pass network has an infinite bandwidth. The use of such a network is seldom justifiable, especially in view of its accompanying shortcomings, just enumerated. In most practical situations, either the systems involved are band-limited or the signals encountered are limited in frequency content. Hence most delay networks are designed to pass certain bands of frequency only. These bands are determined either by other considerations on the system or by the frequency spectra of the signals involved. Two related groups of delay networks that have limited bandwidths and are of engineering importance are the low-pass and the band-pass constant-delay networks.

A low-pass constant-delay network may be considered as a network whose magnitude and phase characteristics are approximations of those shown in Fig. 7.3. Similarly, a band-pass constant-delay network may be regarded as a network whose characteristics are approximations of those shown in Fig. 7.4.

The relationship between the low-pass and the band-pass characteristics in the frequency domain is well known. The method of designing band-pass characteristics from corresponding low-pass characteristics is also well explained in many basic network-theory texts. Although some special techniques are available for synthesizing band-pass filters directly to produce certain specially desirable features in the final networks, for our purposes we may regard the low-pass networks as the only basic type of network or transmission system with which we need to be concerned here.

One simple method of obtaining a low-pass delay characteristic from an all-pass delay characteristic is to take the latter, which has the form

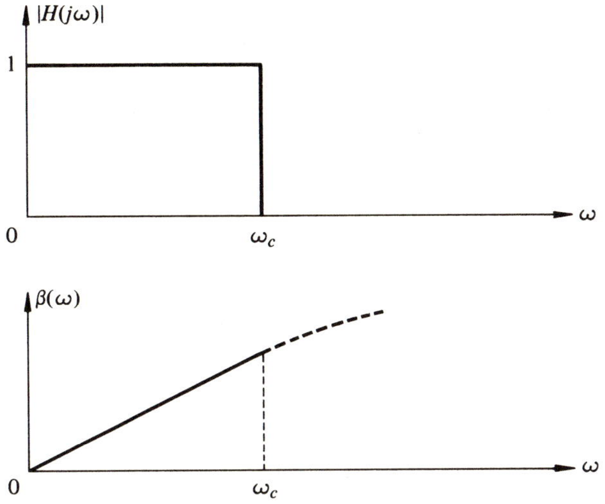

Fig. 7.3 Ideal low-pass constant-delay characteristics.

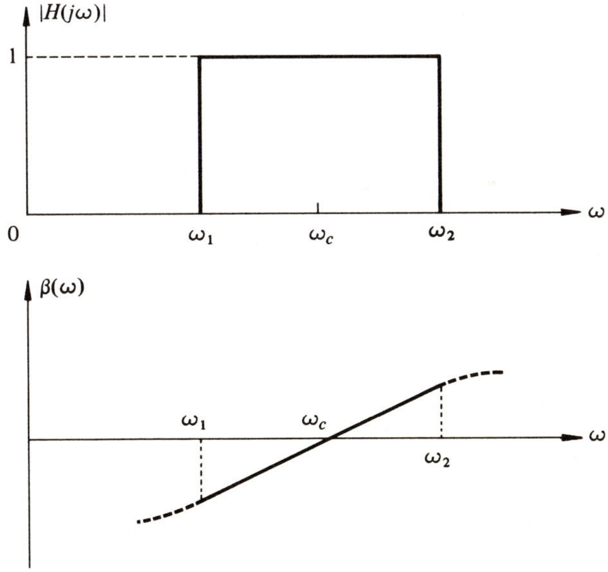

Fig. 7.4 Ideal band-pass constant-delay characteristics.

$$H_{ap}(s) = \frac{P(-s)}{P(s)} = \frac{M(s) - N(s)}{M(s) + N(s)} \tag{7.18}$$

and which has somehow been properly determined to have the desirable delay characteristic, and form a corresponding low-pass function

$$H_{lp}(s) = \frac{1}{P(s)} = \frac{1}{M(s) + N(s)} \tag{7.19}$$

This procedure simply halves the delay (or phase), or

$$\arg\,[H_{lp}(j\omega)] = \tfrac{1}{2}\arg\,[H_{ap}(j\omega)] \tag{7.20}$$

But now $|H_{lp}(j\omega)|$ will be a monotonically decreasing function of ω—hence the low-pass property—because all zeros are located at infinity and $P(s)$ is a Hurwitz polynomial.

In doing so, we exercise no control over $|H_{lp}(j\omega)|$. However, the phase characteristic is obtained readily except for a factor of 2, which can be compensated for by either doubling the bandwidth or constructing two networks in cascade. If the magnitude characteristic obtained this way is not acceptable, certain steps can be taken to modify it without altering the phase characteristic. These steps are explained in Section 7.2.

In the foregoing discussion, our emphases have been on the phase characteristics and, in the case of low-pass networks, the magnitude characteristic. This is necessary because we have better theoretical facilities in designing and controlling these characteristics. The reader should be reminded that these characteristics are not the final aims of our goals here. The final payoff is still in the time domain—how the designed network responds to a prescribed excitation. Even if a pair of magnitude and phase characteristics appear very satisfactory in the frequency domain but their time response is not acceptable, these characteristics are still unsatisfactory. On the other hand, even if their phase and magnitude characteristics appear to be inferior to what we are accustomed to regard as good frequency responses, they may still be acceptable as long as the associated time-domain response is satisfactory.

The frequency-domain characteristics of low-pass constant-delay networks are sometimes an important consideration, especially in band-limited situations. It is therefore important to realize the relationship between time-domain scaling and frequency-domain scaling. This is best explained by citing a theorem in Fourier transform: *If* $F(\omega)$ *is the Fourier transform of* $f(t)$, *then* $(1/a)F(\omega/a)$ *is the Fourier transform of* $f(at)$. Hence the time-domain response of a network may be expanded or contracted merely by a corresponding reverse change in the frequency scale except for a constant gain, which can be adjusted by inserting an amplifier into the system.

7.2 MAGNITUDE MODIFICATION OF DELAY OR PHASE-EQUALIZATION NETWORKS

In the design of frequency-domain filters or equalizers, it is customary to concentrate our effort on the magnitude characteristics. If the phase characteristic of a designed network does not turn out to be acceptable, phase equalization can be effected as the final step before the network becomes acceptable. One convenient procedure is to connect an all-pass section in cascade with the network to introduce the necessary phase equalization without altering the magnitude characteristic.

In many ways the situation encountered in the delay or phase-equalization network design is just the reverse. We design the network to meet the phase requirement by selecting an appropriate all-pass function or alternatively, its corresponding low-pass function, and then hope that the magnitude characteristic of the latter will turn out to be acceptable. Hence it is natural to ask whether or not it is possible to modify the magnitude without altering the phase characteristic in a manner analogous to the way an all-pass section modifies the phase without affecting the magnitude characteristic. This question is answered affirmatively in certain situations by the following analogy.

In the phase correction, an additional factor of the form $P(-s)/P(s)$ can be introduced into any network function without altering the magnitude of the original function. Analogously, in the magnitude correction any even function of s of the form

$$\frac{M_1(s)}{M_2(s)}$$

where $M_1(s)$ and $M_2(s)$ are both even polynomials in s whose j-axis zeros are of even order but whose real-axis zeros and complex zeros may be of odd order, may be added as an additional factor of a network function without altering the phase characteristics. In other words,

$$\arg\left[\frac{M_1(s)}{M_2(s)}\,H(s)\right]_{s=j\omega} = \arg\,[H(s)]_{s=j\omega} \tag{7.21}$$

However, passive realizability and active stability immediately rule out pole distributions of this type. Hence $M_1(s)$ is the only polynomial that can be used for this purpose. Here again the realizability conditions severely limit the extent to which $M_1(s)$ can be chosen, because $M_1(s)$ has all its poles at infinity and the degree of $M_1(s)$ is limited by the number by which the degree of the denominator of $H(s)$ exceeds that of the numerator. In the case of the low-pass characteristic obtained from an all-pass function by dropping the numerator, this limitation is not too severe, because these functions are all-pole functions.

However, in the case of an all-pass network function, no new $M_1(s)$ can be added to the function.

In case the original function does not have a sufficient number of zeros at infinity to allow a sufficiently high degree of $M_1(s)$ for magnitude modification, the following artifice may sometimes be useful.

Suppose the given function is of the form

$$H(s) = \frac{A(-s)}{A(s)}\frac{P(s)}{Q(s)} \tag{7.22}$$

in which $A(-s)/A(s)$ is the all-pass part of this function [$P(-s)$ and $Q(s)$ have no common factors]. The phase of a new function

$$H_1(s) = \frac{1}{A^2(s)}\frac{P(s)}{Q(s)} \tag{7.23}$$

is the same as that of $H(s)$. So is

$$H_2(s) = \frac{M_1(s)}{A^2(s)}\frac{P(s)}{Q(s)} \tag{7.24}$$

Although $|H_1(j\omega)|$ and $|H(j\omega)|$ are different, if we need to modify $|H(j\omega)|$, we might as well modify $|H_1(j\omega)|$. Although the task of modifying $|H_1(j\omega)|$ might be more difficult than that of modifying $|H(j\omega)|$, the additional flexibility provided by a higher degree $M_1(s)$ allowed by $H_1(s)$ may more than compensate for the additional correction needed.

The problem of magnitude correction is thus reduced to one of finding an even polynomial $M_1(s)$ to modify the magnitude characteristic of a given function

$$\frac{1}{D(s)}$$

where $D(s)$ includes all or some of the zeros of $A^2(s)$ and $Q(s)$, such that

$$\left|\frac{M_1(s)P(s)}{A^2(s)Q(s)}\right|_{s=j\omega} = \left|\frac{M_1(s)P(s)}{D(s)Q_1(s)}\right|_{s=j\omega} \tag{7.25}$$

has the desired magnitude characteristic. Suppose the desired magnitude is $|H_d(j\omega)|$; the problem is to find $M_1(s)$ for a given $D(s)$ such that

$$\left|\frac{M_1(j\omega)}{D(j\omega)}\right| = \left|\frac{H_d(j\omega)Q_1(j\omega)}{P(j\omega)}\right| \tag{7.26}$$

A typical application of the magnitude-modification scheme is to introduce j-axis zeros outside the passband to improve the cutoff rate of the low-pass magnitude characteristic. These effects are depicted in Fig. 7.5. Theoretically, these j-axis zeros must be of even order so as not to introduce discontinuities in the phase characteristic at $\omega_1, \omega_2, \ldots$. However, since these poles are usually situated in the region in which the network has high attenuations, the discontinuities usually will have little consequence on the time-domain response of the network. Therefore, in this particular application of the scheme, simple j-axis poles are sometimes used.

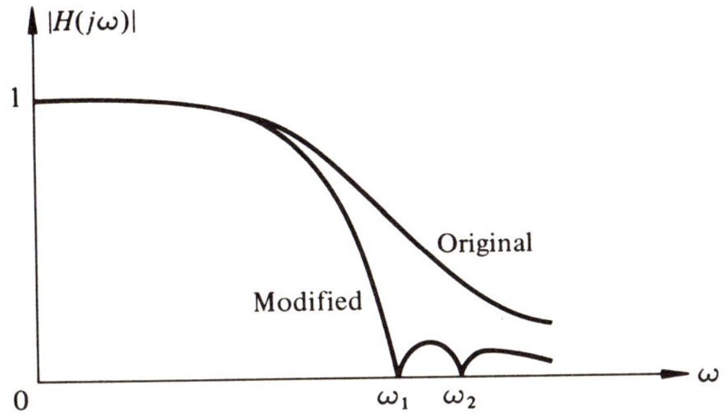

Fig. 7.5 Introduction of j-axis zeros to improve the magnitude characteristic beyond cut-off.

7.3 MAXIMALLY FLAT DELAY FUNCTIONS

A simple and direct approach in finding an

$$H^*(s) = \frac{P(-s)}{P(s)} \tag{7.27}$$

such that it approximates the ideal delay function

$$H(s) = \epsilon^{-T_D s} \tag{7.28}$$

is to equate the transfer function of the all-pass symmetric lattice (7.11) to that of (7.28). For convenience we let $T_D = 2$. (Any other value of T_D may be obtained by a frequency scaling.) We have

$$\frac{1 - z_a}{z + z_a} = \epsilon^{-2s} \tag{7.29}$$

Solving for z_a we get

$$z_a = \tanh s \tag{7.30}$$

By using the continued-fraction expansion of tanh s, we write

$$z_a = \cfrac{1}{\cfrac{1}{s} + \cfrac{1}{\cfrac{3}{s} + \cfrac{1}{\cfrac{5}{s} + \cfrac{1}{\cfrac{7}{s} + \cdots}}}} \tag{7.31}$$

Thus z_a and z_b, which is the dual of z_a, can be realized by the infinite networks of Fig. 7.6.

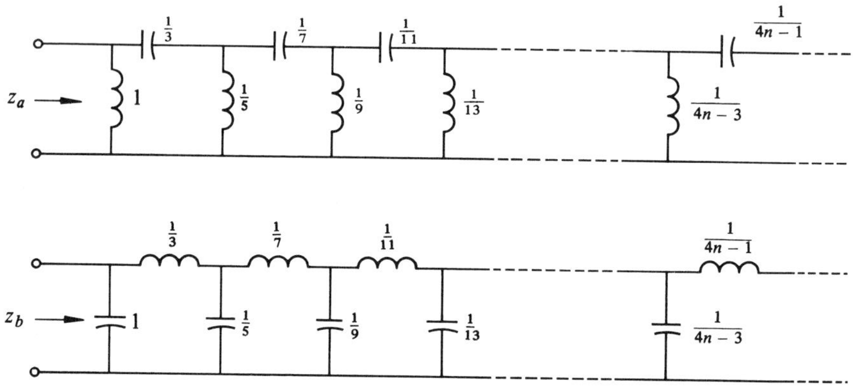

Fig. 7.6 Networks realizing the impedance of (7.31) and its dual.

These infinite networks are, of course, not practical. It is not unrealistic to presume that a good approximation of z_a can be obtained if the networks of Fig. 7.6 are cut off after a number of elements, say after the nth, and either an open circuit or a short circuit placed between the leads. This abbreviation of the network amounts to truncating the continued-fraction expansion (7.31) after the nth term. Call z_{an} or z_{bn} the impedances of the finite networks that include the first n elements of Fig. 7.6. Then

$$z_{an} = \cfrac{1}{\cfrac{1}{s} + \cfrac{1}{\cfrac{3}{s} + \cfrac{1}{\cfrac{5}{s} + \cdots \cfrac{1}{\cfrac{2n-1}{s}}}}} = \frac{M_n(1/s)}{L_n(1/s)} \tag{7.32}$$

Polynomials M_n and L_n are related to the Lommel polynomials [WA3] and may be defined by the recursion formulas

$$M_k\left(\frac{1}{s}\right) = \frac{2k-1}{s} M_{k-1}\left(\frac{1}{s}\right) + M_{k-2}\left(\frac{1}{s}\right) \tag{7.33}$$

$$L_k\left(\frac{1}{s}\right) = \frac{2k-1}{s} L_{k-1}\left(\frac{1}{s}\right) + L_{k-2}\left(\frac{1}{s}\right) \tag{7.34}$$

with the initial values $L_{-1} = 0$, $L_0 = 1$, $M_{-1} = 1$, and $M_0 = 0$. It can be seen that M_k is a polynomial in $1/s$ of degree $k - 1$ and L_k is a polynomial in $1/s$ of degree k. Furthermore, all L's and M's are either odd or even polynomials (that is, every other term missing). Table 7.1 lists z_{an} for values of n up to 6. For higher values of n, z_{an} may be obtained through the Bessel polynomials $y_n(s)$ tabulated in Table 7.3.

TABLE 7.1

Explicit Expressions of z_{an} of (7.32)

n	z_{an}
1	s
2	$\dfrac{3s}{s^2+3}$
3	$\dfrac{s^3+15s}{6s^2+15}$
4	$\dfrac{10s^3+105s}{s^4+45s^2+105}$
5	$\dfrac{s^5+105s^3+945s}{15s^4+420s^2+945}$
6	$\dfrac{21s^5+1260s^3+10{,}395s}{s^6+210s^4+4725s^2+10{,}395}$

Depending on the complexity of the networks used, the all-pass network of Fig. 7.1 will have the transfer function given by

$$H_n(s) = \frac{1 - z_a}{1 + z_a} = \frac{L_n(1/s) - M_n(1/s)}{L_n(1/s) + M_n(1/s)} \tag{7.35}$$

If we let

$$G_n\left(\frac{1}{s}\right) = L_n\left(\frac{1}{s}\right) + M_n\left(\frac{1}{s}\right) \tag{7.36}$$

then $G_n(1/s)$ must satisfy the recursion formula

$$G_k\left(\frac{1}{s}\right) = \frac{2k-1}{s} G_{k-1}\left(\frac{1}{s}\right) + G_{k-2}\left(\frac{1}{s}\right) \tag{7.37}$$

with the initial values $G_{-1} = G_0 = 1$. Furthermore,

$$H_n(s) = (-1)^n \frac{G_n(-1/s)}{G_n(1/s)} \tag{7.38}$$

and $G_n(1/s)$ is a polynomial in $1/s$ of degree n. Hence we may write

$$H_n(s) = (-1)^n \frac{s^n G_n(-1/s)}{s^n G_n(1/s)} = \frac{y_n(-s)}{y_n(s)} \tag{7.39}$$

where $y_n(s) = s^n G_n(1/s)$.

The recursion formula (7.37) and its associated initial values are satisfied by the Bessel polynomials in the variable $1/s$ [KR1]. Thus many well-known relationships for Bessel polynomials are available for the derivation of various forms of $H_n(s)$. Some of these relationships are

$$G_n\left(\frac{1}{s}\right) = \sum_{k=0}^{n} \frac{(n+k)!}{(n-k)!\,k!(2s)^k} \tag{7.40}$$

$$\begin{aligned} y_n(s) &= s^n + a_{n-1}s^{n-1} + a_{n-2}s^{n-2} + \cdots + a_1 s + a_0 \\ &= \sum_{k=0}^{n} \frac{(2n-k)!\,s^k}{2^{n-k}k!(n-k)} \end{aligned} \tag{7.41}$$

$$a_0 = \frac{(2n)!}{2^n n!} = 1\cdot 3\cdot 5 \cdots (2n-1) = (2n-1)!! \tag{7.42}$$

$$y_n(s) = \sum_{k=0}^{n} \frac{(n+k)s^{n-k}}{(n-k)!\,k!\,2^k} \tag{7.43}$$

$$y_n(s) = \sum_{k=0}^{n} \binom{2n-k}{k}(2n-2k-1)!!\,s^k \tag{7.44}$$

If we write [BU1]

$$y_n(s) = \prod_{\nu=1}^{n} (s - s_\nu) \tag{7.45}$$

then

$$\prod_{\nu=1}^{n} \left(\frac{1}{s_\nu}\right) = 1 \tag{7.46}$$

$$\prod_{\nu=1}^{n} \left(\frac{1}{s_\nu}\right)^{2m-1} = 0, \quad \text{for} \quad m = 2, 3, 4, \ldots, n \tag{7.47}$$

It should also be pointed out that $H_n(s)$ of (7.39) is precisely the (n, n) Padé approximant of the function ϵ^{-2s}.

Polynomials $y_n(s)$ for n up to 10 are tabulated in Table 7.2. Zeros of these polynomials are tabulated in Table 7.3. For larger values of n see [OR1].

Low-pass delay network functions can now be formed by writing

TABLE 7.2

Coefficients of the Polynomials $y_n(s) = s^n + a_{n-1}s^{n-1} + a_{n-2}s^{n-2} + \dots + a_1 s + a_0$

n	a_0	a_1	a_2	a_3	a_4	a_5	a_6	a_7	a_8	a_9
1	1									
2	3	3								
3	15	15	6							
4	105	105	45	10						
5	945	945	420	105	15					
6	10,395	10,395	4,725	1,260	210	21				
7	135,135	135,135	62,370	17,325	3,150	378	28			
8	2,027,025	2,027,025	945,945	270,270	51,975	6,930	630	36		
9	34,459,425	34,459,425	16,216,200	4,729,725	945,945	135,135	13,860	990	45	
10	654,729,075	654,729,075	310,134,825	91,891,800	18,918,900	2,837,835	315,315	25,740	1,485	55

TABLE 7.3

Zeros of the Polynomials $y_n(s)$

n	Zeros
1	-1.0000000
2	$-1.5000000 \pm j0.8660254$
3	-2.3221854; $-1.8389073 \pm j1.7543810$
4	$-2.8962106 \pm j0.8672341$; $-2.1037894 \pm j2.6574180$
5	-3.6467386; $-3.3519564 \pm j1.7426614$; $-2.3246743 \pm j3.5710229$
6	$-4.2483594 \pm j0.8675097$; $-3.7357084 \pm j2.6262723$; $-2.5159322 \pm j4.4926730$
7	-4.9717869; $-4.7582905 \pm j1.7392861$; $-4.0701392 \pm j3.5171740$; $-2.6856769 \pm j5.4206941$
8	$-5.5878860 \pm j0.8676144$; $-2.8389840 \pm j6.3539113$; $-4.3682892 \pm j4.4144425$; $-5.2048408 \pm j2.6161751$
9	-6.29070193;$-6.1293679 \pm j1.7378484$; $-5.6044218 \pm j3.4981573$; $-4.6384399 \pm j5.3172717$; $-2.9792608 \pm j7.2914637$
10	$-6.9220449 \pm j0.8676651$; $-3.1089162 \pm j8.2326995$; $-6.6152916 \pm j2.6115683$; $-5.9675282 \pm j4.3849471$; $-4.8862195 \pm j6.2249855$

$$H_n(s) = \frac{k}{y_n(s)} = \frac{k}{m_n(s) + n_n(s)} \tag{7.48}$$

where $m_n(s)$ is the even part and $n_n(s)$ is the odd part of the denominator polynomial. The phase lag of such a function is

$$\beta_n(\omega) = \tan^{-1}\left[\frac{n_n(j\omega)}{jm_n(j\omega)}\right] = \tan^{-1}[z_{an}(j\omega)] \tag{7.49}$$

where $z_{an}(s)$ is the impedance of the z_a network abbreviated after the nth element and is given in (7.32).

The group delay characteristic as defined by

$$T_d(\omega) = \frac{d\beta(\omega)}{d\omega} \tag{7.50}$$

may be obtained for different n from (7.49). Let $X_n(\omega)$ denote the reactance of $z_{an}(j\omega)$, which is purely imiginary. Then we have

$$T_{dn}(\omega) = \frac{1}{1 + X_n^2(\omega)} \frac{dX_n(\omega)}{d\omega} \tag{7.51}$$

$T_{dn}(\omega)$ for n up to 4 are listed in Table 7.4.

TABLE 7.4

Delay Characteristics of Network Functions of (7.48)

n	$T_{dn}(\omega)$
1	$\frac{1}{1 + \omega^2}$
2	$\frac{9 + 3\omega^2}{9 + 3\omega^2 + \omega^4}$
3	$\frac{225 + 45\omega^2 + 6\omega^4}{225 + 45\omega^2 + 6\omega^4 + \omega^6}$
4	$\frac{11{,}025 + 1575\omega^2 + 135\omega^4 + 10\omega^6}{11{,}025 + 1575\omega^2 + 135\omega^4 + 10\omega^6 + \omega^8}$

In Table 7.4 the denominator of $T_{dn}(\omega)$ is simply $m_n^2(j\omega) - n_n^2(j\omega)$. The numerators are simply the corresponding denominators with the highest terms deleted. Thus the delay characteristic $T_{dn}(\omega)$ is maximally flat at $\omega = 0$. Low-pass delay networks obtained this way are known as *maximally flat delay* networks [TH1, TH2]. The reader should not confuse this term with the Butterworth network, which has a *maximally flat magnitude* characteristic.

For analytical studies and computational work, advantage may frequently be taken of the relationships between Bessel polynomials and Bessel functions. For example [KR1],

$$G_n\left(\frac{1}{j\omega}\right) = j^{-n}\sqrt{\frac{\pi\omega}{2}}\,[\epsilon^{j\omega}(-1)^n J_{-n-(1/2)}(\omega) - jJ_{n+(1/2)}(\omega)] \tag{7.52}$$

Thus $H_n(s)$ of (7.48) becomes

$$H_n(j\omega) = \frac{K\epsilon^{-j\omega}}{\omega^{n+1}\sqrt{\pi/2\omega}\,[(-1)^n J_{-n-(1/2)}(\omega) - jJ_{n+(1/2)}(\omega)]} \tag{7.53}$$

the phase lag is

$$\beta_n(\omega) = \omega - \tan^{-1}\left[\frac{J_{n+(1/2)}(\omega)}{(-1)^n J_{-n-(1/2)}(\omega)}\right] \tag{7.54}$$

and the delay function is

$$T_{dn}(\omega) = 1 - \frac{2}{\pi\omega[J^2_{-n-(1/2)}(\omega) + J^2_{n+(1/2)}(\omega)]} \tag{7.55}$$

The Maclaurin series of (7.55) is

$$T_{dn}(\omega) = 1 - \frac{\omega^{2n}}{a_0^2} + \frac{\omega^{2n+2}}{(2n-1)a_0^2} - \frac{2(n-2)\omega^{2n+4}}{(2n-1)^2(2n-3)a_0^2} + \cdots \tag{7.56}$$

The maximally flat property of the delay characteristic associated with the network function of (7.48) is now not only apparent but also analytically established for any order n. Figure 7.7 shows the delay characteristics of (7.48) for values of n up to 10. It should be noted that Fig. 7.7 is also applicable to the all-pass transfer function of (7.39) if their zero-frequency values are adjusted accordingly.

The magnitude function of the low-pass delay function (7.48) can also be obtained readily from (7.53):

$$|H(j\omega)|^2 = \frac{2K^2}{\omega^{2n+1}\pi[J^2_{-n-(1/2)}(\omega) + J^2_{n+(1/2)}(\omega)]} \tag{7.57}$$

And the loss when expressed in decibels (dB) [loss $= L = -20 \log |H(j\omega)|$] is approximately

$$L \cong \frac{10\omega^2}{(2n-1)\ln 10}\,\text{dB} \tag{7.58}$$

and is approached for large n. Equation (7.58) gives the approximate 3-dB bandwidth to be

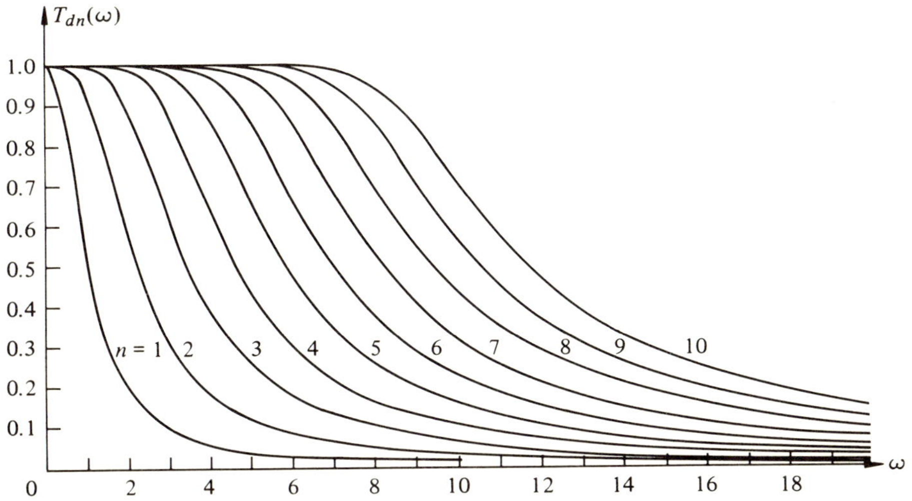

Fig. 7.7 The maximally flat delay characteristics.

$$\omega_{3\text{-dB}} \cong \sqrt{(2n - 1)\ln 2} \tag{7.59}$$

Although the transfer characteristic obtained by using the Bessel polynomials has been shown to be maximally flat for all values of n, the question of whether the polynomials are unique or not has not been established. That is, are there more than one polynomial that gives the maximally flat delay for each n? Here we shall attempt to approach the problem of finding the maximally flat characteristic from the opposite direction. Suppose it is desired that the low-pass network function

$$H(s) = \frac{K}{P(s)} = \frac{K}{m(s) + n(s)} \tag{7.60}$$

possesses a delay that is maximally flat about $\omega = 0$. As was shown for (7.49),

$$\beta(\omega) = \tan^{-1}\left[\frac{n(j\omega)}{jm(j\omega)}\right] \tag{7.61}$$

and thus

$$T_d(\omega) = \left.\frac{m(s)n'(s) - n(s)m'(s)}{m^2(s) - n^2(s)}\right|_{s=j\omega} \tag{7.62}$$

Expression (7.62) is the even part of the rational function

$$\frac{P'(s)}{P(s)} = \frac{m'(s) + n'(s)}{m(s) + n(s)}$$

that crops up frequently in connection with the study of passive networks, especially in the derivation of the positive-real properties of certain network functions. We shall now determine what the values of the coefficients of $P(s)$ should be in order that $T_d(\omega)$ be maximally flat.

For $n = 2$, we let

$$P(s) = s^2 + a_1 s + a_0 \tag{7.63}$$

$$P'(s) = 2s + a_1$$

$$T_{d2}(\omega) = \frac{a_1\omega^2 + a_1 a_0}{\omega^4 + (a_1^2 - 2a_0)\omega^2 + a_0^2} \tag{7.64}$$

For maximal flatness at the origin it is necessary that

$$a_1 = a_1^2 - 2a_0, \qquad a_0 a_1 = a_0^2 \tag{7.65}$$

which requires that

$$a_1 = a_0 = 3 \tag{7.66}$$

These values agree with the coefficients given in Table 7.2 for $y_2(s)$.

For $n = 3$,

$$P(s) = s^3 + a_2 s^2 + a_1 s + a_0 \tag{7.67}$$

$$P'(s) = 3s^2 + 2a_2 s + a_1 \tag{7.68}$$

$$T_{d3}(\omega) = \frac{a_2\omega^4 + (a_1 a_2 - 3a_0)\omega^2 + a_1 a_0}{\omega^6 + (a_2^2 - 2a_1)\omega^4 + (a_1^2 - 2a_0 a_2)\omega^2 + a_0^2} \tag{7.69}$$

For maximal flatness it is necessary that

$$a_2 = a_2^2 - 2a_1, \qquad a_1 a_2 - 3a_0 = a_1^2 - 2a_0 a_2, \qquad a_1 a_0 = a_0^2 \tag{7.70}$$

which yields

$$a_0 = 15, \qquad a_1 = 15, \qquad a_2 = 6 \tag{7.71}$$

These values again agree with the coefficients of $y_3(s)$ in Table 7.2.

This procedure is straightforward and may be extended to higher values of n. However, the complexity of the equations relating the coefficients of $P(s)$ becomes quite complicated as the value of n becomes higher. Hence this method is not very useful for the purpose of obtaining $H_n(s)$ directly. The existence of unique sets of coefficients for $n = 2$ and 3 does suggest that the Bessel polynomials are the unique polynomials to provide maximally flat delay characteristics.

Impulse Responses

The impulse response of a maximally flat delay network function, $h_n(t)$, is difficult to obtain in general terms. For a definite value of n, $h_n(t)$ can be obtained by taking the inverse Laplace transform of $H_n(s)$ given by (7.48), with $y_n(s)$ as tabulated in Table 7.2. Impulse responses for $n = 2$, 5, and 9 are shown in Fig. 7.8.

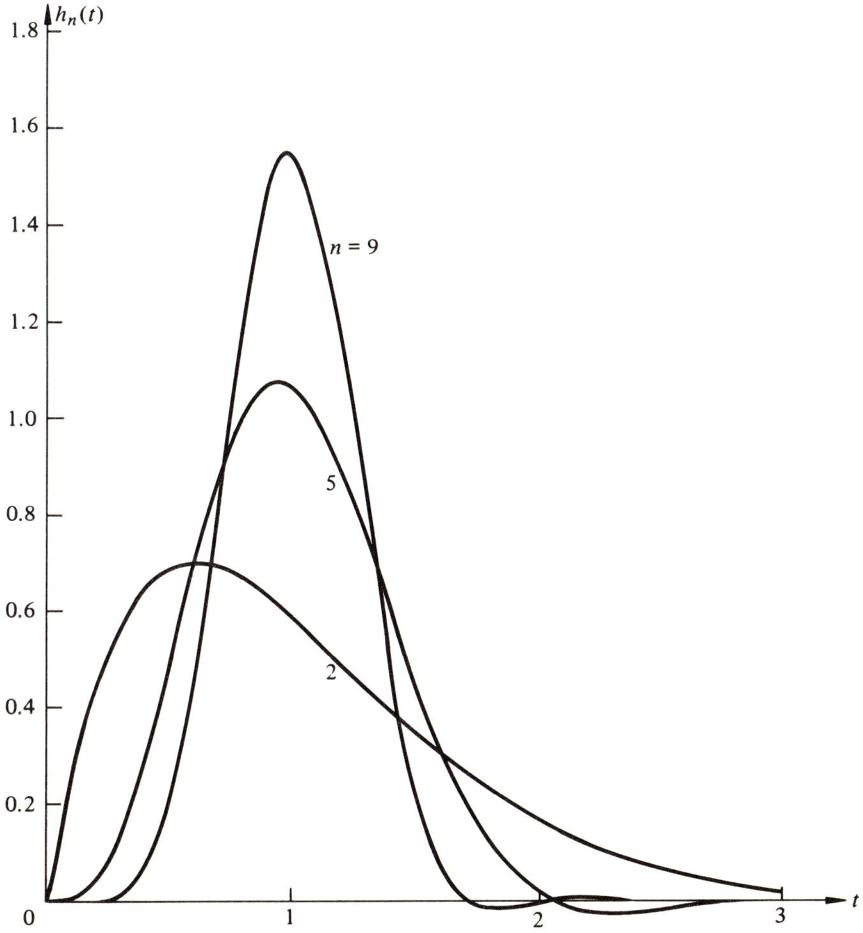

Fig. 7.8 Impulse responses of maximally flat delay networks for $n = 2, 5, 9$.

Networks

Once the transfer function for a maximally flat delay network has been formed, various methods of synthesizing different classes of passive or active networks can be used to arrive at the network. The network for one particular case is especially simple to obtain. This is the case when the low-pass $H_n(s)$ is the transfer impedance of an LC two-port terminated in a resistance (normalized to 1 ohm). Referring to Fig. 7.9, we have

$$H_n(s) = Z_{12}(s) = \frac{E_2}{I_1} \tag{7.72}$$

and

$$Z_{12}(s) = \frac{Kz_{12}}{1 + z_{22}} \tag{7.73}$$

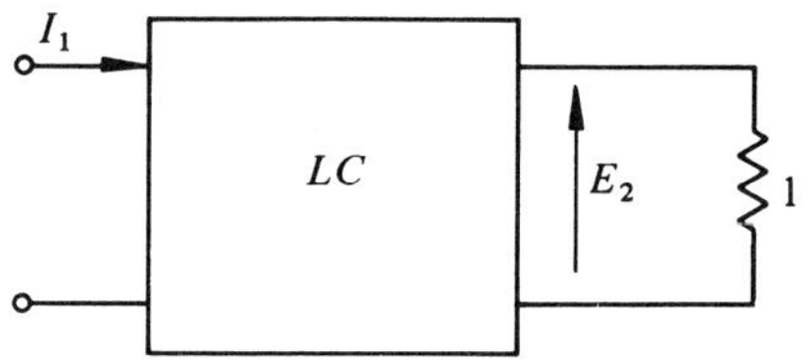

Fig. 7.9 An LC two-port terminated in a 1-ohm resistance.

We may then assign

$$z_{22}(s) = \frac{n(s)}{m(s)}, \qquad z_{12}(s) = \frac{K}{m(s)} \tag{7.74}$$

Hence the LC network may be obtained by expanding z_{22} into a Cauer's ladder with all shunt capacitors and series inductors. In fact, this network is none other than the truncated z_a of Fig. 7.6. We, of course, have no control over the multiplicative constant K.

7.4 MAXIMALLY FLAT COMPROMISE CHARACTERISTICS

The magnitude and phase characteristics of an ideal low-pass constant-delay network are shown in Fig. 7.3. The delay-versus-frequency characteristic of such a network will have the shape shown in Fig. 7.10. In a lumped-network realization of such a characteristic we must accept some kind of deviation from this ideal characteristic.

The low-pass *delay* characteristics obtained by the procedure of Section 7.3 were found to be maximally flat at the origin of the ω axis. Thus these delay characteristics $[T_d(\omega)]$ are optimum in that limited sense. When this is done we must accept the magnitude characteristic $[|H(j\omega)|]$ that results. The only thing we can say about the magnitude characteristics obtained in this manner is that they are generally of the low-pass type.

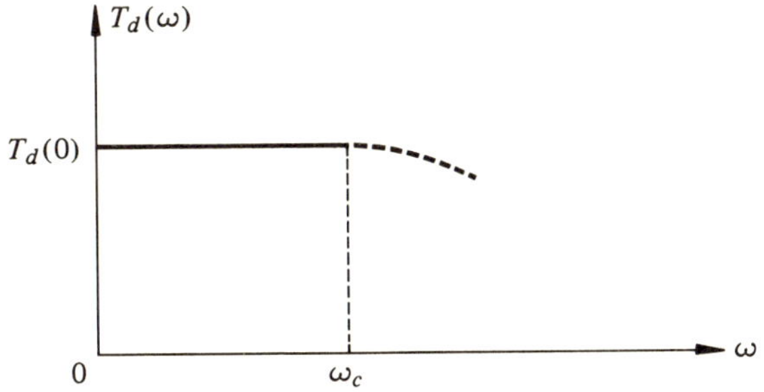

Fig. 7.10 Ideal low-pass delay characteristic.

In this context we might wonder what would be the result if we place emphasis of approximation on the magnitude characteristic and accept the delay that comes with it. When this criterion is applied to the magnitude characteristic, the result is the well-known maximally flat magnitude or Butterworth characteristic [VA5, WE2],

$$H(s) = \frac{1}{B_n(s)} \tag{7.75}$$

where $B_n(s)$ is the Butterworth polynomial whose zeros are given by

$$s_\nu = -\sin\frac{(2\nu + 1)\pi}{2n} + j\cos\frac{(2\nu + 1)\pi}{2n}, \qquad \nu = 0, 1, 2, \ldots, n - 1 \tag{7.76}$$

and are uniformly spaced along the left half of the unit circle centered at the origin in the s plane. The zeros of $B_n(s)$ for n up to 10 are listed in Table 7.5. The corresponding coefficients of $B_n(s)$ are listed in Table 7.6.

Quantitative comparison between the maximally flat delay and the maximally flat magnitude characteristics is made together with comparisons of other delay characteristics later in this chapter. Qualitatively we may say that both the maximally flat delay and the maximally flat magnitude characteristics have important shortcomings in their corresponding time-domain responses because of different emphases in the approximation procedures by which they are obtained. The magnitude characteristic corresponding to a maximally flat delay function tends to attenuate the higher-frequency portion of its passband too rapidly. On the other hand, the delay characteristic corresponding to a maximally flat magnitude function will also roll off too fast in the passband.

Hence we may want, in some situations, to accept the inferior features of each characteristic to a lesser extent while attempting to retain the desirable features of each of these characteristics. When this is done, a new set of characteristics that lie somewhere between the two classes of maximally flat

TABLE 7.5

Zeros of the Butterworth Polynomials $B_n(s)$

$n = 1$	$n = 2$	$n = 3$	$n = 4$	$n = 5$	$n = 6$	$n = 7$	$n = 8$	$n = 9$	$n = 10$
-1.0000000	-0.7071068 $\pm j0.7071068$	-1.0000000	-0.3826834 $\pm j0.9238795$	-1.0000000	-0.2588190 $\pm j0.9659258$	-1.0000000	-0.1950903 $\pm j0.9807853$	-1.0000000	-0.1564345 $\pm j0.9876883$
		-0.5000000 $\pm j0.8660254$	-0.9238795 $\pm j0.3826834$	-0.3090170 $\pm j0.9510565$	-0.7071068 $\pm j0.7071068$	-0.2225209 $\pm j0.9749279$	-0.5555702 $\pm j0.8314696$	-0.1736482 $\pm j0.9848078$	-0.4539905 $\pm j0.8910065$
				-0.8090170 $\pm j0.5877852$	-0.9659258 $\pm j0.2588190$	-0.6234898 $\pm j0.7818315$	-0.8314696 $\pm j0.5555702$	-0.5000000 $\pm j0.8660254$	-0.7071068 $\pm j0.7071068$
						-0.9009689 $\pm j0.4338837$	-0.9807853 $\pm j0.1950903$	-0.7660444 $\pm j0.6427876$	-0.8910065 $\pm j0.4539905$
								-0.9396926 $\pm j0.3420201$	-0.9876883 $\pm j0.1564345$

TABLE 7.6

Coefficients of the Butterworth Polynomials

$B_n(s) = s^n + a_{n-1}s^{n-1} + \dots + a_2s^2 + a_1s + 1$

n	a_1	a_2	a_3	a_4	a_5	a_6	a_7	a_8	a_9
2	1.4142136								
3	2.0000000	2.0000000							
4	2.6131259	3.4142136	2.6131259						
5	3.2360680	5.2360680	5.2360680	3.2360680					
6	3.8637033	7.4641016	9.1416202	7.4641016	3.8637033				
7	4.4939592	10.0978347	14.5917939	14.5917939	10.0978347	4.4939592			
8	5.1258309	13.1370712	21.8461510	25.6883559	21.8461510	13.1370712	5.1258309		
9	5.7587705	16.5817187	31.1634375	41.9863857	41.9863857	31.1634375	16.5817187	5.7587705	
10	6.3924532	20.4317291	42.8020611	64.8823963	74.2334292	64.8823963	42.8020611	20.4317291	6.3924532

characteristics are obtained. These new characteristics shall be called the maximally flat compromise characteristics.

This compromise may be done by returning to the original approximation procedure but allotting the available degrees of freedom partly to the delay and partly to the magnitude characteristics instead of wholly to one or the other. A simpler approach would be to achieve the compromise by interpolating the two sets of poles. This scheme has been studied more fully and will be discussed in detail here.

Before a meaningful interpolating scheme can be developed, it is necessary to observe the frequency-domain scales in the two classes of characteristics. As was mentioned earlier, the zeros of $B_n(s)$ lie on a unit circle. Because of this, the a_0 of each $B_n(s)$ is unity regardless of n. This is not true with the $y_n(s)$ in Table 7.2. A brief study of the locations of the zeros of $y_n(s)$ in Table 7.3 will reveal that these zeros are almost all outside the unity circle. Hence the frequency scales of these two polynomials are very different. It is useful to adjust either of the two scales so they will be comparable to each other. In this connection it would be convenient to normalize $y_n(s)$ such that the new a_0 will be unity. This is equivalent to replacing $y_n(s)$ by

$$Y_n(s) = y_n\left(\frac{s}{\sqrt[n]{a_0}}\right) = s^n + c_{n-1}s^{n-1} + \cdots + c_1 s + 1 \tag{7.77}$$

Zeros of $Y_n(s)$ are tabulated in Table 7.7.

This normalization procedure amounts to a change in the frequency-domain scale such that the geometric mean distance of all zeros from the origin is unity. The zeros of $Y_7(s)$ and $B_7(s)$ are depicted in Fig. 7.11.

This normalization has another interesting by-product worth mentioning here. The fact that both a_0 of $B_n(s)$ and c_0 of $Y_n(s)$ are unity implies that both

$$\frac{1}{B_n(s)}$$

and

$$\frac{1}{Y_n(s)}$$

when considered as transfer functions, have unity dc ($s = 0$) gain.

After this normalization has been applied to the poles of maximally flat delay functions, an interpolation between their poles and those of the maximally flat magnitude functions can be performed in many different ways. The scheme used here is to pair off the zeros, one from each set starting from the ones nearest the negative real axis. Let $z_b = r_b\underline{/\phi_b}$ represent a zero of $B_n(s)$ and

TABLE 7.7

Zero of $Y_n(s)$

n	Zeros
2	$-0.8660254 \pm j0.5000000$
3	-0.9416000
	$-0.7456404 \pm j0.7113666$
4	$-0.6572112 \pm j0.8301614$
	$-0.9047588 \pm j0.2709187$
5	-0.9264420
	$-0.5905759 \pm j0.9072068$
	$-0.8515536 \pm j0.4427175$
6	$-0.5385527 \pm j0.9616877$
	$-0.7996542 \pm j0.5621717$
	$-0.9093907 \pm j0.1856964$
7	-0.9194872
	$-0.4966917 \pm j1.0025085$
	$-0.7527356 \pm j0.6504696$
	$-0.8800029 \pm j0.3216653$
8	$-0.4621741 \pm j1.0343887$
	$-0.7111382 \pm j0.7186517$
	$-0.8473251 \pm j0.4259017$
	$-0.9096831 \pm j0.1412438$
9	-0.9154953
	$-0.4331413 \pm j1.0600692$
	$-0.6743705 \pm j0.7730542$
	$-0.8148017 \pm j0.5085814$
	$-0.8911212 \pm j0.2526580$
10	$-0.4101120 \pm j1.0860148$
	$-0.6445646 \pm j0.8211676$
	$-0.7608222 \pm j0.5784394$
	$-0.8726548 \pm j0.3445045$
	$-0.9131201 \pm j0.1144579$

$z_y = r_y\underline{/\phi_y}$ represent the zero of $Y_n(s)$ from the same pair. Then the intermediate zero is given by

$$z_m = r_m\underline{/\phi_m} \tag{7.78}$$

where

$$r_m = r_y^m \tag{7.79}$$

$$\phi_m = \phi_b - m(\phi_b - \phi_y) \tag{7.80}$$

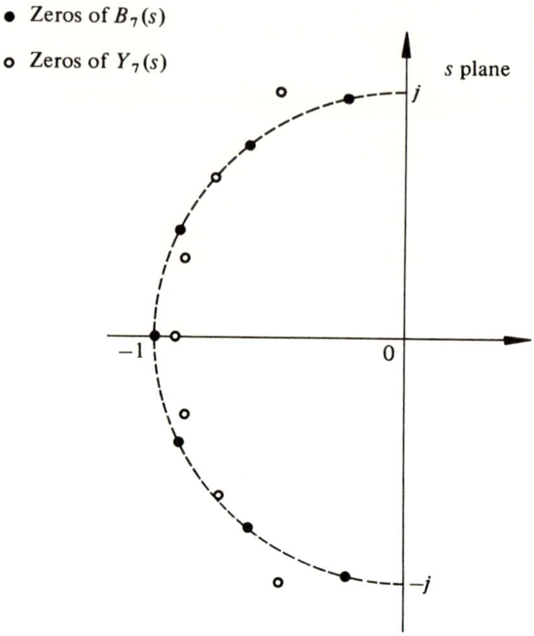

Fig. 7.11 Zeros of $B_7(s)$ and $Y_7(s)$.

and m is an interpolating parameter that usually may assume any positive value between 0 and 1. When $m = 0$, $z_m = z_b$ and $r_m = 1$. When $m = 1$, $z_m = z_y$. Thus as m is varied from 0 to 1, the angle of the vector joining the origin and z_m varies linearly while its magnitude varies logarithmically from z_b to z_y; the

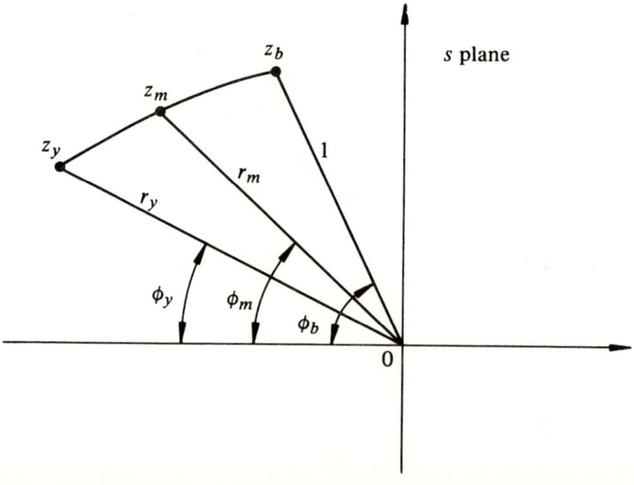

Fig. 7.12 Interpolation scheme for compromise network functions.

transition is depicted in Fig. 7.12. This interpolation scheme has the advantage that the zeros of the intermediate polynomial also have the property that their geometric mean distance to the origin remains unity.

Zeros of the denominator of the maximally flat compromise network function are listed in Table 7.8 for several values of m.

TABLE 7.8(A)

Poles of Maximally Flat Compromise Network Functions
($n = 2$)

m	Poles
−0.2	$-0.6691306 \pm j0.7431448$
−0.1	$-0.6883546 \pm j0.7253744$
0	$-0.7071068 \pm j0.7071068$
0.1	$-0.7253744 \pm j0.6883546$
0.2	$-0.7431448 \pm j0.6691306$
0.3	$-0.7604060 \pm j0.6494480$
0.4	$-0.7771460 \pm j0.6293204$
0.5	$-0.7933533 \pm j0.6087614$
0.6	$-0.8090170 \pm j0.5877853$
0.7	$-0.8241262 \pm j0.5664062$
0.8	$-0.8386706 \pm j0.5446390$
0.9	$-0.8526402 \pm j0.5224986$
1.0	$-0.8660254 \pm j0.5000000$
1.1	$-0.8788171 \pm j0.4771588$
1.2	$-0.8910065 \pm j0.4539905$

TABLE 7.8(B)

Poles of Maximally Flat Compromise Network Functions
$(n = 3)$

m	Poles
−0.2	−1.0121076 −0.4470959 ± j0.8877739
−0.1	−1.0060356 −0.4736633 ± j0.8772934
0	−1.0000000 −0.5000000 ± j0.8660254
0.1	−0.9940006 −0.5260829 ± j0.8539745
0.2	−0.9880372 −0.5518889 ± j0.8411458
0.3	−0.9821096 −0.5773951 ± j0.8275453
0.4	−0.9762175 −0.6025786 ± j0.8131795
0.5	−0.9703608 −0.6274166 ± j0.7980557
0.6	−0.9645392 −0.6518865 ± j0.7821818
0.7	−0.9587525 −0.6759660 ± j0.7655664
0.8	−0.9530006 −0.6996328 ± j0.7482187
0.9	−0.9472832 −0.7228649 ± j0.7301486
1.0	−0.9416000 −0.7456404 ± j0.7113666
1.1	−0.9359510 −0.7679378 ± j0.6918841
1.2	−0.9303359 −0.7897358 ± j0.6717127

TABLE 7.8(C)

Poles of Maximally Flat Compromise Network Functions
$(n = 4)$

m	Poles
-0.2	$-0.3271895 \pm j0.9329230$ $-0.9264298 \pm j0.4060200$
-0.1	$-0.3549198 \pm j0.9287984$ $-0.9252104 \pm j0.3943119$
0	$-0.3826834 \pm j0.9238795$ $-0.9238795 \pm j0.3826834$
0.1	$-0.4104585 \pm j0.9181612$ $-0.9224387 \pm j0.3711353$
0.2	$-0.4382228 \pm j0.9116387$ $-0.9208893 \pm j0.3596682$
0.3	$-0.4659539 \pm j0.9043082$ $-0.9192327 \pm j0.3482827$
0.4	$-0.4936292 \pm j0.8961662$ $-0.9174706 \pm j0.3369794$
0.5	$-0.5212256 \pm j0.8872099$ $-0.9156041 \pm j0.3257589$
0.6	$-0.5487201 \pm j0.8774372$ $-0.9136349 \pm j0.3146219$
0.7	$-0.5760894 \pm j0.8668466$ $-0.9115643 \pm j0.3035688$
0.8	$-0.6033101 \pm j0.8554371$ $-0.9093939 \pm j0.2926002$
0.9	$-0.6303586 \pm j0.8432086$ $-0.9071249 \pm j0.2817167$
1.0	$-0.6572112 \pm j0.8301614$ $-0.9047588 \pm j0.2709187$
1.1	$-0.6838439 \pm j0.8162967$ $-0.9022971 \pm j0.2602068$
1.2	$-0.7102330 \pm j0.8016161$ $-0.8997412 \pm j0.2495813$

TABLE 7.8(D)

Poles of Maximally Flat Compromise Network Functions ($n = 5$)

m	Poles
−0.2	−1.0153981 −0.2545375 ± j0.9507887 −0.7976842 ± j0.6166553
−0.1	−1.0076696 −0.2816675 ± j0.9512802 −0.8034697 ± j0.6022328
0	−1.0000000 −0.3090170 ± j0.9510565 −0.8090170 ± j0.5877852
0.1	−0.9923887 −0.3365687 ± j0.9501065 −0.8143268 ± j0.5733162
0.2	−0.9848354 −0.3643050 ± j0.9484191 −0.8193997 ± j0.5588293
0.3	−0.9773395 −0.3922076 ± j0.9459841 −0.8242364 ± j0.5443282
0.4	−0.9699008 −0.4202580 ± j0.9427911 −0.8288378 ± j0.5298164
0.5	−0.9625186 −0.4484373 ± j0.9388306 −0.8332047 ± j0.5152975
0.6	−0.9551926 −0.4767261 ± j0.9340934 −0.8373379 ± j0.5007749
0.7	−0.9479224 −0.5051047 ± j0.9285704 −0.8412384 ± j0.4862522
0.8	−0.9407075 −0.5335530 ± j0.9222536 −0.8449072 ± j0.4717328
0.9	−0.9335475 −0.5620503 ± j0.9151348 −0.8483452 ± j0.4572201
1.0	−0.9264420 −0.5905759 ± j0.9072068 −0.8515536 ± j0.4427175
1.1	−0.9193906 −0.6191085 ± j0.8984626 −0.8545335 ± j0.4282284
1.2	−0.9123929 −0.6476266 ± j0.8888959 −0.8572860 ± j0.4137561

TABLE 7.8(E)

Poles of Maximally Flat Compromise Network Functions ($n = 6$)

m	Poles
-0.2	$-0.2064173 \pm j0.9587546$ $-0.6853878 \pm j0.7344335$ $-0.9771933 \pm j0.2745255$
-0.1	$-0.2324461 \pm j0.9626486$ $-0.6963776 \pm j0.7208483$ $-0.9715623 \pm j0.2666260$
0	$-0.2588190 \pm j0.9659258$ $-0.7071068 \pm j0.7071068$ $-0.9659258 \pm j0.2588190$
0.1	$-0.2855239 \pm j0.9685719$ $-0.7175733 \pm j0.6932137$ $-0.9602846 \pm j0.2511040$
0.2	$-0.3125481 \pm j0.9705724$ $-0.7277751 \pm j0.6791740$ $-0.9546391 \pm j0.2434801$
0.3	$-0.3398784 \pm j0.9719133$ $-0.7377103 \pm j0.6649926$ $-0.9489899 \pm j0.2359469$
0.4	$-0.3675011 \pm j0.9725806$ $-0.7473772 \pm j0.6506744$ $-0.9433377 \pm j0.2285037$
0.5	$-0.3954021 \pm j0.9725606$ $-0.7567741 \pm j0.6362244$ $-0.9376830 \pm j0.2211497$
0.6	$-0.4235667 \pm j0.9718398$ $-0.7658993 \pm j0.6216476$ $-0.9320263 \pm j0.2138843$
0.7	$-0.4519799 \pm j0.9704050$ $-0.7747513 \pm j0.6069489$ $-0.9263682 \pm j0.2067070$
0.8	$-0.4806261 \pm j0.9682432$ $-0.7833286 \pm j0.5921334$ $-0.9207092 \pm j0.1996170$
0.9	$-0.5094892 \pm j0.9653416$ $-0.7916300 \pm j0.5772060$ $-0.9150498 \pm j0.1926137$
1.0	$-0.5385527 \pm j0.9616877$ $-0.7996542 \pm j0.5621717$ $-0.9093907 \pm j0.1856964$
1.1	$-0.5677996 \pm j0.9572695$ $-0.8073999 \pm j0.5470357$ $-0.9037322 \pm j0.1788646$
1.2	$-0.5972125 \pm j0.9520750$ $-0.8148662 \pm j0.5318029$ $-0.8980749 \pm j0.1721175$

TABLE 7.8(F)

Poles of Maximally Flat Compromise Network Functions
$(n = 7)$

m	Poles
−0.2	−1.0169295
	−0.1724368 ± j0.9624729
	−0.5947639 ± j0.8051859
	−0.9039578 ± j0.4574419
−0.1	−1.0084292
	−0.1972649 ± j0.9689605
	−0.6092385 ± j0.7936385
	−0.9025164 ± j0.4456171
0	−1.0000000
	−0.2225209 ± j0.9749279
	−0.6234898 ± j0.7818315
	−0.9009689 ± j0.4338837
0.1	−0.9916412
	−0.2481971 ± j0.9803594
	−0.6375130 ± j0.7697692
	−0.8993168 ± j0.4222422
0.2	−0.9833523
	−0.2742854 ± j0.9852389
	−0.6513036 ± j0.7574559
	−0.8975618 ± j0.4106929
0.3	−0.9751327
	−0.3007771 ± j0.9895506
	−0.6648571 ± j0.7448962
	−0.8957054 ± j0.3992363
0.4	−0.9669817
	−0.3276632 ± j0.9932787
	−0.6781692 ± j0.7320946
	−0.8937492 ± j0.3878728
0.5	−0.9588989
	−0.3549341 ± j0.9964073
	−0.6912354 ± j0.7190557
	−0.8916948 ± j0.3766027
0.6	−0.9508837
	−0.3825797 ± j0.9989206
	−0.7040517 ± j0.7057843
	−0.8895436 ± j0.3654264
0.7	−0.9429355
	−0.4105895 ± j1.0008029
	−0.7166138 ± j0.6922851
	−0.8872973 ± j0.3543442

TABLE 7.8(F) Continued

m	Poles
0.8	-0.9350537
	$-0.4389526 \pm j1.0020387$
	$-0.7289178 \pm j0.6785629$
	$-0.8849574 \pm j0.3433564$
0.9	-0.9272377
	$-0.4676573 \pm j1.0026124$
	$-0.7409596 \pm j0.6646228$
	$-0.8825254 \pm j0.3324633$
1.0	-0.9194872
	$-0.4966917 \pm j1.0025085$
	$-0.7527356 \pm j0.6504696$
	$-0.8800029 \pm j0.3216653$
1.1	-0.9118014
	$-0.5260433 \pm j1.0017118$
	$-0.7642418 \pm j0.6361085$
	$-0.8773914 \pm j0.3109625$
1.2	-0.9041798
	$-0.5556990 \pm j1.0002070$
	$-0.7754746 \pm j0.6215446$
	$-0.8746924 \pm j0.3003552$

TABLE 7.8(G)

Poles of Maximally Flat Compromise Network Functions
$(n = 8)$

m	Poles
−0.2	−0.1472773 ± j0.9641613
	−0.5222475 ± j0.8502235
	−0.8262376 ± j0.5820432
	−0.9954378 ± j0.2067770
−0.1	−0.1709415 ± j0.9726890
	−0.5389893 ± j0.8410109
	−0.8289421 ± j0.5687877
	−0.9880936 ± j0.2008940
0	−0.1950903 ± j0.9807853
	−0.5555702 ± j0.8314696
	−0.8314696 ± j0.5555702
	−0.9807853 ± j0.1950903
0.1	−0.2197197 ± j0.9884339
	−0.5719837 ± j0.8216024
	−0.8338215 ± j0.5423926
	−0.9735129 ± j0.1893652
0.2	−0.2448256 ± j0.9956184
	−0.5882235 ± j0.8114122
	−0.8359991 ± j0.5292569
	−0.9662765 ± j0.1837178
0.3	−0.2704033 ± j1.0023224
	−0.6042832 ± j0.8009021
	−0.8380039 ± j0.5161650
	−0.9590762 ± j0.1781473
0.4	−0.2964477 ± j1.0085292
	−0.6201566 ± j0.7900751
	−0.8398375 ± j0.5031187
	−0.9519118 ± j0.1726531
0.5	−0.3229533 ± j1.0142220
	−0.6358373 ± j0.7789345
	−0.8415012 ± j0.4901199
	−0.9447835 ± j0.1672342
0.6	−0.3499141 ± j1.0193843
	−0.6513193 ± j0.7674838
	−0.8429966 ± j0.4771705
	−0.9376912 ± j0.1618899
0.7	−0.3773237 ± j1.0239990
	−0.6665964 ± j0.7557265
	−0.8443251 ± j0.4642722
	−0.9306351 ± j0.1566195

TABLE 7.8(G) Continued

m	Poles
0.8	$-0.4051752 \pm j1.0280493$
	$-0.6816626 \pm j0.7436662$
	$-0.8454884 \pm j0.4514268$
	$-0.9236150 \pm j0.1514222$
0.9	$-0.4334613 \pm j1.0315182$
	$-0.6965118 \pm j0.7313066$
	$-0.8464878 \pm j0.4386361$
	$-0.9166310 \pm j0.1462972$
1.0	$-0.4621741 \pm j1.0343887$
	$-0.7111382 \pm j0.7186517$
	$-0.8473251 \pm j0.4259017$
	$-0.9096831 \pm j0.1412438$
1.1	$-0.4913052 \pm j1.0366438$
	$-0.7255359 \pm j0.7057055$
	$-0.8480017 \pm j0.4132255$
	$-0.9027714 \pm j0.1362612$
1.2	$-0.5208461 \pm j1.0382664$
	$-0.7396991 \pm j0.6924719$
	$-0.8485192 \pm j0.4006089$
	$-0.8958958 \pm j0.1313488$

TABLE 7.8(H)

Poles of Maximally Flat Compromise Network Functions
($n = 9$)

m	Poles
−0.2	−1.0178148
	−0.1279626 ± *j*0.9648089
	−0.4637050 ± *j*0.8802370
	−0.7537836 ± *j*0.6693745
	−0.9490455 ± *j*0.3611549
−0.1	−1.0088681
	−0.1505439 ± *j*0.9749842
	−0.4818982 ± *j*0.8733161
	−0.7600195 ± *j*0.6560972
	−0.9443860 ± *j*0.3515338
0	−1.0000000
	−0.1736482 ± *j*0.9848078
	−0.5000000 ± *j*0.8660254
	−0.7660444 ± *j*0.6427876
	−0.9396926 ± *j*0.3420201
0.1	−0.9912099
	−0.1972749 ± *j*0.9942634
	−0.5180031 ± *j*0.8583659
	−0.7718588 ± *j*0.6294488
	−0.9349665 ± *j*0.3326131
0.2	−0.9824970
	−0.2214230 ± *j*1.0033349
	−0.5359001 ± *j*0.8503387
	−0.7774630 ± *j*0.6160837
	−0.9302086 ± *j*0.3233124
0.3	−0.9738607
	−0.2460912 ± *j*1.0120059
	−0.5536836 ± *j*0.8419448
	−0.7828577 ± *j*0.6026956
	−0.9254200 ± *j*0.3141172
0.4	−0.9653003
	−0.2712779 ± *j*1.0202598
	−0.5713462 ± *j*0.8331857
	−0.7880433 ± *j*0.5892873
	−0.9206017 ± *j*0.3050270
0.5	−0.9568152
	−0.2969809 ± *j*1.0280797
	−0.5888805 ± *j*0.8240628
	−0.7930204 ± *j*0.5758618
	−0.9157545 ± *j*0.2960412

TABLE 7.8(H) Continued

m	Poles
0.6	-0.9484046
	$-0.3231978 \pm j1.0354486$
	$-0.6062792 \pm j0.8145778$
	$-0.7977896 \pm j0.5624223$
	$-0.9108796 \pm j0.2871593$
0.7	-0.9400680
	$-0.3499258 \pm j1.0423495$
	$-0.6235348 \pm j0.8047325$
	$-0.8023517 \pm j0.5489715$
	$-0.9059779 \pm j0.2783806$
0.8	-0.9318047
	$-0.3771616 \pm j1.0487649$
	$-0.6406401 \pm j0.7945287$
	$-0.8067072 \pm j0.5355125$
	$-0.9010503 \pm j0.2697046$
0.9	-0.9236140
	$-0.4049015 \pm j1.0546773$
	$-0.6575878 \pm j0.7839685$
	$-0.8108569 \pm j0.5220481$
	$-0.8960978 \pm j0.2611305$
1.0	-0.9154953
	$-0.4331413 \pm j1.0600692$
	$-0.6743705 \pm j0.7730542$
	$-0.8148017 \pm j0.5085814$
	$-0.8911212 \pm j0.2526580$
1.1	-0.9074480
	$-0.4618766 \pm j1.0649227$
	$-0.6909810 \pm j0.7617881$
	$-0.8185422 \pm j0.4951150$
	$-0.8861216 \pm j0.2442862$
1.2	-0.8994714
	$-0.4911023 \pm j1.0692199$
	$-0.7074120 \pm j0.7501727$
	$-0.8220795 \pm j0.4816519$
	$-0.8810998 \pm j0.2360147$

TABLE 7.8(I)

Poles of Maximally Flat Compromise Network Functions
$(n = 10)$

m	Poles
-0.2	$-0.1126086 \pm j0.9640521$
	$-0.4154539 \pm j0.9001944$
	$-0.6939632 \pm j0.7325892$
	$-0.8935261 \pm j0.4769262$
	$-1.0031883 \pm j0.1655564$
-0.1	$-0.1342376 \pm j0.9760056$
	$-0.4347254 \pm j0.8957975$
	$-0.7006361 \pm j0.7198636$
	$-0.8923169 \pm j0.4654158$
	$-0.9954135 \pm j0.1609640$
0.0	$-0.1564345 \pm j0.9876883$
	$-0.4539905 \pm j0.8910065$
	$-0.7071068 \pm j0.7071068$
	$-0.8910065 \pm j0.4539905$
	$-0.9876883 \pm j0.1564345$
0.1	$-0.1792022 \pm j0.9990846$
	$-0.4732415 \pm j0.8858200$
	$-0.7133758 \pm j0.6943215$
	$-0.8895964 \pm j0.4426508$
	$-0.9800126 \pm j0.1519674$
0.2	$-0.2025433 \pm j1.0101784$
	$-0.4924706 \pm j0.8802366$
	$-0.7194436 \pm j0.6815108$
	$-0.8880880 \pm j0.4313971$
	$-0.9723860 \pm j0.1475619$
0.3	$-0.2264599 \pm j1.0209532$
	$-0.5116700 \pm j0.8742551$
	$-0.7253109 \pm j0.6686772$
	$-0.8864828 \pm j0.4202297$
	$-0.9648085 \pm j0.1432173$
0.4	$-0.2509539 \pm j1.0313926$
	$-0.5308318 \pm j0.8678744$
	$-0.7309782 \pm j0.6558238$
	$-0.8847822 \pm j0.4091491$
	$-0.9572798 \pm j0.1389331$
0.5	$-0.2760271 \pm j1.0414794$
	$-0.5499480 \pm j0.8610937$
	$-0.7364462 \pm j0.6429532$
	$-0.8829876 \pm j0.3981555$
	$-0.9497996 \pm j0.1347085$

TABLE 7.8(I) Continued

m	Poles
0.6	$-0.3016806 \pm j1.0511966$
	$-0.5690107 \pm j0.8539121$
	$-0.7417156 \pm j0.6300683$
	$-0.8811005 \pm j0.3872493$
	$-0.9423678 \pm j0.1305429$
0.7	$-0.3279154 \pm j1.0605267$
	$-0.5880118 \pm j0.8463291$
	$-0.7467872 \pm j0.6171717$
	$-0.8791222 \pm j0.3764308$
	$-0.9349842 \pm j0.1264357$
0.8	$-0.3547322 \pm j1.0694518$
	$-0.6069433 \pm j0.8383442$
	$-0.7516616 \pm j0.6042663$
	$-0.8770542 \pm j0.3657003$
	$-0.9276485 \pm j0.1223862$
0.9	$-0.3821311 \pm j1.0779540$
	$-0.6257969 \pm j0.8299571$
	$-0.7563397 \pm j0.5913546$
	$-0.8748980 \pm j0.3550581$
	$-0.9203605 \pm j0.1183938$
1.0	$-0.4101120 \pm j1.0860148$
	$-0.6445646 \pm j0.8211676$
	$-0.7608222 \pm j0.5784394$
	$-0.8726548 \pm j0.3445045$
	$-0.9131201 \pm j0.1144579$
1.1	$-0.4386744 \pm j1.0936157$
	$-0.6632381 \pm j0.8119759$
	$-0.7651101 \pm j0.5655233$
	$-0.8703261 \pm j0.3340397$
	$-0.9059269 \pm j0.1105777$
1.2	$-0.4678175 \pm j1.1007379$
	$-0.6818092 \pm j0.8023820$
	$-0.7692042 \pm j0.5526089$
	$-0.8679133 \pm j0.3236639$
	$-0.8987809 \pm j0.1067529$

7.5 EQUIRIPPLE DELAY NETWORK FUNCTION

The reader is undoubtedly familiar with the equiripple-magnitude characteristics in the design of low-pass filters. These magnitude characteristics are frequently known as Chebyshev low-pass characteristics because they can be readily expressed in terms of Chebyshev polynomials. The equiripple and maximally flat magnitude characteristics complement each other in offering a wider choice to different engineering requirements in low-pass filters.

Analogously, we may ask ourselves if it is possible to find a network function such that its delay characteristic has an equiripple character to complement the maximally flat delay characteristic explained in Section 7.3. This question is answered affirmatively in this section, although the mathematics involved is not as neat and straightforward as in the equiripple magnitude case [UL1].

1. Further Properties of the Delay Function of an All-Pass Network

Before investigating the necessary relationship for an all-pass network to have an equiripple delay characteristic, it is expedient to explain some basic properties of the delay function $T_d(\omega)$ of an all-pass function in addition to those discussed in Section 7.1. Let the all-pass network function be

$$H(s) = \frac{Q(-s)}{Q(s)} = \frac{M(s) - N(s)}{M(s) + N(s)} \tag{7.81}$$

where $Q(s) = M(s) + N(s)$ is a Hurwitz polynomial of degree n.

The phase-lag function of the $H(s)$ of (7.81) may be expressed as

$$\beta(\omega) = 2\tan^{-1}\left[\frac{N(j\omega)}{jM(j\omega)}\right] \tag{7.82}$$

Hence the delay function of $H(s)$ is

$$T_d(\omega) = \frac{d\beta(\omega)}{d\omega} = \frac{2[M(j\omega)N'(j\omega) - N(j\omega)M'(j\omega)]}{M^2(j\omega) - N^2(j\omega)} \tag{7.83}$$

But

$$\frac{H'(s)}{H(s)} = \frac{2[N(s)M'(s) - M(s)N'(s)]}{M^2(s) - N^2(s)} \tag{7.84}$$

Hence

$$T_d(\omega) = -\frac{H'(s)}{H(s)}\bigg|_{s=j\omega} \tag{7.85}$$

$$= -\frac{d}{ds}[\ln H(s)]_{s=j\omega} \tag{7.86}$$

For convenience let us designate the function of s in (7.86) $T(s)$, or

$$T(s) = -\frac{d}{ds}[\ln H(s)] = -\frac{H'(s)}{H(s)} = T_d\left(\frac{s}{j}\right) \tag{7.87}$$

Now suppose

$$Q(s) = \prod_{\nu=1}^{n} (s - s_\nu) \tag{7.88}$$

$$H(s) = \prod_{\nu=1}^{n} \frac{-s - s_\nu}{s - s_\nu} \tag{7.89}$$

Then, from (7.86) and (7.87), $T(s)$ will have poles at $\pm s_\nu$ and may be written as

$$T(s) = \sum_{\nu=1}^{n} \frac{k_\nu}{s - s_\nu} + \sum_{\nu=1}^{n} \frac{k_{-\nu}}{-s - s_\nu} \tag{7.90}$$

From (7.85) it can be easily shown that

$$\left.\begin{aligned} k_\nu &= 1 \\ k_{-\nu} &= 1 \end{aligned}\right\} \quad \nu = 1, 2, \ldots, n \tag{7.91}$$

Hence

$$T(s) = \sum_{\nu=1}^{n} \left(\frac{1}{s - s_\nu} - \frac{1}{s + s_\nu}\right) \tag{7.92}$$

From (7.92) the following properties for the $T(s)$ function of an all-pass network can be observed:

1. $T(s)$ is an even function of s of order $2n$.
2. $T(j\omega)$ has at most n maxima on the ω axis.
3. If $T(j\omega)$ has n maxima on the positive ω axis, it decreases monotonically for ω beyond the outermost maximum.
4. $T(s)$ has a double zero at $s = \infty$.
5. Poles of $T(s)$ must occur in quadrantal symmetry. Those in the left halfplane must have residues of $+1$ and those in the right halfplane -1.

2. Determination of Pole Location of All-Pass Functions to Give Equiripple Delay

Suppose now we form the function

$$T(s) = T_0 - \delta T_R(s) \tag{7.93}$$

where T_0 is the dc delay, δ the ripple factor, and $T_R(s)$ the Chebyshev rational function described in Appendix C. Explicitly $T_R(s)$ is given by

$$T_R(s) = \frac{1}{2}\left(\prod_{\nu=1}^{n} \frac{z - z_\nu}{-z - z_\nu} + \prod_{\nu=1}^{n} \frac{-z - z_\nu}{z - z_\nu}\right) \tag{7.94}$$

and

$$z^2 = \frac{s^2 + 1}{s^2}$$

$$z_\nu^2 = \frac{s_\nu^2 + 1}{s_\nu^2}, \qquad \nu = 1, 2, \ldots, n$$

or

$$T_R(s) = \frac{1}{2} \frac{\prod_{\nu=1}^{n} (s_\nu\sqrt{s^2 + 1} - s\sqrt{s_\nu^2 + 1})^2 + \prod_{\nu=1}^{n} (s_\nu\sqrt{s^2 + 1} + s\sqrt{s_\nu^2 + 1})^2}{\prod_{\nu=1}^{n} (s^2 - s_\nu^2)},$$

$$\operatorname{Re}(s_\nu) < 0 \tag{7.95}$$

$T(s)$ thus defined obviously satisfies conditions 1 through 4. Furthermore, the variation of $T(j\omega) = T_d(\omega)$ oscillates between $T_0 + \delta$ and $T_0 - \delta$ in the range $-1 < \omega < 1$ and has exactly n maxima. An example of $T(j\omega)$ for $n = 5$ is shown in Fig. 7.13.

As far as these four conditions are concerned, s_ν are completely arbitrary. That is, any set of n values of s_ν [$\operatorname{Re}(s_\nu) < 0$] will give $T(j\omega)$ the equiripple character. But in order for the $T(j\omega)$ to be the delay of an all-pass network, s_ν must be chosen to satisfy condition 5 as well. Thus, for any particular value of ν, say μ,

$$\operatorname{res}[T(s), s_\mu] = 1, \qquad \mu = 1, 2, \ldots, n \tag{7.96}$$

Since

$$\operatorname{res}[T(s), s_\mu] = -\delta \operatorname{res}[T_R(s), s_\mu] \tag{7.97}$$

it is necessary that

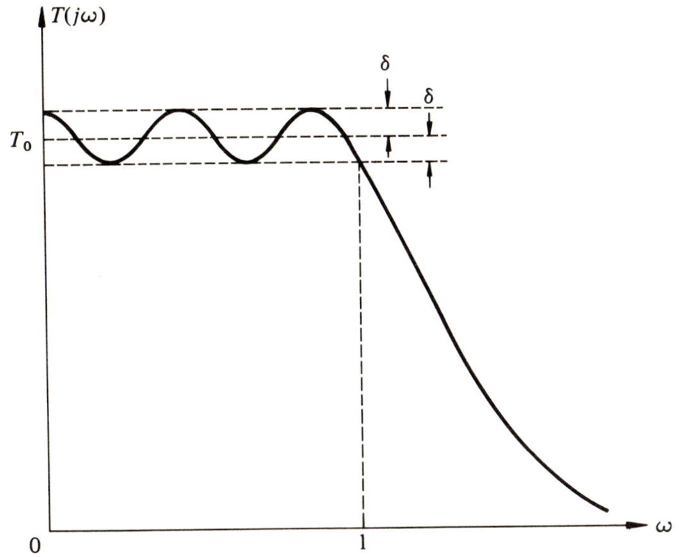

Fig. 7.13 An equiripple delay characteristic ($n = 5$).

$$\operatorname{res}\,[T_R(s), s_\mu] = -\frac{1}{\delta} \tag{7.98}$$

Equation (7.98) is equivalent to

$$\left\{\frac{d}{dz}\left[\frac{1}{\frac{1}{2}\left(\prod_{\nu=1}^{n}\frac{z - z_\nu}{-z - z_\nu} + \prod_{\nu=1}^{n}\frac{-z - z_\nu}{z - z_\nu}\right)}\right]\right\}_{z=z_\mu} \times \left(\frac{dz}{ds}\right)_{s=s_\mu} = -\delta \tag{7.99}$$

But

$$\frac{dz}{ds} = -\frac{(z^2 - 1)^{3/2}}{z} \tag{7.100}$$

Suppose now that we denote the function of (7.94) as

$$W(z) = \frac{1}{2}\left(\prod_{\nu=1}^{n}\frac{z - z_\nu}{-z - z_\nu} + \prod_{\nu=1}^{n}\frac{-z - z_\nu}{z - z_\nu}\right) \tag{7.101}$$

Then

$$\left\{\frac{d}{dz}\left[\frac{1}{\frac{1}{2}\left(\prod_{\nu=1}^{n}\frac{z - z_\nu}{-z - z_\nu} + \prod_{\nu=1}^{n}\frac{-z - z_\nu}{z - z_\nu}\right)}\right]\right\}_{z=z_\mu} = \frac{1}{\operatorname{res}\,[W(z), z_\mu]} \tag{7.102}$$

Since z_μ is not a pole of the first term of $W(z)$ of (7.101), it does not contribute anything to the residue there. Hence

$$\begin{aligned} \text{res}\,[W(z), z_\mu] &= \frac{1}{2}\lim_{z \to z_\mu}\left[(z - z_\nu)\prod_{\nu=1}^{n}\frac{-z - z_\nu}{z - z_\nu}\right] \\ &= \frac{1}{2}\left(-2z_\mu \prod_{\substack{\nu=1\\ \mu\neq\nu}}^{n}\frac{-z_\mu - z_\nu}{z_\mu - z_\nu}\right) \end{aligned} \tag{7.103}$$

and (7.99) becomes

$$\left(-\frac{1}{z_\mu}\prod_{\substack{\nu=1\\ \mu\neq\nu}}^{n}\frac{z_\mu - z_\nu}{-z_\mu - z_\nu}\right)\left[\frac{-(z_\mu^2 - 1)^{3/2}}{z_\mu}\right] = -\delta \tag{7.104}$$

or

$$\frac{(z_\mu^2 - 1)^{3/2}}{z_\mu^2} + \delta\prod_{\substack{\nu=1\\ \mu\neq\nu}}^{n}\frac{-z_\mu - z_\nu}{z_\mu - z_\nu} = 0 \tag{7.105}$$

Thus, for a given order n and a ripple factor δ, n equations of the form of (7.105) may be obtained. When a set n values of z_ν is found such that these n equations are satisfied, condition 5 will also be fulfilled. The following properties can be observed for this set of equations.

1. When $n = 1$, (7.105) degenerates into

$$\frac{(z_1^2 - 1)^{3/2}}{z_1^2} + \delta = 0 \tag{7.106}$$

2. For each pair of complex poles, two equations that are conjugate of each other are obtained. Of course, only one of these equations need be used. (Actually the real and the imaginary parts of the equation give two separate real equations.)
3. Two important questions regarding the set of n equations formed according to (7.105) are: (a) Does a set of z_ν exists for such a group of equations? (b) Are these z_ν uniquely determined by these equations? The answers to both these questions are affirmative [UL1].
4. Once the existence and the uniqueness of z_ν have been established, the set of equations formed this way can be used to determine the set of poles that will give the equiripple delay characteristic for an all-pass network.
5. Equation (7.105) is completely specified once δ and n are given. Although it appears that T_0 is another undetermined parameter, this is not the case. Actually, n and δ also determine T_0. From (7.93), as $s \to \infty$, $T(s) \to 0$. Hence

$$T_0 = \lim_{s \to \infty} [\delta T_R(s)] = \lim_{z \to -1} [\delta T_R(s)] \tag{7.107}$$

From (7.94)

$$T_0 = \frac{\delta}{2} \left(\prod_{\nu=1}^{n} \frac{z_\nu + 1}{z_\nu - 1} + \prod_{\nu=1}^{n} \frac{z_\nu - 1}{z_\nu + 1} \right) \tag{7.108}$$

6. Except for small values of n, the set of equations formed this way are nonlinear and irrational and are not amenable to algebraic solution. Some form of numerical method must be employed.

For example, for $n = 1$, z_1 is real. Equation (7.106) becomes

$$s_1^3 + s_1 + \frac{1}{\delta} = 0 \tag{7.109}$$

For $n = 2$, a pair of conjugate poles are expected. Let $z_1 = x_1 + jy_1$, $z_2 = x_1 - jy_1$. Equation (7.105) gives

$$\frac{(x_1^2 - y_1^2 - 1 + j2x_1y_1)^{3/2}}{x_1^2 - y_1^2 - 1} + \delta \frac{x_1}{jy_1} = 0 \tag{7.110}$$

Equation (7.110) may be solved for x_1 and y_1 for any given δ.

For $n > 2$, the set of equations must be solved numerically. Ulbrich and Piloty [UL1] employed Newton's method of successive approximation for several variables to obtain pole positions in the s plane for different values of δ for n up to 10. These results are tabulated in Table 7.9 with the poles normalized such that their distances from the origin have a geometric mean of unity.

TABLE 7.9(A)

Poles for Equiripple Delay Network Functions
($\delta = 0.01$)

n	Poles
2	$-0.831579 \pm j0.555407$
3	-0.842525 $-0.688133 \pm j0.844619$
4	$-0.759277 \pm j0.353301$ $-0.583811 \pm j1.041654$
5	-0.717921 $-0.677831 \pm j0.609074$ $-0.507213 \pm j1.191671$
6	$-0.660560 \pm j0.272383$ $-0.608665 \pm j0.805860$ $-0.448577 \pm j1.311234$
7	-0.620769 $-0.606323 \pm j0.489799$ $-0.551522 \pm j0.963651$ $-0.402511 \pm j1.410259$
8	$-0.578560 \pm j0.224433$ $-0.558373 \pm j0.667697$ $-0.503899 \pm j1.093474$ $-0.365235 \pm j1.493682$
9	-0.545656 $-0.539305 \pm j0.412300$ $-0.516840 \pm j0.816395$ $-0.463920 \pm j1.202831$ $-0.334526 \pm j1.565545$
10	$-0.513339 \pm j0.191555$ $-0.503980 \pm j0.571769$ $-0.480778 \pm j0.942640$ $-0.429865 \pm j1.296281$ $-0.308715 \pm j1.627986$

TABLE 7.9(B)

Poles for Equiripple Delay Network Functions
($\delta = 0.02$)

n	Poles
2	$-0.819538 \pm j0.573024$
3	-0.819726
	$-0.674865 \pm j0.874344$
4	$-0.733575 \pm j0.367060$
	$-0.570501 \pm j1.077363$
5	-0.688657
	$-0.652885 \pm j0.631528$
	$-0.494778 \pm j1.230907$
6	$-0.631198 \pm j0.281884$
	$-0.585137 \pm j0.833354$
	$-0.436940 \pm j1.351778$
7	-0.590837
	$-0.578276 \pm j0.505622$
	$-0.529699 \pm j0.994272$
	$-0.391740 \pm j1.451417$
8	$-0.549404 \pm j0.231107$
	$-0.531898 \pm j0.687617$
	$-0.483612 \pm j1.125749$
	$-0.355192 \pm j1.534603$
9	-0.516948
	$-0.511495 \pm j0.423675$
	$-0.492027 \pm j0.839094$
	$-0.445084 \pm j1.236086$
	$-0.325175 \pm j1.606071$
10	$-0.485624 \pm j0.196426$
	$-0.477598 \pm j0.586430$
	$-0.457490 \pm j0.967095$
	$-0.412283 \pm j1.329862$
	$-0.299949 \pm j1.667749$

TABLE 7.9(C)

Poles for Equiripple Delay Network Functions
($\delta = 0.03$)

n	Poles
2	$-0.810473 \pm j0.585776$
3	-0.803999
	$-0.665657 \pm j0.894809$
4	$-0.716405 \pm j0.376096$
	$-0.561489 \pm j1.100993$
5	-0.669685
	$-0.636591 \pm j0.646000$
	$-0.486546 \pm j1.256432$
6	$-0.612437 \pm j0.287860$
	$-0.569930 \pm j0.850777$
	$-0.429311 \pm j1.377710$
7	-0.571996
	$-0.560544 \pm j0.515486$
	$-0.515727 \pm j1.013519$
	$-0.384755 \pm j1.477524$
8	$-0.531197 \pm j0.235215$
	$-0.515256 \pm j0.699924$
	$-0.470690 \pm j1.145853$
	$-0.348711 \pm j1.560310$
9	-0.499179
	$-0.494240 \pm j0.430641$
	$-0.476511 \pm j0.853055$
	$-0.433147 \pm j1.256706$
	$-0.319179 \pm j1.631409$
10	$-0.468557 \pm j0.199387$
	$-0.461292 \pm j0.595355$
	$-0.442972 \pm j0.982053$
	$-0.401175 \pm j1.350564$
	$-0.294349 \pm j1.692455$

TABLE 7.9(D)

Poles for Equiripple Delay Network Functions
($\delta = 0.04$)

n	Poles
2	$-0.802898 \pm j0.596117$
3	-0.791619
	$-0.658368 \pm j0.910926$
4	$-0.703147 \pm j0.382983$
	$-0.554455 \pm j1.119115$
5	-0.655328
	$-0.624193 \pm j0.656912$
	$-0.480216 \pm j1.275816$
6	$-0.598361 \pm j0.292295$
	$-0.558432 \pm j0.863766$
	$-0.423478 \pm j1.397179$
7	-0.557999
	$-0.547333 \pm j0.522769$
	$-0.505225 \pm j1.027805$
	$-0.379451 \pm j1.497032$
8	$-0.517739 \pm j0.238220$
	$-0.502900 \pm j0.708952$
	$-0.461006 \pm j1.160685$
	$-0.343808 \pm j1.579396$
9	-0.486117
	$-0.481536 \pm j0.435726$
	$-0.465029 \pm j0.863274$
	$-0.424231 \pm j1.271883$
	$-0.314662 \pm j1.650168$
10	$-0.456051 \pm j0.201536$
	$-0.449311 \pm j0.601846$
	$-0.432247 \pm j0.992962$
	$-0.392894 \pm j1.365742$
	$-0.290138 \pm j1.710669$

TABLE 7.9(E)

Poles for Equiripple Delay Network Functions
($\delta = 0.05$)

n	Poles
2	$-0.796256 \pm j0.604959$
3	-0.781255
	$-0.652234 \pm j0.924436$
4	$-0.692201 \pm j0.388608$
	$-0.548593 \pm j1.133997$
5	-0.643652
	$-0.614067 \pm j0.665763$
	$-0.475002 \pm j1.291632$
6	$-0.586987 \pm j0.295845$
	$-0.549080 \pm j0.874211$
	$-0.418693 \pm j1.412921$
7	-0.546772
	$-0.536710 \pm j0.528584$
	$-0.496722 \pm j1.039260$
	$-0.375123 \pm j1.512760$
8	$-0.506981 \pm j0.240603$
	$-0.492988 \pm j0.716126$
	$-0.453182 \pm j1.172520$
	$-0.339816 \pm j1.594703$
9	-0.475721
	$-0.471412 \pm j0.439753$
	$-0.455841 \pm j0.871385$
	$-0.417044 \pm j1.283974$
	$-0.310995 \pm j1.665188$
10	$-0.446117 \pm j0.203232$
	$-0.439780 \pm j0.606969$
	$-0.423675 \pm j1.001592$
	$-0.386229 \pm j1.377798$
	$-0.286725 \pm j1.725206$

TABLE 7.9(F)

Poles for Equiripple Delay Network Functions
($\delta = 0.06$)

n	Poles
2	$-0.790269 \pm j0.612760$
3	-0.772261
	$-0.646885 \pm j0.936183$
4	$-0.682800 \pm j0.393393$
	$-0.543518 \pm j1.146718$
5	-0.633750
	$-0.605447 \pm j0.673256$
	$-0.470530 \pm j1.305089$
6	$-0.577385 \pm j0.298821$
	$-0.541144 \pm j0.882988$
	$-0.414600 \pm j1.426215$
7	-0.537351
	$-0.527782 \pm j0.533445$
	$-0.489533 \pm j1.048868$
	$-0.371439 \pm j1.526015$
8	$-0.497978 \pm j0.242586$
	$-0.484672 \pm j0.722099$
	$-0.446576 \pm j1.182408$
	$-0.336424 \pm j1.607548$
9	-0.467050
	$-0.462960 \pm j0.443097$
	$-0.448148 \pm j0.878130$
	$-0.410991 \pm j1.294069$
	$-0.307886 \pm j1.677778$
10	$-0.437847 \pm j0.204636$
	$-0.431831 \pm j0.611213$
	$-0.416505 \pm j1.008753$
	$-0.380619 \pm j1.387836$
	$-0.283837 \pm j1.737353$

TABLE 7.9(G)

Poles for Equiripple Delay Network Functions
($\delta = 0.07$)

n	Poles
2	$-0.784772 \pm j0.619784$
3	-0.764268
	$-0.642110 \pm j0.946645$
4	$-0.674516 \pm j0.397575$
	$-0.539011 \pm j1.157881$
5	-0.625117
	$-0.597907 \pm j0.679780$
	$-0.466591 \pm j1.316859$
6	$-0.569046 \pm j0.301386$
	$-0.534220 \pm j0.890581$
	$-0.411004 \pm j1.437767$
7	-0.529210
	$-0.520055 \pm j0.537633$
	$-0.483280 \pm j1.057171$
	$-0.368213 \pm j1.537513$
8	$-0.490215 \pm j0.244286$
	$-0.477484 \pm j0.727226$
	$-0.440835 \pm j1.190920$
	$-0.333458 \pm j1.618649$
9	-0.459593
	$-0.455687 \pm j0.445963$
	$-0.441509 \pm j0.883920$
	$-0.405741 \pm j1.302757$
	$-0.305177 \pm j1.688650$
10	$-0.430745 \pm j0.205835$
	$-0.424998 \pm j0.614842$
	$-0.410321 \pm j1.014884$
	$-0.375756 \pm j1.396453$
	$-0.281320 \pm j1.747815$

TABLE 7.9(H)

Poles for Equiripple Delay Network Functions
($\delta = 0.08$)

n	Poles
2	$-0.779661 \pm j0.626202$
3	-0.757043
	$-0.637775 \pm j0.956123$
4	$-0.667082 \pm j0.401299$
	$-0.534939 \pm j1.167861$
5	-0.617439
	$-0.591184 \pm j0.685573$
	$-0.463057 \pm j1.327358$
6	$-0.561654 \pm j0.303648$
	$-0.528058 \pm j0.897287$
	$-0.407784 \pm j1.448010$
7	-0.522025
	$-0.513225 \pm j0.541319$
	$-0.477729 \pm j1.064497$
	$-0.365336 \pm j1.547699$
8	$-0.483376 \pm j0.245775$
	$-0.471139 \pm j0.731724$
	$-0.435746 \pm j1.198410$
	$-0.330814 \pm j1.628449$
9	-0.453042
	$-0.449291 \pm j0.448476$
	$-0.435658 \pm j0.888998$
	$-0.401093 \pm j1.310396$
	$-0.302765 \pm j1.698242$
10	$-0.424511 \pm j0.206884$
	$-0.418993 \pm j0.618015$
	$-0.404874 \pm j1.020251$
	$-0.371452 \pm j1.404017$
	$-0.279082 \pm j1.757028$

TABLE 7.9(I)

Poles for Equiripple Delay Network Functions
($\delta = 0.09$)

n	Poles
2	$-0.774862 \pm j0.632131$
3	-0.750432
	$-0.633791 \pm j0.964819$
4	$-0.660319 \pm j0.404662$
	$-0.531209 \pm j1.176910$
5	-0.610511
	$-0.585101 \pm j0.690797$
	$-0.459841 \pm j1.336858$
6	$-0.555002 \pm j0.305672$
	$-0.522490 \pm j0.903301$
	$-0.404858 \pm j1.457229$
7	-0.515584
	$-0.507095 \pm j0.544616$
	$-0.472726 \pm j1.071066$
	$-0.362728 \pm j1.556863$
8	$-0.477254 \pm j0.247102$
	$-0.465448 \pm j0.735737$
	$-0.431164 \pm j1.205104$
	$-0.328421 \pm j1.637236$
9	-0.447189
	$-0.443575 \pm j0.450713$
	$-0.430418 \pm j0.893529$
	$-0.396912 \pm j1.317224$
	$-0.300589 \pm j1.706843$
10	$-0.418948 \pm j0.207815$
	$-0.413630 \pm j0.620837$
	$-0.399999 \pm j1.025030$
	$-0.367585 \pm j1.410762$
	$-0.277063 \pm j1.765268$

TABLE 7.9(J)

Poles for Equiripple Delay Network Functions
($\delta = 0.1$)

n	Poles
2	$-0.770323 \pm j0.637654$
3	-0.744321
	$-0.630095 \pm j0.972875$
4	$-0.654099 \pm j0.407734$
	$-0.527761 \pm j1.185204$
5	-0.604187
	$-0.579535 \pm j0.695560$
	$-0.456883 \pm j1.345554$
6	$-0.548944 \pm j0.307504$
	$-0.517405 \pm j0.908762$
	$-0.402170 \pm j1.465626$
7	-0.509738
	$-0.501526 \pm j0.547603$
	$-0.468167 \pm j1.077027$
	$-0.360341 \pm j1.565204$
8	$-0.472460 \pm j0.248698$
	$-0.450404 \pm j0.740546$
	$-0.427672 \pm j1.213102$
	$-0.326754 \pm j1.647844$
9	-0.441895
	$-0.438403 \pm j0.452733$
	$-0.425666 \pm j0.897621$
	$-0.393108 \pm j1.323404$
	$-0.298599 \pm j1.714647$
10	$-0.413921 \pm j0.208654$
	$-0.408779 \pm j0\ 623379$
	$-0.395579 \pm j1.029338$
	$-0.364067 \pm j1.416857$
	$-0.275218 \pm j1.772730$

TABLE 7.9(K)

Poles for Equiripple Delay Network Functions
($\delta = 0.2$)

n	Poles
2	$-0.733607 \pm j0.679574$
3	-0.699132
	$-0.602261 \pm j1.033261$
4	$-0.608883 \pm j0.429450$
	$-0.502031 \pm j1.244680$
5	-0.559488
	$-0.539835 \pm j0.729181$
	$-0.435314 \pm j1.407806$
6	$-0.506451 \pm j0.320108$
	$-0.481276 \pm j0.946554$
	$-0.382646 \pm j1.524523$
7	-0.469265
	$-0.462814 \pm j0.568156$
	$-0.436035 \pm j1.118336$
	$-0.343188 \pm j1.623725$
8	$-0.433451 \pm j0.256426$
	$-0.424465 \pm j0.764012$
	$-0.397620 \pm j1.252677$
	$-0.310536 \pm j1.700492$
9	-0.405667
	$-0.402928 \pm j0.466459$
	$-0.392868 \pm j0.925505$
	$-0.366480 \pm j1.365815$
	$-0.284435 \pm j1.768778$
10	$-0.379607 \pm j0.214306$
	$-0.375570 \pm j0.640516$
	$-0.365108 \pm j1.058472$
	$-0.339467 \pm j1.458361$
	$-0.262107 \pm j1.824062$

TABLE 7.9(L)

Poles for Equiripple Delay Network Functions
($\delta = 0.3$)

n	Poles
2	$-0.705261 \pm j0.708947$
3	-0.668281
	$-0.582681 \pm j1.075574$
4	$-0.578605 \pm j0.443349$
	$-0.484092 \pm j1.283620$
5	-0.530712
	$-0.513939 \pm j0.750836$
	$-0.420747 \pm j1.448779$
6	$-0.479337 \pm j0.327897$
	$-0.457798 \pm j0.970137$
	$-0.369505 \pm j1.562037$
7	-0.443907
	$-0.438428 \pm j0.580944$
	$-0.415396 \pm j1.144293$
	$-0.331834 \pm j1.661185$
8	$-0.409605 \pm j0.261366$
	$-0.401955 \pm j0.779054$
	$-0.378787 \pm j1.278267$
	$-0.300167 \pm j1.735170$
9	-0.383313
	$-0.380986 \pm j0.474859$
	$-0.372392 \pm j0.942629$
	$-0.349517 \pm j1.392105$
	$-0.275172 \pm j1.802879$
10	$-0.358503 \pm j0.217710$
	$-0.355063 \pm j0.650858$
	$-0.346107 \pm j1.076123$
	$-0.323814 \pm j1.483751$
	$-0.253547 \pm j1.855950$

TABLE 7.9(M)

Poles for Equiripple Delay Network Functions
($\delta = 0.4$)

n	Poles
2	$-0.681247 \pm j0.732054$
3	-0.644206 $-0.567021 \pm j1.109408$
4	$-0.555207 \pm j0.453717$ $-0.469793 \pm j1.313161$
5	-0.509051 $-0.494271 \pm j0.767192$ $-0.409384 \pm j1.480201$
6	$-0.459016 \pm j0.333595$ $-0.439988 \pm j0.987496$ $-0.359269 \pm j1.590062$
7	-0.425134 $-0.420311 \pm j0.590393$ $-0.399862 \pm j1.163578$ $-0.323088 \pm j1.689388$
8	$-0.391994 \pm j0.264948$ $-0.385247 \pm j0.789981$ $-0.364615 \pm j1.296978$ $-0.292186 \pm j1.760850$
9	-0.366921 $-0.364867 \pm j0.480998$ $-0.357263 \pm j0.955168$ $-0.336815 \pm j1.411468$ $-0.268095 \pm j1.828285$
10	$-0.343051 \pm j0.220168$ $-0.340010 \pm j0.658323$ $-0.332067 \pm j1.088895$ $-0.312089 \pm j1.502238$ $-0.247010 \pm j1.879430$

TABLE 7.9(N)

Poles for Equiripple Delay Network Functions
($\delta = 0.5$)

n	Poles
2	$-0.660026 \pm j0.751243$
3	-0.624204
	$-0.553736 \pm j1.138164$
4	$-0.535882 \pm j0.462024$
	$-0.457677 \pm j1.337158$
5	-0.491515
	$-0.478239 \pm j0.780495$
	$-0.399914 \pm j1.506066$
6	$-0.442607 \pm j0.338104$
	$-0.425471 \pm j1.001292$
	$-0.350739 \pm j1.612604$
7	-0.410116
	$-0.405781 \pm j0.597951$
	$-0.387278 \pm j1.179068$
	$-0.315868 \pm j1.712275$
8	$-0.377925 \pm j0.267766$
	$-0.371848 \pm j0.798592$
	$-0.353130 \pm j1.311786$
	$-0.285597 \pm j1.781386$
9	-0.353896
	$-0.352044 \pm j0.485868$
	$-0.345174 \pm j0.965132$
	$-0.326560 \pm j1.426915$
	$-0.262286 \pm j1.848737$
10	$-0.330783 \pm j0.222093$
	$-0.328038 \pm j0.664181$
	$-0.320848 \pm j1.098937$
	$-0.302622 \pm j1.516835$
	$-0.241644 \pm j1.898135$

7.6 DELAY FUNCTIONS BASED ON A DELAYED APPROXIMATE IMPULSE

It is shown in Appendix B and explained in Section 6.5 that a series of two-unit impulses spaced T seconds apart as illustrated in Fig. 6.23 may be approximated by the truncated finite series

$$\hat{i}_{2n}(t) = \frac{4}{T} \sum_{k=1}^{n} \frac{\sin\left[\frac{(2k-1)}{2n}\pi\right]}{\frac{(2k-1)}{2n}\pi} \cos\left[(2k-1)\frac{\pi}{T}t\right] \tag{7.111}$$

Now, let us form a new function $g_n(t)$ which is $\hat{i}_{2n}(t)$ delayed by τ seconds, $\tau < T$; or

$$g_n(t) = \hat{i}_{2n}(t-\tau) \tag{7.112}$$

If that part of $g_n(t)$ for $t > T$ is cut off, an approximate two-unit impulse delayed by τ seconds should result. This latter feat may be accomplished by convolving the delayed $\hat{i}_{2n}(i)$—that is, $\hat{i}_{2n}(t-\tau)$—with the function $w(t)$ of Fig. 6.22 in which

$$w(t) = u_0(t) + u_0(t-T) \tag{7.113}$$

with T replaced by τ. In other words, the function

$$\hat{h}(t) = \tfrac{1}{2}w(t) \circledast g_n(t) \cong u_0(t-\tau) \tag{7.114}$$

will be approximately another unit impulse centered at $t = \tau$, and $\hat{h}(t)$ of (7.114) may be considered as an approximate impulse response of a constant-delay network.

In the frequency domain, $\mathscr{L}[w(t)]$ may be approximated by

$$\hat{W}(s) = \frac{2}{1 + \hat{I}_{2n}(s)} \cong \mathscr{L}[w(t)] \tag{7.115}$$

as was explained for (6.126).

The function $g_n(t)$ of (7.112) is already Laplace transformable. Hence a network function to produce an approximate constant-delay characteristic is formed by finding

$$H^*(s) = \tfrac{1}{2}\hat{W}_n(s)G_n(s) = \frac{G_n(s)}{1 + \hat{I}_{2n}(s)} \tag{7.116}$$

For example, suppose that it is desired to find an $H^*(s)$ for

$$n = 4, \qquad T = 2\pi, \qquad \tau = 2.1082$$

[This value of τ is the third zero of $\hat{\imath}_8(t)$ to the left of the origin.] From (7.111) and (7.112) we have

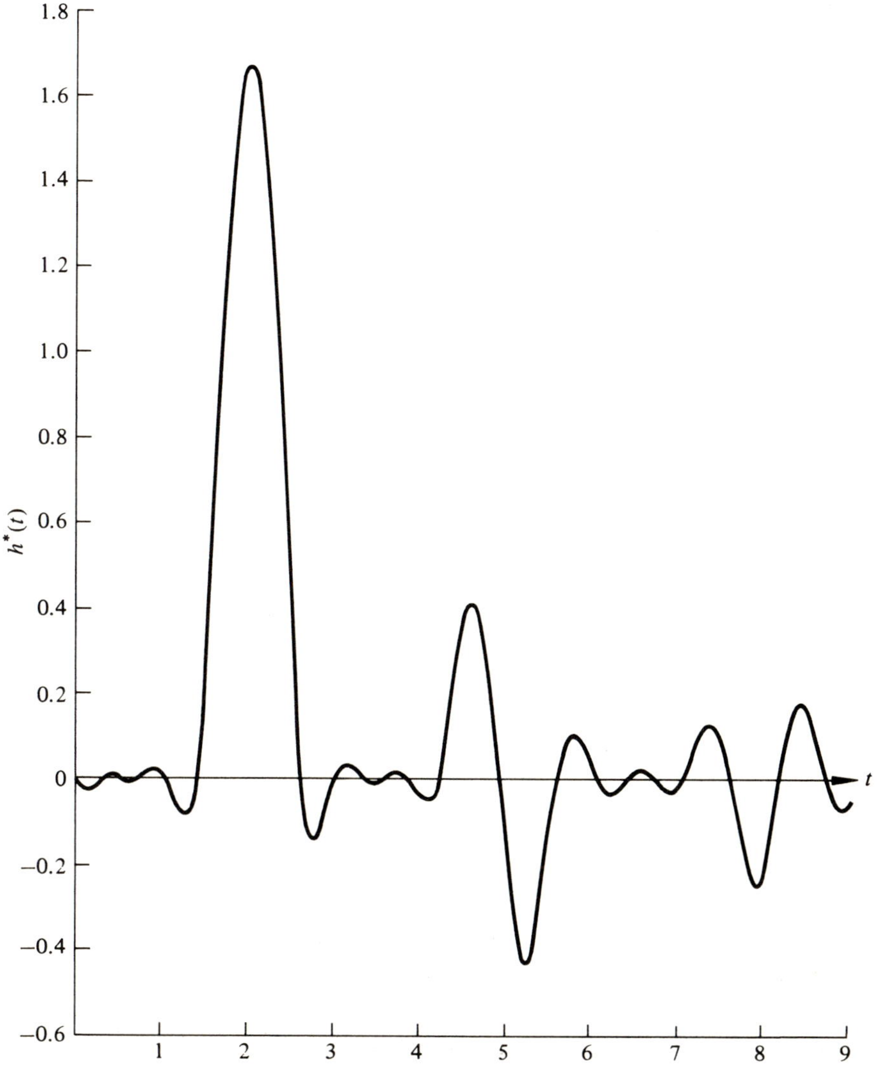

Fig. 7.14 The $h^*(t)$ of (7.121).

$$g_4(t) = \hat{\imath}_8(t - 2.1082)$$
$$= \frac{4}{\pi}\{0.97450 \cos(t - 2.1082) + 0.78421 \cos[3(t - 2.1082)]$$
$$+ 0.47053 \cos[5(t - 2.1082)] + 0.13921 \cos[7(t - 2.1082)]\} \tag{7.117}$$

The Laplace transform of $g_4(t)$ is

$$G_4(s) = \frac{\begin{array}{r}-0.0035961s^7 - 0.49170s^6 + 2.9788s^5 - 25.679s^4 \\ - 63.095s^3 + 1095.2s^2 - 5919.7s + 10,943\end{array}}{(s^2 + 1)(s^2 + 9)(s^2 + 25)(s^2 + 49)} \tag{7.118}$$

and

$$\hat{I}_8(s) = \mathcal{L}^{-1}\left[\frac{4}{\pi}[0.97450 \cos t + 0.78421 \cos 3t\right.$$
$$\left. + 0.47053 \cos 5t + 0.13921 \cos 7t]\right]$$
$$= \frac{3.0157s^7 + 219.43s^5 + 3988s^3 + 15207s}{(s^2 + 1)(s^2 + 9)(s^2 + 25)(s^2 + 49)} \tag{7.119}$$

Thus

$$H^*(s) = \frac{\begin{array}{r}-0.0035961s^7 - 0.49170s^6 + 2.9788s^5 - 25.679s^4 \\ - 63.095s^3 + 1095.2s^2 - 5919.7s + 10,943\end{array}}{\begin{array}{r}s^8 + 3.0157s^7 + 84s^6 + 219.43s^5 + 1974s^4 \\ + 3988s^3 + 12,916s + 15,207s + 11,025\end{array}} \tag{7.120}$$

The inverse Laplace transform of (7.120) of the function is

$$\begin{aligned} h^*(t) = &-2.6037\epsilon^{-0.7690t} \cos(0.8034t + 75.91^\circ) \\ &+1.4336\epsilon^{-0.4421t} \cos(2.7530t + 32.37^\circ) \\ &-0.6508\epsilon^{-0.2280t} \cos(4.8577t - 31.41^\circ) \\ &-0.1697\epsilon^{-0.0687t} \cos(6.9624t + 81.46^\circ) \end{aligned} \tag{7.121}$$

The variation of this $h^*(t)$ is shown in Fig. 7.14.

7.7 DELAY FUNCTIONS BASED ON THE INFINITE-PRODUCT EXPANSIONS OF THE HYPERBOLIC FUNCTIONS

The method of obtaining the maximally flat delay functions of Section 7.3 is equivalent to expanding the function

$$H(s) = \epsilon^{-T_D s} \tag{7.122}$$

into the ratio of two infinite series, which are, in turn, truncated to leave a realizable rational function. Specifically, the process may be reformulated in the following way. First, we set $T_D = 2$ to simplify the notation. Then the all-pass function is written as

$$H_{ap}(s) = \epsilon^{-2s} = \frac{\cosh s - \sinh s}{\cosh s + \sinh s} \tag{7.123}$$

Then we replace each hyperbolic function by its power-series expansion,

$$\cosh s = 1 + \frac{s^2}{2!} + \frac{s^4}{4!} + \frac{s^6}{6!} + \cdots \tag{7.124}$$

$$\sinh s = s + \frac{s^3}{3!} + \frac{s^5}{5!} + \frac{s^7}{7!} + \cdots \tag{7.125}$$

After the truncation of these series, the all-pass rational-function approximation is obtained and it turned out to be maximally flat.

In the low-pass case, a similar process is applied to

$$H_{lp}(s) = \epsilon^{-s} = \frac{1}{\cosh s + \sinh s} \tag{7.126}$$

An interesting alternative may be applied to this scheme by replacing the infinite series of the hyperbolic functions with their infinite-product expansion [AB2],

$$\cosh s = \left(1 + \frac{4s^2}{\pi^2}\right)\left(1 + \frac{4s^2}{9\pi^2}\right)\left(1 + \frac{4s^2}{25\pi^2}\right)\cdots \tag{7.127}$$

$$\sinh s = s\left(1 + \frac{s^2}{\pi^2}\right)\left(1 + \frac{s^2}{4\pi^2}\right)\left(1 + \frac{s^2}{9\pi^2}\right)\cdots \tag{7.128}$$

These expansions clearly indicate the equally spaced zeros of these functions.

Now, if these expansions are truncated and substituted into (7.126), another approximation for the low-pass delay function is obtained. Here

$$H^*_{lp}(s) = \frac{1}{\prod_{k=1}^{N}\left(1 + \frac{4s^2}{(2k-1)^2\pi^2}\right) + s\prod_{k=1}^{M}\left(1 + \frac{s^2}{k^2\pi^2}\right)} \tag{7.129}$$

in which the upper limits of the indices do not have to be the same. It is easily seen that the denominator of (7.129) is always Hurwitz, as the zeros of its even and odd parts always alternate (they are also mutually equally spaced) as long as the degrees of these two parts differ by one. Hence $N - M$ must be either 0 or 1 in (7.129). A plot of the phase function of (7.129) for $N = M = 4$ is shown in Fig. 7.15.

It is seen that the delay function obtained this way oscillates about 1 second but not in the equiripple manner. It has been found that two additional pa-

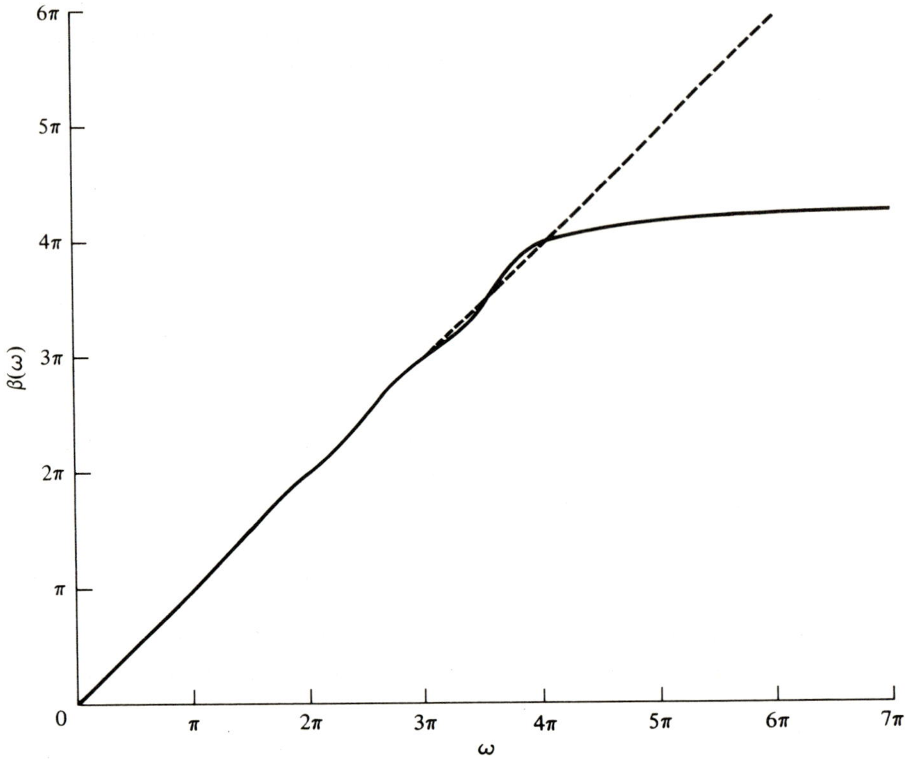

Fig. 7.15 The phase characteristic of the network function of (7.129) for $M = N = 4$.

rameters, k_1 and k_2, may be introduced artificially to offer more flexibility in the design of low-pass constant-delay networks. We make

$$H_{\mathrm{lp}}^{*}(s) = \frac{1}{k_1 \prod_{k=1}^{N}\left(1 + \frac{4s^2}{(2k-1)^2\pi^2}\right) + k_2 s \prod_{k=1}^{M}\left(1 + \frac{s^2}{k^2\pi^2}\right)} \tag{7.130}$$

By properly choosing k_1 and k_2, the amount of oscillation can be controlled to some extent [VA1].

7.8 LOW-PASS CONSTANT-DELAY FUNCTIONS BASED ON AN APPROXIMATION OF THE GAUSSIAN-RESPONSE CHARACTERISTIC

If a transfer function $H(s)$ has neither poles nor zeros in the right halfplane and is analytic at infinity, then its attenuation and phase functions can be related through the equation [BO1, p. 307]

$$\beta(\omega_c) = \frac{2\omega_c}{\pi}\int_0^\infty \frac{\alpha(\omega) - \alpha(\omega_c)}{\omega^2 - \omega_c^2}\, d\omega \tag{7.131}$$

where

$$\ln\,[H(j\omega)] = \alpha(\omega) + j\beta(\omega) \tag{7.132}$$

and both $\alpha(\omega)$ and $\beta(\omega)$ are real functions of ω. Thus the phase function at ω_c is a function of $\alpha(\omega)$ for all values of ω.

Now suppose that this $H(s)$ is of the low-pass type and its attenuation and phase characteristic have the form shown in Fig. 7.16. We may assume that

$$\alpha(\omega) - \alpha(\omega_c) \cong 0, \qquad \text{for} \quad \omega, \omega_c \ll \omega_h \tag{7.133}$$

$$\left.\begin{aligned} \omega^2 - \omega_c^2 &\cong \omega^2 \\ \alpha(\omega) - \alpha(\omega_c) &\cong \alpha(\omega) \end{aligned}\right\} \quad \text{for} \quad \omega > \omega_h \tag{7.134}$$

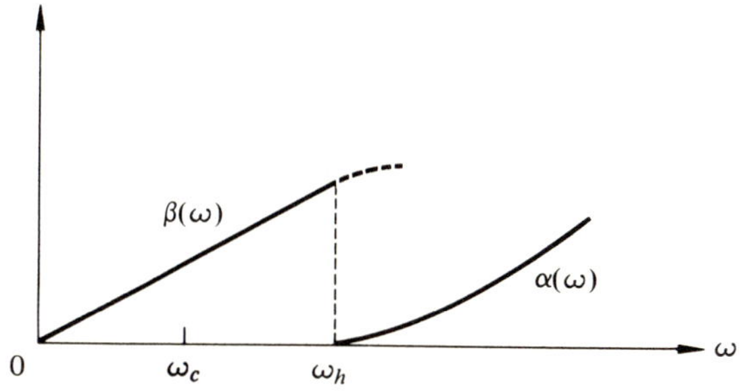

Fig. 7.16 Phase and attenuation characteristics satisfying (7.133) and (7.134).

and (7.131) becomes

$$\frac{\beta(\omega_c)}{\omega_c} \cong \frac{1}{\pi}\int_0^\infty \frac{\alpha(\omega)}{\omega^2}\, d\omega \tag{7.135}$$

Hence if

$$\alpha(\omega) \propto \omega^2 \tag{7.136}$$

then

$$\frac{\beta(\omega_c)}{\omega_c} \cong \text{constant} \tag{7.137}$$

for $\omega_c < \omega_h$. Filters whose attenuation characteristic follow (7.136) are thus seen to have nearly a constant delay for low frequencies and may be considered an approximate low-pass constant-delay network. The magnitude characteristic corresponding to (7.136) is

TABLE 7.10

Poles for Approximate Gaussian Network Functions

n	Poles
2	$-0.9238795 \pm j0.3826834$
3	-0.9372044
	$-0.8524403 \pm j0.5833941$
4	$-0.9110191 \pm j0.2204342$
	$-0.7938767 \pm j0.7127431$
5	-0.9148936
	$-0.8777058 \pm j0.3704086$
	$-0.7457514 \pm j0.8051058$
6	$-0.9016732 \pm j0.1552831$
	$-0.8445992 \pm j0.4814055$
	$-0.7055238 \pm j0.8753226$
7	-0.9034763
	$-0.8824275 \pm j0.2730820$
	$-0.8136408 \pm j0.5679828$
	$-0.6713112 \pm j0.9310244$
8	$-0.8955111 \pm j0.1199754$
	$-0.8613025 \pm j0.3666768$
	$-0.7852392 \pm j0.6380106$
	$-0.6417683 \pm j0.9765989$
9	-0.8965570
	$-0.8829773 \pm j0.2166663$
	$-0.8400318 \pm j0.4434746$
	$-0.7593204 \pm j0.6961901$
	$-0.6159247 \pm j1.0147744$
10	$-0.8912373 \pm j0.0977934$
	$-0.8683463 \pm j0.2969130$
	$-0.7356647 \pm j0.7455338$
	$-0.8193693 \pm j0.5080159$
	$-0.5930664 \pm j1.0473497$

$$|H(\omega)| = K\epsilon^{-(\omega^2)} \tag{7.138}$$

and filters designed on this basis are known as Gaussian filters.

Magnitude characteristics of the type (7.138) are, of course, not realizable with lumped networks. A simple approximation of this $|H(\omega)|$ is obtained if we write

$$|H(\omega)|^2 = \frac{K}{\epsilon^{2(\omega^2)}} \tag{7.139}$$

and then replace the denominator by a truncated infinite series of $\epsilon^{2(\omega^2)}$, or

$$|H_n^*(\omega)|^2 = \frac{K}{1 + 2\omega^2 + \frac{2^2}{2!}\omega^4 + \frac{2^3}{3!}\omega^6 + \cdots + \frac{2^n}{n!}\omega^{2n}} \tag{7.140}$$

Now the poles of the corresponding $H_n^*(s)$ can be obtained by finding the poles of $|H_n^*(s/j)|^2$ and then keeping only those in the left halfplane. Poles obtained this way after their distances to the origin have been normalized such that their geometric mean is unity are tabulated in Table 7.10.

7.9 EMPIRICAL METHODS

The foregoing sections embody several useful techniques for deriving the approximate low-pass (or all-pass) constant-delay characteristic through various mathematical artifices and engineering reasonings. Although some of the characteristics found by these techniques have proved to be adequate for various applications, each method seems to render certain advantages while possessing other disadvantages. This is partly because there are no easily formulated criteria to compare these networks. It is also because of the indirect relationship between the time-domain and the frequency-domain behavior of a network function, which makes it extremely difficult to settle the question of optimizing the time response of a network. Most of the foregoing methods of approximating the ideal delay characteristics by a finite network have some theoretical basis for their reaching the particular pole distributions in the frequency domain. This particular feature of these methods makes them scholastically attractive and intellectually gratifying in the way these conclusions are reached.

From the user's point of view, the final payoff of these undertakings is in the performance of the product but not in the way the product is manufactured. This motive has led engineers to try their ingenuity by applying their experience and judgment in attempting to arrive at a good constant-delay characteristic by empirically distributing poles in the frequency domain.

A few of the suggested pole distributions are described below. The rationales of some of the suggested distributions are included. In specifying each pole distribution it is convenient to describe (1) the contour on which the poles are situated and (2) the spacing along the contour by which the poles are placed.

1. Linear Contour

It is easy to show that if an infinite array of poles is placed uniformly along a straight line parallel to the $j\omega$ axis as shown in Fig. 7.17, the delay-versus-frequency characteristic will have an equiripple behavior. This property is particularly apparent if the notion of potential analogy is used. Since the phase function $\beta(\omega)$ may be identified with the stream function when each pole is

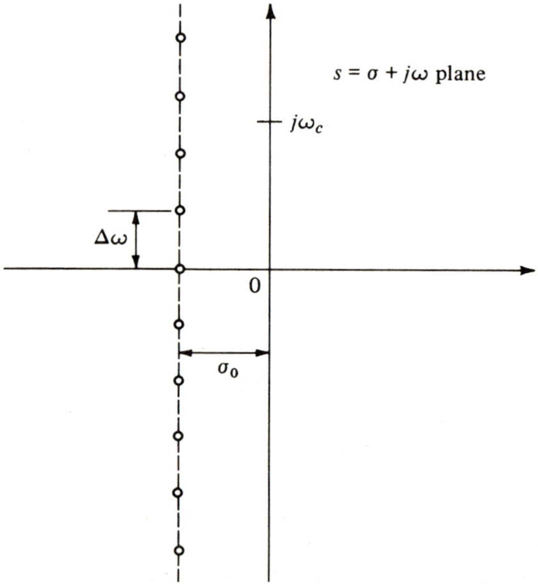

Fig. 7.17 Uniformly spaced poles on a linear contour.

identified with a unit line charge, $\beta(\omega)$ is a monotonically increasing function of ω. Furthermore, the rate of increase has a cyclic pattern every $\Delta\omega$. Hence $T_d(\omega) = d\beta(\omega)/d\omega$ will be equiripple.

For a practical network function, only a finite number of poles can be included. The simplest scheme is to leave out all poles beyond the passband ($\omega > \omega_c$). The deletion of poles for $\omega > \omega_c$ will no doubt disturb somewhat the equiripple character of $T_d(\omega)$. However, it is not too much to expect that for $\omega \ll \omega_c$, $T_d(\omega)$ may remain fairly close to equiripple provided the number of poles is sufficiently large. On the other hand, for large ω the phase shift tends to a constant and the delay to zero. Also, a smaller σ_0 will give a larger ripple in $T_d(\omega)$ but the in-band equiripple character will remain intact when high-frequency poles are discarded.

This empirical scheme is generally not very satisfactory for small numbers of poles.

2. Parabolic Contour

One point of view we may assume in seeking new and, possibly, better pole distributions for a low-pass delay network is to take any one of the existing distributions and find out how the path of the $j\omega$ axis can be altered so that the delay along the new path might be an improvement over the original. If we know this desirable new path, then we might find a conformal transformation to map the original $j\omega$ axis onto the new path. Conversely, the application of the inverse transform will map the new path onto the $j\omega$ axis and also give the new pole positions.

Alternatively, instead of altering the $j\omega$ axis, an equivalent step will be to alter the contour on which the poles lie. It is based on this rationale that some of the methods in this section are developed.

The conformal transformation [RA3]

$$s = b - \frac{4(b+1)}{(p-1)^2} \tag{7.141}$$

maps the unit circle on the p plane onto a parabola on the s plane. Table 7.11 indicates several key image points on these two planes.

TABLE 7.11

Images of Transformation (7.141)

$s = \sigma + j\omega$	$p = u + jv$
$-3b - 4$	0
-1	-1
∞	1
$b \mp j2(b+1)$	$\pm j$
$\omega^2 = 4(b+1)(\sigma+1)$	$u^2 + v^2 = 1$

The last pair of entries in Table 7.11 says that the unit circle $|p| = 1$ is mapped onto a parabola which is symmetric about the real axis with its vertex at $-1 + j0$ and focus at $b + j0$ on the s plane. Of particular interest is the case when $b = 0$. The relative image points are listed in Table 7.12.

TABLE 7.12

Images of Transformation $s = -4/(p-1)^2$

$s = \sigma + j\omega$	$p = u + jv$
∞	1
4	0
-1	-1
$\pm j2$	$\pm j$
$\pm j$	$(\sqrt{2}+1) \pm j\sqrt{2}$
0	∞
$\omega^2 = 4(\sigma + 1)$	$u^2 + v^2 = 1$

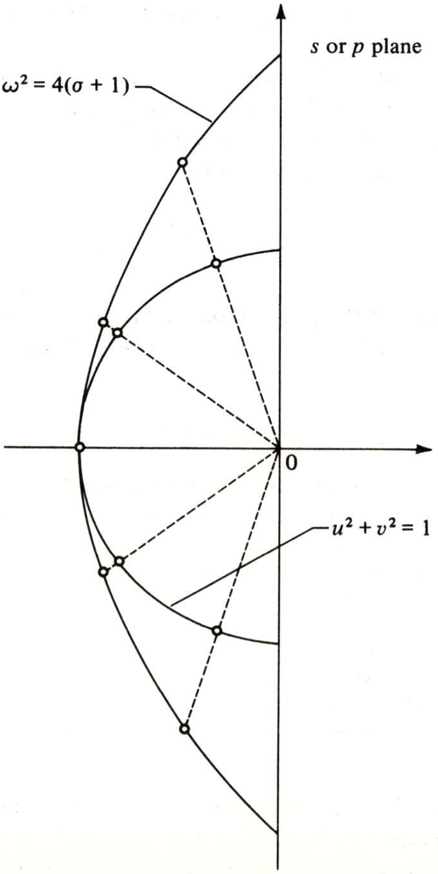

Fig. 7.18 Equal-angle pole distribution on a circle in the p plane and on a parabola in the s plane.

For this particular value of b, points on the unit circle $u^2 + v^2 = 1$ are mapped onto the parabola in such a way that

$$\left|\frac{u}{v}\right| = \left|\frac{\sigma}{\omega}\right| \tag{7.142}$$

Hence corresponding points on these two contours lie on the same radial line when the two planes are superimposed on each other. Figure 7.18 illustrates such a relationship for the case of five poles. For other values of b, this simple pole-to-pole mapping no longer holds.

Armed with transformation (7.141), we may now take different sets of known poles and proceed to find their images with various values of b. The transient responses of these new network functions can then be studied and evaluated.

Alternatively, we may regard the transformation at hand as a way of generating the new contour for the new poles and we are free to arbitrarily space the poles on these contours. The following two systems have been proposed and found useful and convenient.

1. Equal-angle spacing [MU2]. The poles are located at the intersections of the parabola and a series of uniformly spaced radial lines. This system is similar to the manner in which poles are located for the maximally flat magnitude functions except that the latter has a circular contour. This system of pole spacing is illustrated in Fig. 7.19 for even and odd degrees.

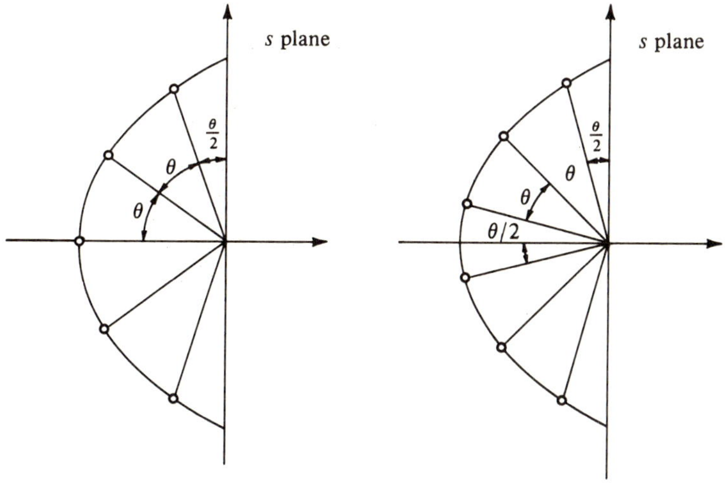

Fig. 7.19 Equal-angle pole distribution on a parabolic contour.

The study of typical transient responses of functions with poles generated this way has revealed that a value of 1.25 for b is a good compromise for most applications of the constant-delay network. Poles

TABLE 7.13(A)

Poles on a Parabolic Contour with Equal-Angle Spacing
($b = 0$)

n	Poles
2	$-0.7071068 \pm j0.7071068$
3	-0.8254818
	$-0.5503212 \pm j0.9531843$
4	$-0.4514036 \pm j1.0897975$
	$-0.7832234 \pm j0.3244217$
5	-0.8108326
	$-0.7252307 \pm j0.5269109$
	$-0.3828232 \pm j1.1782087$
6	$-0.7942801 \pm j0.2128267$
	$-0.6696091 \pm j0.6696091$
	$-0.3323762 \pm j1.2404450$
7	-0.8067524
	$-0.7647245 \pm j0.3682719$
	$-0.6196551 \pm j0.7770229$
	$-0.2936871 \pm j1.2867271$
8	$-0.7979366 \pm j0.1587195$
	$-0.7316081 \pm j0.4888449$
	$-0.5755476 \pm j0.8613678$
	$-0.2630673 \pm j1.3225286$
9	-0.8050668
	$-0.7800346 \pm j0.2839100$
	$-0.6984162 \pm j0.5860407$
	$-0.5367112 \pm j0.9296111$
	$-0.2382288 \pm j1.3510626$
10	$-0.7995922 \pm j0.1266430$
	$-0.7582017 \pm j0.3863231$
	$-0.6665324 \pm j0.6665324$
	$-0.5024376 \pm j0.9860893$
	$-0.2176749 \pm j1.3743452$

TABLE 7.13(B)

Poles on a Parabolic Contour with Equal-Angle Spacing
($b = 1.25$)

n	Poles
2	$-0.7071068 \pm j0.7071068$
3	-0.7363603
	$-0.5826729 \pm j1.0092191$
4	$-0.4988344 \pm j1.2042927$
	$-0.7087591 \pm j0.2935776$
5	-0.7155502
	$-0.6778842 \pm j0.4925117$
	$-0.4359779 \pm j1.3418020$
6	$-0.3868485 \pm j1.4437382$
	$-0.6468075 \pm j0.6468075$
	$-0.7064949 \pm j0.1893047$
7	-0.7100155
	$-0.6926059 \pm j0.3335414$
	$-0.6164934 \pm j0.7730583$
	$-0.3474254 \pm j1.5221699$
8	$-0.3151384 \pm j1.5843077$
	$-0.5874768 \pm j0.8792212$
	$-0.6766298 \pm j0.4521096$
	$-0.7055891 \pm j0.1403504$
9	-0.7077628
	$-0.6976408 \pm j0.2539205$
	$-0.6596632 \pm j0.5535232$
	$-0.5600441 \pm j0.9700248$
	$-0.2882437 \pm j1.6347111$
10	$-0.2655150 \pm j1.6763959$
	$-0.5343142 \pm j1.0486508$
	$-0.6422868 \pm j0.6422868$
	$-0.6878105 \pm j0.3504570$
	$-0.7051497 \pm j0.1116847$

TABLE 7.13(C)

Poles on a Parabolic Contour with Equal-Angle Spacing
($b = 3$)

n	Poles
2	$-0.7071068 \pm j0.7071068$
3	-0.6960557
	$-0.5993052 \pm j1.0380270$
4	$-0.5275673 \pm j1.2736601$
	$-0.6701579 \pm j0.2775885$
5	-0.6687526
	$-0.6480353 \pm j0.4708252$
	$-0.4717465 \pm j1.4518863$
6	$-0.4261954 \pm j1.5905830$
	$-0.6272267 \pm j0.6272267$
	$-0.6612896 \pm j0.1771920$
7	-0.6615807
	$-0.6522596 \pm j0.3141117$
	$-0.6068717 \pm j0.7609930$
	$-0.3882408 \pm j1.7009943$
8	$-0.3561822 \pm j1.7906489$
	$-0.5868230 \pm j0.8782426$
	$-0.6424251 \pm j0.4292548$
	$-0.6582535 \pm j0.1309348$
9	-0.6587209
	$-0.6533554 \pm j0.2378019$
	$-0.6320346 \pm j0.5303400$
	$-0.5671598 \pm j0.9823496$
	$-0.3288013 \pm j1.8647248$
10	$-0.3051852 \pm j1.9268637$
	$-0.5480080 \pm j1.0755264$
	$-0.6212357 \pm j0.6212357$
	$-0.6475571 \pm j0.3299468$
	$-0.6568709 \pm j0.1040381$

of this distribution for $b = 0$, 1.25, and 2.5 are listed in Table 7.13 for n up to 10. These poles have been normalized such that their geometric mean distance from the origin is unity.

2. Equal-frequency spacing [RA3]. The distribution of this set of poles is such that they are equally spaced in the vertical direction in the complex-frequency plane. This system is illustrated in Fig. 7.20 for both an even and an odd order.

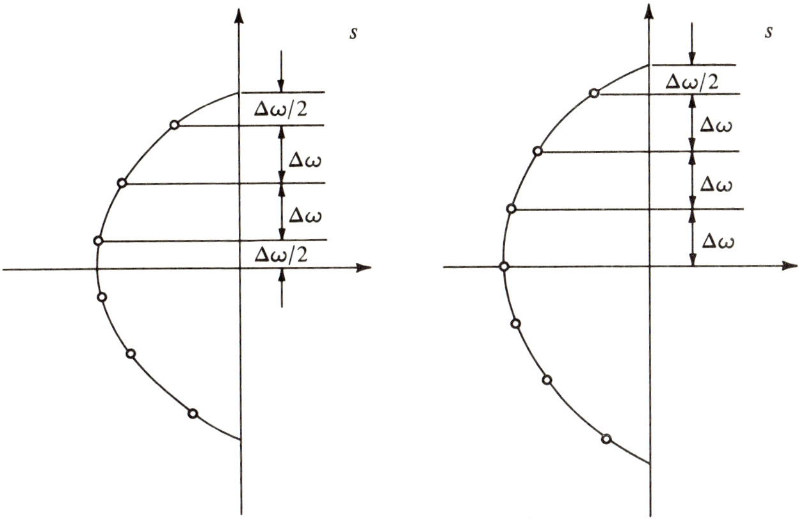

Fig. 7.20 Equal-frequency pole distribution on a parabolic contour.

3. Elliptic-Circular Contours

It is well known that the equal-angle spacing of poles along a circular contour results in a maximally flat magnitude function. The poles of a maximally flat delay function do not lie on any simple contour. But an inspection of the pole positions of the latter (Table 7.3) will reveal that they are nearly equally distributed vertically. These observations had led to another suggestion that poles be located on a circular contour with the equal-frequency spacing [SC1].

Further variations may be obtained by transforming a circular contour into an ellipse. This can be accomplished by the conformal transformation [DA1]

$$s = kp - \frac{1}{kp} \tag{7.143}$$

where $k > 1$ is a real constant. Substituting $p = u + jv$ and $s = \sigma + j\omega$ into (7.143) and separating real and imaginary parts yield

$$\sigma = ku - \frac{u}{k(u^2 + v^2)} \tag{7.144}$$

$$\omega = kv + \frac{v}{k(u^2 + v^2)} \tag{7.145}$$

For points on the unit circle $u^2 + v^2 = 1$, (7.144) and (7.145) become

$$\sigma = ku - \frac{u}{k} = \left(k - \frac{1}{k}\right)u \tag{7.146}$$

$$\omega = kv + \frac{v}{k} = \left(k + \frac{1}{k}\right)v \tag{7.147}$$

Equations (7.146) and (7.147) show that the unit circle $u^2 + v^2 = 1$ is mapped onto the ellipse

$$\left(\frac{\sigma}{k - (1/k)}\right)^2 + \left(\frac{\omega}{k + (1/k)}\right)^2 = 1 \tag{7.148}$$

in the s plane. This ellipse is shown in Fig. 7.21.

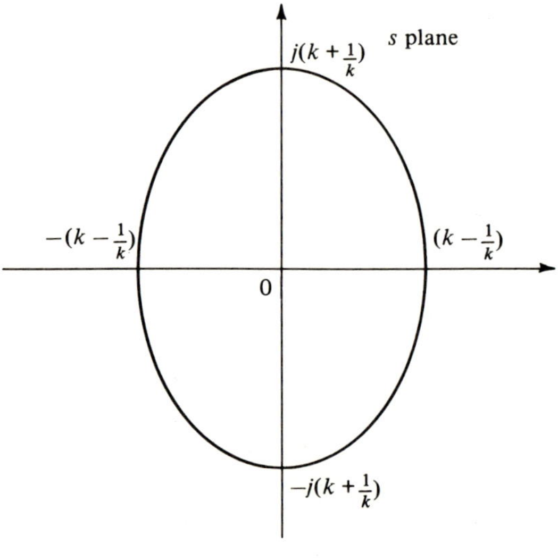

Fig. 7.21 Elliptic image of a circular contour in the p plane.

It is convenient to describe the shape of the transformed ellipse of Fig. 7.21 by the ratio of its minor axis (real-axis intersect) to its major axis (imaginary-axis intersect). This ratio is the eccentricity of the ellipse and will be denoted by a. Then

$$a = \frac{k - (1/k)}{k + (1/k)} = \frac{k^2 - 1}{k^2 + 1} \tag{7.149}$$

For $k = 1, a = 0$, the ellipse becomes that part of the straight line that lies on the j axis between $\pm j2$. For $k \gg 1, a \to 1$, the ellipse approaches a circle.

After the elliptic contour has been chosen by selecting a value for a, the poles may again be spaced along the contour in various ways. Two systems comparable to the parabolic contour can be formulated.

1. Equal-angle spacing. This system places poles on the ellipse where the contour intersects with the same equally spaced radially lines of Fig. 7.19 for the parabolic contour. When $k \to \infty$ and $a = 1$, this system reduces to the maximally flat magnitude case.
2. Equal-frequency spacing [SC3, KU2]. This system is again parallel to the equal-frequency spacing for the parabolic contour. Poles are placed on the contour such that they are uniformly spaced in the vertical direction as shown in Fig. 7.20.

An interesting property of transformation (7.143) may be pointed out by referring to (7.146) and (7.147). From these equations it is seen that points on the unit circle are moved proportionately in both the vertical and the horizontal directions, each by a different proportionality constant. Hence this system may be regarded as the direct modification of the circular equal-frequency spacing system described earlier.

Another interesting point is that transformation (7.143) is the very transformation that had been used to relate the poles of the maximally flat magnitude characteristic and those of the equiripple magnitude characteristic. Specifically, (7.146) and (7.147) are the same relationships that are used to transform the poles of the function

$$\frac{1}{B_n(s)}$$

defined in (7.75) into those of the function that produce the magnitude characteristic

$$\frac{1}{1 + \epsilon^2 T_n^2(\omega)}$$

where $T_n(\omega)$ is the normalized Chebyshev polynomial of order n. The value of k is related to the ripple parameter ϵ through the relationship [WE1]

$$k = \left(\sqrt{\frac{1}{\epsilon^2} + 1} + \frac{1}{\epsilon}\right)^{1/n} \tag{7.150}$$

Since the circular equal-frequency pole distribution approximates the constant-delay characteristic in somewhat the maximally flat manner, it may be suspected that the elliptic equal-frequency distribution will render a delay characteristic that fluctuates about a constant value. This is indeed the case, as revealed by actual detailed study of this class of delay networks. A typical shape of delay characteristic of this class of functions is shown in Fig. 7.22.

TABLE 7.14(A)

Poles on an Elliptic Contour with Equal-Frequency Spacing
($a = 0.4$)

n	Poles
2	$-0.5694948 \pm j0.8219949$
3	-0.6694329
	$-0.4989659 \pm j1.1157216$
4	$-0.6396476 \pm j0.4128907$
	$-0.4369625 \pm j1.2386722$
5	-0.6593851
	$-0.6043364 \pm j0.6593851$
	$-0.3956311 \pm j1.3187702$
6	$-0.6488908 \pm j0.2742064$
	$-0.5699273 \pm j0.8226193$
	$-0.3637759 \pm j1.3710321$
7	-0.6574549
	$-0.6300488 \pm j0.4696106$
	$-0.5395415 \pm j0.9392212$
	$-0.3386410 \pm j1.4088319$
8	$-0.6518550 \pm j0.2053150$
	$-0.6090628 \pm j0.6159451$
	$-0.5128768 \pm j1.0265752$
	$-0.3180727 \pm j1.4372053$
9	-0.6567089
	$-0.6402886 \pm j0.3648383$
	$-0.5882841 \pm j0.7296766$
	$-0.4894819 \pm j1.0945148$
	$-0.3008534 \pm j1.4593531$
10	$-0.6532025 \pm j0.1641233$
	$-0.6262546 \pm j0.4923699$
	$-0.5685398 \pm j0.8206165$
	$-0.4688300 \pm j1.1488632$
	$-0.2861588 \pm j1.4771098$

TABLE 7.14(B)

Poles on an Elliptic Contour with Equal-Frequency Spacing
($a = 0.5$)

n	Poles
2	$-0.6546537 \pm j0.7559289$
3	-0.7539474
	$-0.5619592 \pm j1.0052633$
4	$-0.7244195 \pm j0.3740886$
	$-0.4948727 \pm j1.1222658$
5	-0.7462272
	$-0.6839285 \pm j0.5969818$
	$-0.4477363 \pm j1.1939636$
6	$-0.7346481 \pm j0.2483564$
	$-0.6452488 \pm j0.7450691$
	$-0.4118525 \pm j1.2417819$
7	-0.7443859
	$-0.7133561 \pm j0.4253634$
	$-0.6108816 \pm j0.8507268$
	$-0.3834174 \pm j1.2760901$
8	$-0.7381058 \pm j0.1859852$
	$-0.6896515 \pm j0.5579555$
	$-0.5807385 \pm j0.9299258$
	$-0.3601587 \pm j1.3018962$
9	-0.7436359
	$-0.7250420 \pm j0.3305048$
	$-0.6661538 \pm j0.6610097$
	$-0.5542735 \pm j0.9915145$
	$-0.3406766 \pm j1.3220193$
10	$-0.7396914 \pm j0.1486836$
	$-0.7091754 \pm j0.4460507$
	$-0.6438188 \pm j0.7434179$
	$-0.5309065 \pm j1.0407850$
	$-0.3240483 \pm j1.3381522$

TABLE 7.14(C)

Poles on an Elliptic Contour with Equal-Frequency Spacing
($a = 0.6$)

n	Poles
2	$-0.7205767 \pm j0.6933752$
3	-0.8235798
	$-0.6138601 \pm j0.9150886$
4	$-0.7930284 \pm j0.3412651$
	$-0.5417416 \pm j1.0237953$
5	-0.8170797
	$-0.7488659 \pm j0.5447198$
	$-0.4902478 \pm j1.0894396$
6	$-0.8045963 \pm j0.2266693$
	$-0.7066851 \pm j0.6800080$
	$-0.4510663 \pm j1.1333467$
7	-0.8153667
	$-0.7813780 \pm j0.3882698$
	$-0.6691321 \pm j0.7765397$
	$-0.4199780 \pm j1.1648095$
8	$-0.8085580 \pm j0.1697812$
	$-0.7554787 \pm j0.5093436$
	$-0.6361700 \pm j0.8489061$
	$-0.3945359 \pm j1.1884685$
9	-0.8146643
	$-0.7942945 \pm j0.3017275$
	$-0.7297815 \pm j0.6034550$
	$-0.6072149 \pm j0.9051826$
	$-0.3732163 \pm j1.2069101$
10	$-0.8103777 \pm j0.1357434$
	$-0.7769455 \pm j0.4072301$
	$-0.7053432 \pm j0.6787169$
	$-0.5816409 \pm j0.9502036$
	$-0.3550150 \pm j1.2216903$

TABLE 7.14(D)

Poles on an Elliptic Contour with Equal-Frequency Spacing
($a = 0.7$)

n	Poles
2	$-0.7714543 \pm j0.6362848$
3	-0.8809670
	$-0.6566340 \pm j0.8390162$
4	$-0.8493844 \pm j0.3133001$
	$-0.5802400 \pm j0.9399004$
5	-0.8755468
	$-0.8024519 \pm j0.5003125$
	$-0.5253281 \pm j1.0006250$
6	$-0.8623977 \pm j0.2082455$
	$-0.7574526 \pm j0.6247364$
	$-0.4834704 \pm j1.0412273$
7	-0.8740803
	$-0.8376441 \pm j0.3567675$
	$-0.7173155 \pm j0.7135349$
	$-0.4502202 \pm j1.0703024$
8	$-0.8668707 \pm j0.1560221$
	$-0.8099634 \pm j0.4680662$
	$-0.6820502 \pm j0.7801103$
	$-0.4229896 \pm j1.0921544$
9	-0.8734793
	$-0.8516388 \pm j0.2772950$
	$-0.7824683 \pm j0.5545900$
	$-0.6510530 \pm j0.8318850$
	$-0.4001608 \pm j1.1091800$
10	$-0.8689272 \pm j0.1247578$
	$-0.8330796 \pm j0.3742734$
	$-0.7563040 \pm j0.6237891$
	$-0.6236643 \pm j0.8733047$
	$-0.3806647 \pm j1.1228203$

TABLE 7.14(E)

Poles on an Elliptic Contour with Equal-Frequency Spacing
($a = 0.8$)

n	Poles
2	$-0.8108849 \pm j0.5852057$
3	-0.9283178
	$-0.6919272 \pm j0.7735981$
4	$-0.8961623 \pm j0.2892351$
	$-0.6121954 \pm j0.8677054$
5	-0.9242872
	$-0.8471232 \pm j0.4621436$
	$-0.5545723 \pm j0.9242872$
6	$-0.9106854 \pm j0.1924174$
	$-0.7998642 \pm j0.5772522$
	$-0.5105411 \pm j0.9620871$
7	-0.9231927
	$-0.8847093 \pm j0.3297117$
	$-0.7576198 \pm j0.6594234$
	$-0.4755170 \pm j0.9891351$
8	$-0.9156878 \pm j0.1442073$
	$-0.8555758 \pm j0.4326218$
	$-0.7204593 \pm j0.7210364$
	$-0.4468099 \pm j1.0094509$
9	-0.9227443
	$-0.8996721 \pm j0.2563179$
	$-0.8266003 \pm j0.5126357$
	$-0.6877730 \pm j0.7689536$
	$-0.4227303 \pm j1.0252715$
10	$-0.9179894 \pm j0.1153268$
	$-0.8801177 \pm j0.3459803$
	$-0.7990072 \pm j0.5766338$
	$-0.6588782 \pm j0.8072873$
	$-0.4021582 \pm j1.0379408$

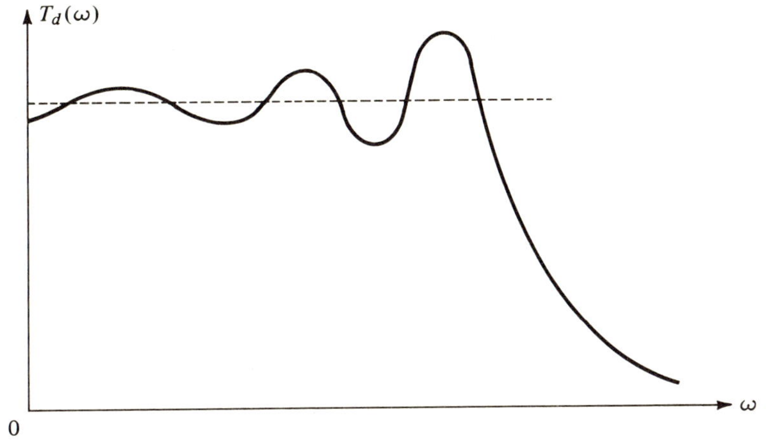

Fig. 7.22 Typical delay characteristic of network functions with elliptically distributed poles.

Experience has shown that to produce a generally good delay characteristic with this approach, the eccentricity value should be chosen to be a moderate one. Typically, it should fall in the range between 0.4 and 0.8. Normalized poles of this class of network functions are listed in Table 7.14 for several values of a in this range.

4. Catenary Contour

The empirical methods of specifying the pole locations of a constant-delay characteristic are largely based on combinations of the engineer's experience, intuition, or ability to reason, and on mathematical convenience. Most of these distributions are justified not by the way they are arrived at (although some form of logic in arriving at these distributions is sometimes helpful) but by the final result—the time-domain response of networks obtained this way. Therefore, it does not matter how systematic or reasonable these contours are arrived at so long as the resulting time-domain response is acceptable.

One of the more useful pole distributions that is arbitrarily created uses the contour that is a catenary passing through $s = -1 + j0$ and is defined by the equation [GH2]

$$\sigma = \frac{1}{b-1}(\cosh \omega - b) \tag{7.151}$$

When the poles are placed in the equal-angle manner on this contour this system of pole distribution has been found to give results comparable to parabolic and elliptic-circular contours. Normalized poles for this class of functions for several selected values of b are listed in Table 7.15.

TABLE 7.15(A)

Poles on a Catenary Contour with Equal-Angle Spacing
$(b = 2)$

n	Poles
2	$-0.7071068 \pm j0.7071068$
3	-0.9542071
	$-0.5118570 \pm j0.8865624$
4	$-0.8772877 \pm j0.3633845$
	$-0.4030073 \pm j0.9729458$
5	-0.9446932
	$-0.7728479 \pm j0.5615068$
	$-0.3328132 \pm j1.0242938$
6	$-0.9112087 \pm j0.2441576$
	$-0.6839193 \pm j0.6839193$
	$-0.2836627 \pm j1.0586437$
7	-0.9419630
	$-0.8513270 \pm j0.4099775$
	$-0.6118466 \pm j0.7672313$
	$-0.2472602 \pm j1.0833176$
8	$-0.9233174 \pm j0.1836592$
	$-0.7895439 \pm j0.5275563$
	$-0.5531602 \pm j0.8278628$
	$-0.2191880 \pm j1.1019325$
9	-0.9408301
	$-0.8851313 \pm j0.3221615$
	$-0.7326990 \pm j0.6148074$
	$-0.5046813 \pm j0.8741337$
	$-0.1968674 \pm j1.1164907$
10	$-0.9289670 \pm j0.1471339$
	$-0.8411450 \pm j0.4285848$
	$-0.6821101 \pm j0.6821101$
	$-0.4640265 \pm j0.9107032$
	$-0.1786885 \pm j1.1281950$

TABLE 7.15(B)

Poles on a Catenary Contour with Equal-Angle Spacing
($b = 2.5$)

n	Poles
2	$-0.7071068 \pm j0.7071068$
3	-0.8829647
	$-0.5321062 \pm j0.9216350$
4	$-0.8304267 \pm j0.3439740$
	$-0.4257491 \pm j1.0278492$
5	-0.8727649
	$-0.7542536 \pm j0.5479973$
	$-0.3547922 \pm j1.0919380$
6	$-0.8509667 \pm j0.2280158$
	$-0.6830693 \pm j0.6830693$
	$-0.3041219 \pm j1.1349984$
7	-0.8698960
	$-0.8105284 \pm j0.3903299$
	$-0.6213231 \pm j0.7791145$
	$-0.2661280 \pm j1.1659830$
8	$-0.8579859 \pm j0.1706640$
	$-0.7658372 \pm j0.5117161$
	$-0.5685898 \pm j0.8509547$
	$-0.2365807 \pm j1.1893717$
9	-0.8687077
	$-0.8330704 \pm j0.3032128$
	$-0.7220512 \pm j0.6058729$
	$-0.5235144 \pm j0.9067536$
	$-0.2129437 \pm j1.2076638$
10	$-0.8612045 \pm j0.1364014$
	$-0.8029172 \pm j0.4091067$
	$-0.6810168 \pm j0.6810168$
	$-0.4847480 \pm j0.9513715$
	$-0.1936039 \pm j1.2223670$

TABLE 7.15(C)

Poles on a Catenary Contour with Equal-Angle Spacing
($b = 5$)

n	Poles
2	$-0.7071068 \pm j0.7071068$
3	-0.7601599
	$-0.5734790 \pm j0.9932948$
4	$-0.7361321 \pm j0.3049159$
	$-0.4802853 \pm j1.1595112$
5	-0.7483725
	$-0.7030713 \pm j0.5108112$
	$-0.4110378 \pm j1.2650442$
6	$-0.7399070 \pm j0.1982575$
	$-0.6669560 \pm j0.6669560$
	$-0.3582207 \pm j1.3368979$
7	-0.7453954
	$-0.7246001 \pm j0.3489490$
	$-0.6304054 \pm j0.7905034$
	$-0.3169556 \pm j1.3886730$
8	$-0.7409919 \pm j0.1473924$
	$-0.7060262 \pm j0.4717516$
	$-0.5949384 \pm j0.8903882$
	$-0.2839807 \pm j1.4276672$
9	-0.7441362
	$-0.7320818 \pm j0.2664560$
	$-0.6856999 \pm j0.5753705$
	$-0.5613905 \pm j0.9723568$
	$-0.2570958 \pm j1.4580629$
10	$-0.7414445 \pm j0.1174333$
	$-0.7206438 \pm j0.3671863$
	$-0.6644900 \pm j0.6644900$
	$-0.5301477 \pm j1.0404735$
	$-0.2347904 \pm j1.4824084$

7.10 DELAY FUNCTIONS WITH EQUIRIPPLE MAGNITUDE CHARACTERISTICS AND EQUIRIPPLE STEP OR IMPULSE RESPONSES—A NUMERICAL APPROACH

In Section 7.9 various techniques for formulating the low-pass constant-delay network function were presented. These network functions were obtained with little or no regard to their frequency-domain magnitude characteristics. When the frequency scale is not an important consideration, the frequency scale can be varied to produce a corresponding time-scale change, as explained in Section 7.1. In a band-limited transmission, we no longer have this freedom. We must produce the best transient response with the restriction that any signal component outside the passband is nearly completely suppressed. In such a situation, the magnitude characteristic and the transient response must be carefully examined simultaneously.

In Fig. 7.23 there are shown a typical magnitude characteristic and the step response of a network. (The ripples in either characteristic may or may not be present.) The frequency characteristic is assumed to be normalized such that its dc gain is unity. One of the main features of these diagrams is that the magnitude $|H(j\omega)|$ beyond the frequency ω_c is no greater than a certain amount, say q_f, or

$$|H(j\omega)| \leq q_f, \qquad \text{for } \omega > \omega_c$$

The step response $g(t)$ usually resembles a ramp function in its main rising portion. Outside of this main portion two regions may be specified by a quantity q_t such that $g(t)$ does not go beyond these regions, as shown shaded in Fig. 7.23. Although in many situations the oscillation may not take place, this description will still be applicable. When $q_t \ll 1$, $g(t)$ inside these shaded regions may be considered sufficiently close to 0 or 1 that the deviation from these values is due to imperfect response of the network and may be regarded as parasitic or noise effect. Between $t = t_1$ and $t = t_2$, $g(t)$ is nearly a straight line and may be regarded as a delayed step function with a finite rise time equal to

$$T_R = t_2 - t_1 \tag{7.152}$$

For a given network function $H(s)$, the product

$$M = \omega_c T_R$$

is invariant with respect to frequency or time scaling.

For many engineering applications, for a given q_f and q_t, it is desirable to

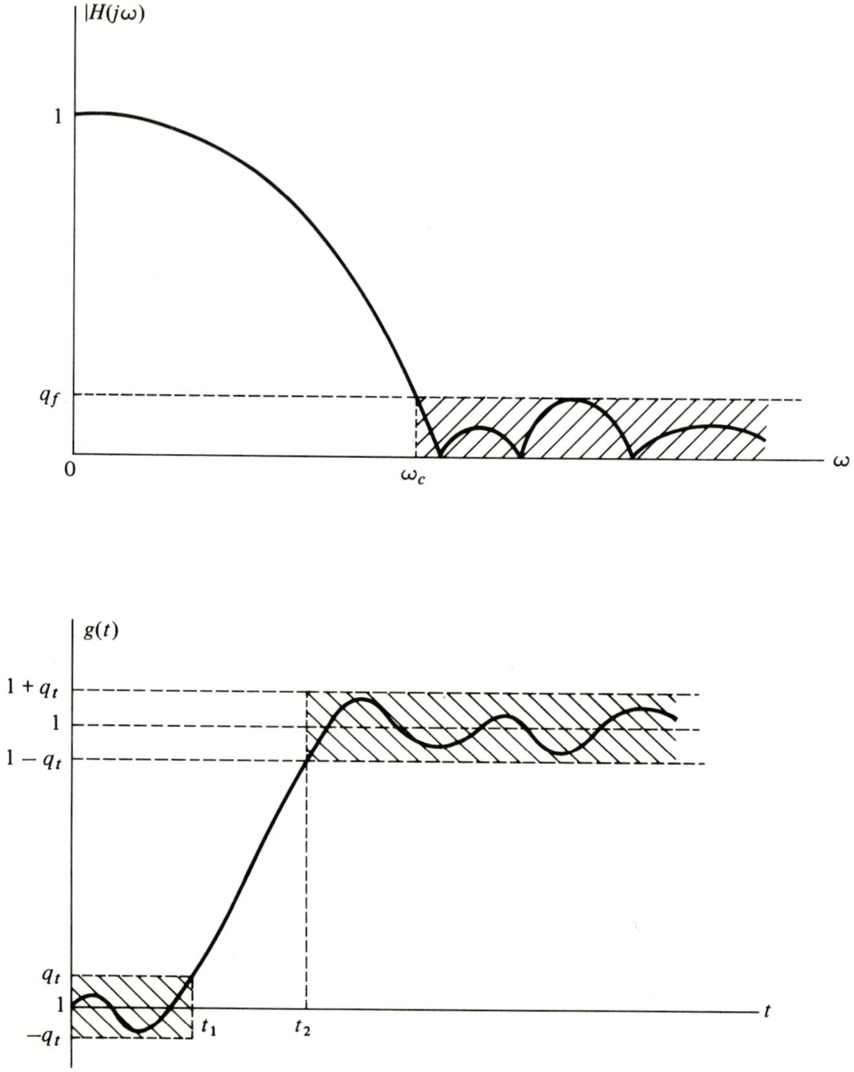

Fig. 7.23 Typical magnitude and step response characteristics.

minimize both ω_c and T_R. In other words, the quantity M should be made as small as possible. The problem of analytically performing this minimization appears to be extremely difficult if not impossible. However, approximate results have been obtained by using a combination of numerical experiment, computational optimization, and a rather bold conjecture.

The network function used in obtaining these results are of the form

$$H(s) = \frac{\prod_{\mu=1}^{m} (s^2 + \omega_\mu^2)}{\prod_{\nu=1}^{n} (s - p_\nu)} \tag{7.153}$$

where $s = \pm j\omega_\mu$, $\omega_\mu > \omega_c$, are j-axis transmission zeros and the p_ν are poles of $H(s)$. Thus, instead of limiting ourselves to all-pole functions, j-axis zeros are admitted. These zeros do not alter the delay characteristic in the passband, and they may be helpful in increasing the cutoff rate of the magnitude characteristic, thus improving the gain distribution over the frequency axis.

After examining a large number of network functions [JE1, JE2, JE3, JE4], it was surmised that for a given set of m, n, q_f, and q_t, M appears to be very nearly minimum if (1) the ripples of the magnitude outside $\omega = \omega_c$ are made equal, and (2) the ripples of $g(t)$ for $t < t_1$ and $t > t_2$ are made equal.

These two conditions can be satisfied by the following derivation. First, both the magnitude characteristic $|H(j\omega)|$ and the step response $g(t)$ may be regarded as functions of $m + n + 1$ variables—t or ω, ω_μ, and p_ν. Thus we may write the magnitude characteristic as

$$|H(j\omega)| = H(\omega, \omega_1, {}_2, \ldots, \omega_m, p_1, p_2, \ldots, p_n) \tag{7.154}$$

and the step response

$$g(t) = g(t, \omega_1, \omega_2, \ldots, \omega_m, p_1, p_2, \ldots, p_n) \tag{7.155}$$

For simplicity we shall consider only the case where all time-domain ripples occur after $t = t_2$. For other situations the procedure is equally applicable with a slight modification. If we denote the points along the ω axis at which the extrema for $|H(j\omega)|$ occurs by $\omega_{e\mu}$, $\mu = 1, 2, \ldots, m$, and the points on the t axis at which the extrema for $g(t)$ occur by $t_{e\nu}$, $\nu = 1, 2, \ldots, n$, these points must satisfy

$$\left[\frac{\partial}{\partial\omega} H(\omega, \omega_1, \omega_2, \ldots, \omega_m, p_1, p_2, \ldots, p_n)\right]_{\omega=\omega_{e\mu}} = 0, \qquad \mu = 1, 2, \ldots, m \tag{7.156}$$

and

$$\left[\frac{\partial}{\partial t} g(t, \omega_1, \omega_2, \ldots, \omega_m, p_1, p_2, \ldots, p_n)\right]_{t=t_{e\nu}} = 0, \qquad \nu = 1, 2, \ldots, n-1 \tag{7.157}$$

This set of notation is illustrated in Fig. 7.24, for $n = 6$ and $m = 3$.

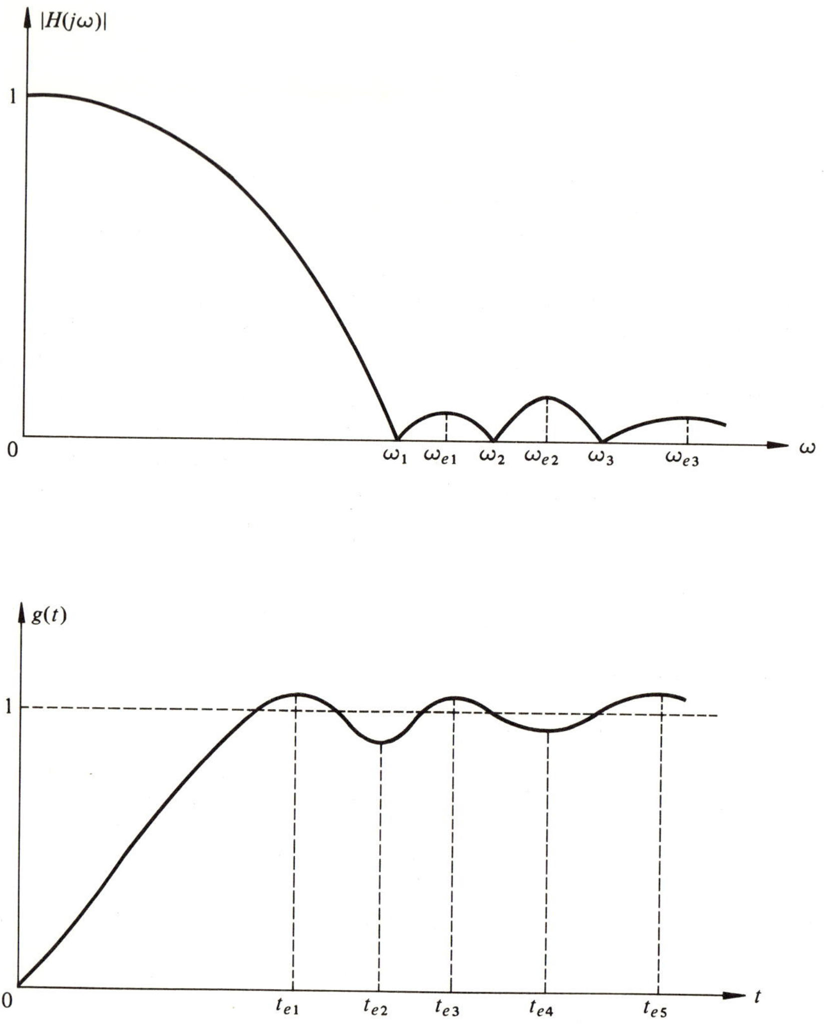

Fig. 7.24 Notation used in formulating (7.156) and (7.157).

In addition, now we also require that the functional values of $|H(j\omega)|$ and $g(t)$ be made to equal the prescribed quantities,

$$
\begin{aligned}
|H(\omega_{e1}, \omega_1, \omega_2, \ldots, \omega_m, p_1, p_2, \ldots, p_n)| &= q_f \\
|H(\omega_{e2}, \omega_1, \omega_2, \ldots, \omega_m, p_1, p_2, \ldots, p_n)| &= q_f \\
&\vdots \\
|H(\omega_{em}, \omega_1, \omega_2, \ldots, \omega_m, p_1, p_2, \ldots, p_n)| &= q_f
\end{aligned}
\tag{7.158}
$$

TABLE 7.16(B)

Poles and Zeros of Delay Networks with Equiripple Step Response $n = 5$, $m = 2$

q_t	q_f	Poles	Zeros
0.001	0.001	-0.70469 $-0.64672 \pm j0.60469$ $-0.48999 \pm j1.25307$	$\pm j2.88577$
0.001	0.005	-0.70664 $-0.64558 \pm j0.60677$ $-0.47201 \pm j1.25702$	$\pm j2.11932$
0.001	0.010	-0.70719 $-0.64468 \pm j0.60740$ $-0.46057 \pm j1.26106$	$\pm j1.86132$
0.001	0.020	-0.70677 $-0.64306 \pm j0.60719$ $-0.44724 \pm j1.26840$	$\pm j1.63901$
0.005	0.001	-0.65571 $-0.57310 \pm j0.65988$ $-0.34214 \pm j1.37091$	$\pm j2.91454$
0.005	0.005	-0.66158 $-0.57479 \pm j0.66175$ $-0.31779 \pm j1.36616$	$\pm j2.16075$
0.005	0.010	-0.66515 $-0.57640 \pm j0.66218$ $-0.30093 \pm j1.36386$	$\pm j1.91060$
0.005	0.020	-0.66923 $-0.57894 \pm j0.66147$ $-0.27838 \pm j1.36245$	$\pm j1.69906$
0.010	0.001	-0.63741 $-0.54213 \pm j0.68189$ $-0.26838 \pm j1.41255$	$\pm j2.92570$
0.010	0.005	-0.64636 $-0.54639 \pm j0.68363$ $-0.24283 \pm j1.40039$	$\pm j2.17594$
0.010	0.010	-0.65245 $-0.55021 \pm j0.68408$ $-0.22499 \pm j1.39214$	$\pm j1.92768$
0.010	0.020	-0.66069 $-0.55655 \pm j0.68363$ $-0.20064 \pm j1.38111$	$\pm j1.71774$
0.020	0.001	-0.62468 $-0.51624 \pm j0.70114$ $-0.19170 \pm j1.44045$	$\pm j2.93414$
0.020	0.005	-0.63784 $-0.52413 \pm j0.70272$ $-0.16736 \pm j1.41845$	$\pm j2.18625$
0.020	0.010	-0.64728 $-0.53087 \pm j0.70332$ $-0.15071 \pm j1.40249$	$\pm j1.93804$
0.020	0.020	-0.66077 $-0.54174 \pm j0.70351$ $-0.12853 \pm j1.37950$	$\pm j1.72650$

TABLE 7.16(A)

Poles and Zeros of Delay Networks with Equiripple Step Response

$q_t = q_f = 0.01$

n	m	Poles	Zeros
2	0	$-0.82608 \pm j0.56355$	
3	0	-0.75377 $-0.61352 \pm j0.97481$	
3	2	-0.75697 $-0.59994 \pm j0.98037$	$\pm j3.38633$
4	0	$-0.64814 \pm j0.39114$ $-0.43247 \pm j1.24817$	
4	2	$-0.65505 \pm j0.39601$ $-0.38978 \pm j1.24608$	$\pm j2.30020$
5	0	-0.62750 $-0.53927 \pm j0.67878$ $-0.29536 \pm j1.42591$	
5	2	-0.65245 $-0.55021 \pm j0.68408$ $-0.22499 \pm j1.39214$	$\pm j1.92768$
5	4	-0.68502 $-0.56702 \pm j0.69220$ $-0.16003 \pm j1.34077$	$\pm j1.63332$ $\pm j2.38431$
6	0	$-0.57474 \pm j0.29010$ $-0.44900 \pm j0.89491$ $-0.19617 \pm j1.53890$	
6	2	$-0.61822 \pm j0.29212$ $-0.46900 \pm j0.88876$ $-0.11442 \pm j1.45085$	$\pm j1.75865$
6	4	$-0.68804 \pm j0.29811$ $-0.50121 \pm j0.89140$ $-0.04720 \pm j1.30322$	$\pm j1.44689$ $\pm j1.82958$
7	0	-0.56913 $-0.51775 \pm j0.51792$ $-0.38063 \pm j1.05624$ $-0.12630 \pm j1.60722$	
7	4	-0.74038 $-0.67343 \pm j0.51660$ $-0.44758 \pm j1.01223$ $-0.01040 \pm j1.23714$	$\pm j1.33498$ $\pm j1.58864$
8	0	$-0.54511 \pm j0.22892$ $-0.46966 \pm j0.69852$ $-0.33096 \pm j1.17516$ $-0.07823 \pm j1.64404$	
9	0	-0.54986 $-0.51802 \pm j0.41505$ $-0.43295 \pm j0.84029$ $-0.29599 \pm j1.26138$ $-0.04613 \pm j1.65822$	

$$
\begin{aligned}
g(0, \omega_1, \omega_2, \ldots, \omega_m, p_1, p_2, \ldots, p_n) &= 0 \\
g(t_{e1}, \omega_1, \omega_2, \ldots, \omega_m, p_1, p_2, \ldots, p_n) &= 1 + q_t \\
g(t_{e2}, \omega_1, \omega_2, \ldots, \omega_m, p_1, p_2, \ldots, p_n) &= 1 - q_t \qquad (7.159) \\
&\vdots \\
g(t_{e(n-1)}, \omega_1, \omega_2, \ldots, \omega_m, p_1, p_2, \ldots, p_n) &= 1 + (-1)^{n-1} q_t
\end{aligned}
$$

Now (7.156), (7.157), (7.158), and (7.159) collectively may be regarded as a set of $2m + 2n - 1$ equations that governs the values of the equal number of unknowns. Hence for a given set of m, n, q_f, and q_t, the parameters ω_1, ω_2, $\ldots$, ω_m, p_1, p_2, $\ldots$, p_n (as well as ω_{e1}, ω_{e2}, $\ldots$, ω_{em}, t_{e1}, t_{e2}, $\ldots$, $t_{e(n-1)}$) may be solved from these equations.

However, equations in (7.156), (7.157), (7.158), and (7.159) are nonlinear. Hence some numerical techniques must be used. Jess and Schüssler used the analog and trial-and-error techniques to obtain the initial set of parameters [JE4] and then the Newton–Raphson iteration technique to refine the final solution. Poles and zeros for a number of the network functions found this way are tabulated in Table 7.16. These poles and zeros have been normalized such that the geometric mean distance of all poles from the origin is unity.

An example of the step response of a network designed this way is found in Fig. 7.38, for which $n = 5$, $M = 0$, and $q_t = 0.01$.

7.11 QUANTITATIVE COMPARISONS OF CONSTANT-DELAY NETWORK FUNCTIONS

The various schemes of designing constant-delay networks presented so far in this chapter are only several of the better-known methods of forming an approximate delay network function. Many other methods, some of them slight variations and others compromises of two or more of the ones presented here, are also available. A natural question at this point is: Which of these methods gives the best delay function? This question cannot be answered in general terms because the criterion for the optimum approximation varies from application to application.

In most of the situations in which delay networks are used, it is desirable to produce a replica of the excitation as nearly proportional to the excitation as possible. But the best approximation for one excitation need not render a similarly good approximation for another excitation if the two excitations have different wave shapes. In these special situations, where a critical network for a specific application is desired, the delay network function must be evaluated for that particular wave shape.

For the purpose of general information, it is desirable to have some guideline in evaluating the performance of the various delay functions when the wave shape of the excitation is either unknown or variable. It has been a common practice to compare the responses of the various delay networks under a standard excitation. This standard excitation is usually either a step function or an impulse function. Of the two functions, the step response is more suitable when the network designed is subjected to time-varying signals. This is partly because in many time-varying signals, the transmission of the low-frequency part of the frequency spectrum is the most important, and a step function is rich in this part of the spectrum. Furthermore, a sudden jump in a signal is, in effect, a step function.

On the other hand, the impulse function seldom occurs in actual signals. However, the impulse response of a network is important because it is convenient to use for many theoretical studies and derivations.

We shall assume that our purpose is to design networks to be used in the time domain and that the step response is of primary importance. We shall let $g(t)$ denote the step response of a network whose dc gain has been normalized to unity. In other words, $g(t)$ tends to unity as t approaches infinity. Of course.

$$g(t) = \int_0^t h(t)\,dt \tag{7.160}$$

Figure 7.25 shows several typical step responses of lumped delay networks. It can be easily seen that it is extremely difficult to compare these vastly different responses. Nevertheless, the following quantities have been commonly used to characterize or compare the relative qualities of the various delay networks.

1. Delay time (T_D). This is a measure of the time lapse between the time the step is applied and the time a step-like response has been essentially established. In a network designed mainly for causing a delay, this quantity is usually maximized.
2. Rise time (T_R). This is the approximate time that the response takes in rising from a near-zero value to nearly its final value. In a band-limited transmission system, this quantity is usually minimized.
3. Delay-to-rise ratio. This is the ratio between the two quantities described in 1 and 2. The delay-to-rise ratio is independent of any scaling in the frequency domain. For most of the developments of constant-delay networks, this is the most significant quantity in measuring the quality of a network. This quantity is usually maximized, because a large delay-to-rise ratio usually indicates either a long delay or a short rise time [KU2].
4. Overshoot (γ). This is the amount by which the maximum $g(t)$ exceeds its final value (unity). It is usually expressed in per-unit or percent values. For certain applications, overshoot must be kept below certain level, say, 5 or 10 percent.

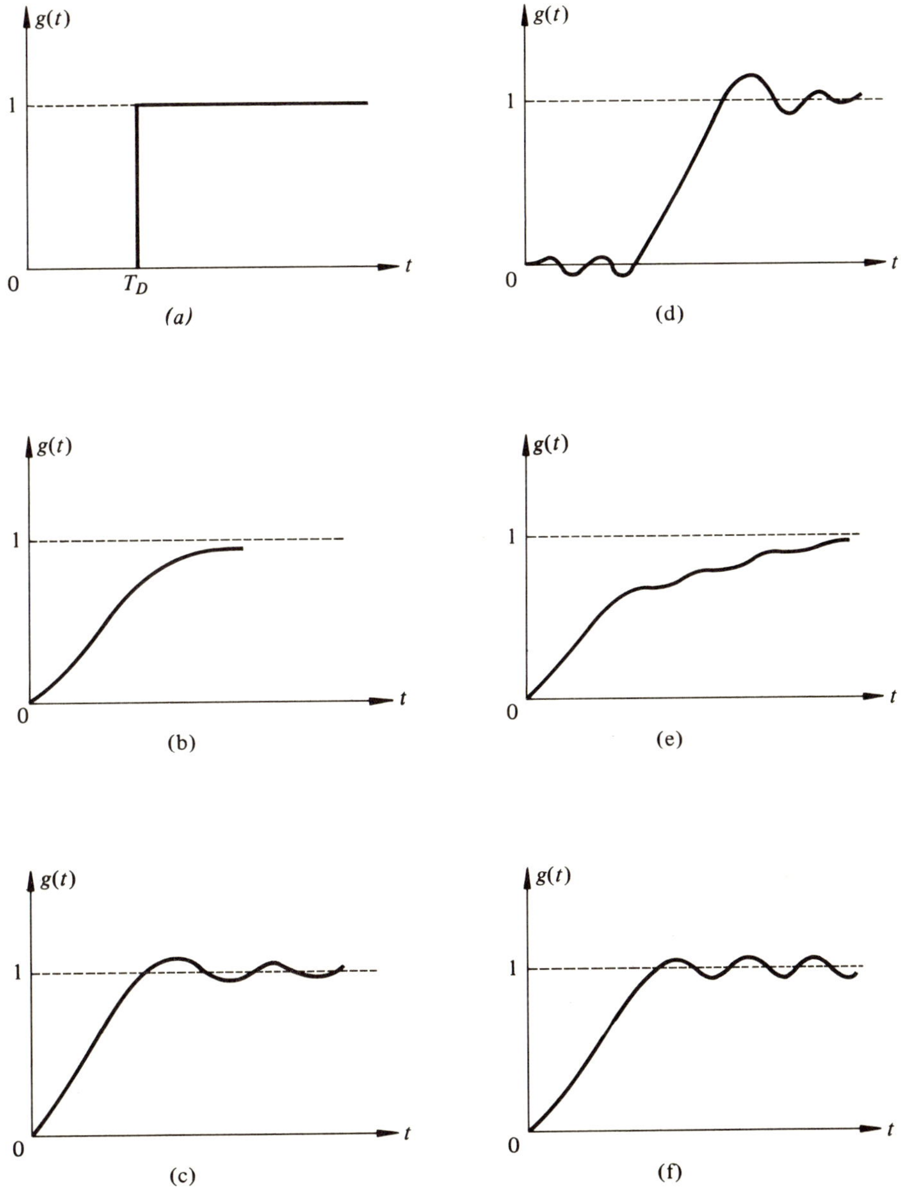

Fig. 7.25 The step response of an ideal delay network and some typical step responses of lumped constant-delay networks.

5. Undershoot. This is the maximum amount that $g(t)$ goes below its final value after the first time the final value has been reached.
6. Sag. This is proportional to the initial slope of $g(t)$ at $t = 0^+$ and has units of volts (or amperes) per second. This quantity is a measure of how much the flat top of the output droops when the network is excited by a square-wave input [GH1].
7. Frequency-domain characteristics are sometimes described as an adjoint in characterizing a delay network. Pertinent quantities are the 3-dB bandwidth, delay bandwidth, peak magnitude, and the frequency at which this peak occurs.

Of the seven quantities, the last four are the least controversial. The first two—the delay and rise times—are very much subject to choice and interpretation. A summary of several definitions of rise and delay times is given in the following two subsections. Thereafter, a new set of definition for the delay and rise times, which is more significant and universally applicable in making comparison of step responses with different wave shapes, is introduced.

It should be noted in passing that other quantities are sometimes used for evaluating a step response. These quantities are usually of interest only for certain specialized applications. For example, in video amplifiers, it is desirable that the response be symmetric about the point where $g(t) = 0.5$ [CO1]. In that case this symmetry is a primary consideration in the design.

1. Traditional Definitions of the Delay Time

The following are some of the existing definitions of the delay time. Some of the implications and possible ambiguities are pointed out with each definition.

(*a*) The time at which $g(t)$ reaches a certain percentage of its final value. Most typical of these percentages is 50 percent. Under this definition, all step responses of Fig. 7.26 will have the same delay time.

This delay time can also be defined in terms of the corresponding impulse response. Delay time is the value of T_D such that

$$\int_0^{T_D} h(t)\,dt = \int_{T_D}^{\infty} h(t)\,dt = 0.5 \tag{7.161}$$

is satisfied.

The 50 percent delay time is by far the most commonly used quantity for indicating the delay produced by a network. Delay time defined this way for various delay functions obtained in Section 7.10 are tabulated in Section 7.12.

(*b*) The design delay time. Frequently, in the procedures of arriving at a constant-delay network function, a parameter is introduced. The nature of this parameter is such that it may logically be considered as the

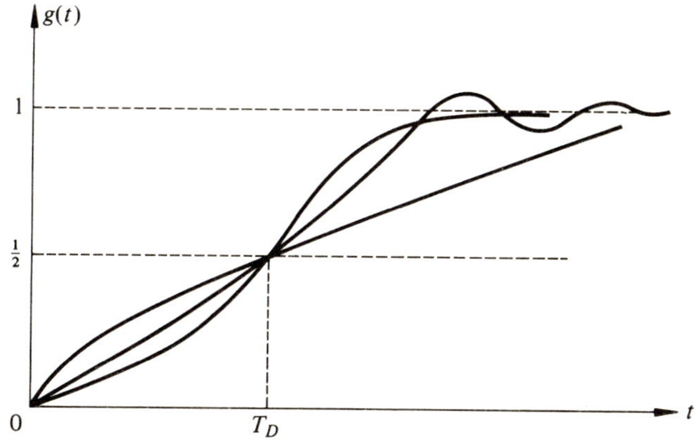

Fig. 7.26 Step responses having the same 50% delay time.

delay time we are trying to achieve. One example of such a parameter is the T_0 of (7.93). Another example is the quantity τ of (7.112). Still another example is the approximate average slop of the $\beta(\omega)$ characteristic (1 second in Fig. 7.15).

(*c*) The dc delay time. This is the slope of the $\beta(\omega)$ characteristic or the value of $T_d(\omega)$ at $\omega = 0$. This quantity usually gives a good indication of the delay of a low-pass network. For a band-pass network, the center-frequency delay may be substituted for the dc delay for this definition.

(*d*) The time at which the slope of $g(t)$ is maximum after the initial ringing (Fig. 7.25d) but before the final ringing. The reasoning for this definition is that this delay time corresponds to the time the maximum of the large peak of the impulse response occurs. This point is usually close to where the impulse response may be considered to be located.

(*e*) The center of gravity of the impulse response [EL1]. Mathematically this delay time is given by

$$T_D = \frac{\int_0^\infty th(t)\,dt}{\int_0^\infty h(t)\,dt} \tag{7.162}$$

This definition has the advantage that T_D calculated this way does not depend on the behavior of $g(t)$ at or in the vicinity of a single point. Thus the variation of $g(t)$ at all values of t affects the values T_D to some extend. Another advantage of this definition is that it is easy to compute. For a given transfer function

$$H(s) = \frac{1 + a_1 s + a_2 s^2 + \ldots + a_m s^m}{1 + b_1 s + b_2 s^2 + \ldots + b_n s^n} \tag{7.163}$$

in which $n \geq m$, it can be shown

$$T_D = b_1 - a_1 \tag{7.164}$$

The drawback of this definition is that when $h(t)$ goes negative at any time, it tends to offset its positive portions. For example, the two impulse responses of Fig. 7.27 will have the same T_D. To remedy this we can square $h(t)$ and redefine T_D as

$$T_D = \frac{\int_0^\infty t h^2(t)\, dt}{\int_0^\infty h^2(t)\, dt} \tag{7.165}$$

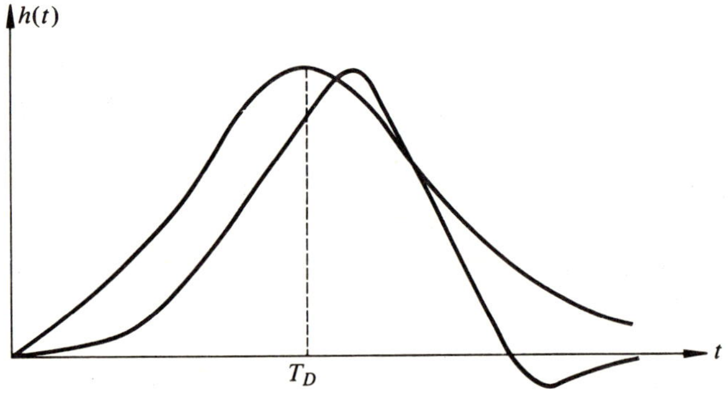

Fig. 7.27 Two impulse responses having the same center of gravity.

(*f*) The arithmetical mean of t_2 and t_1 defined in the following subsection (Fig. 7.28). Loosely speaking, t_1 is the instant at which $g(t)$ begins to rise and t_2 is the time at which $g(t)$ has practically attained its final value. Hence $T_D = \frac{1}{2}(t_2 - t_1)$ is approximately the mean position of the front of the rising part of the step response.

2. Traditional Definitions of the Rise Time

Other than the delay time, rise time is the most important parameter of a step response. In many applications this is the only significant parameter in measuring the performance of a delay network. The rise time is frequently defined as the time interval between two points on the step response t_1 and t_2. The question is: How are t_1 and t_2 selected? Some definitions use the slope of the step response at a certain point.

(*a*) Preselected levels of g_1 and g_2. Here t_1 and t_2 are defined as the moments at which $g(t)$ reaches certain predetermined levels regardless of the wave shape of the step response, as depicted in Fig. 7.28. Traditionally these levels are fixed at $g_1 = 0.1$ and $g_2 = 0.9$, although other levels have also been used for special purposes. This definition has very little theoretical basis. But it is a very practical way of giving a first-order approximate indication of the time the response takes to reach approximately its final value. The 10 to 90 percent rise times of all delay functions presented in Section 7.10 are tabulated in Section 7.12.

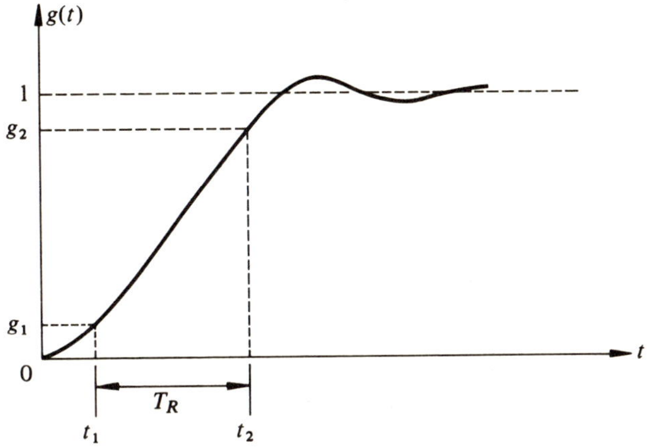

Fig. 7.28 Rise time defined as the time interval between two fixed levels.

The 10 to 90 percent rise time has been so widely accepted for such a long time that it is difficult to ignore and revolutionize. However, it suffers the disadvantages that two vastly different responses may have the identical rise time (Fig. 7.29), whereas two very similar responses

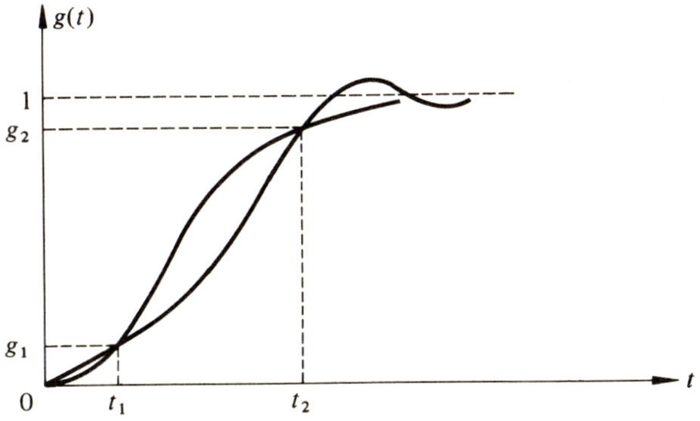

Fig. 7.29 Two step responses having the same rise time between two preselected levels.

may have very different rise times (Fig. 7.30). Alternatively, one may say that the rise time defined this way may sometimes be changed widely with a very slight adjustment in the levels g_1 and g_2.

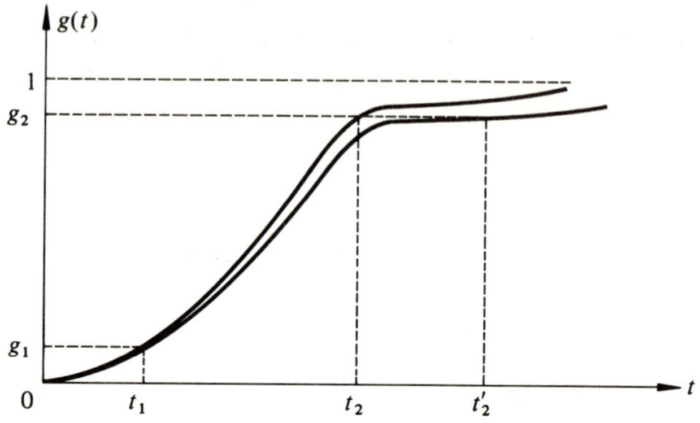

Fig. 7.30 Two similar responses having vastly different rise times between two preselected levels.

For certain classes of filters it may be more meaningful to choose $g_1 = 0$ and $g_2 = 1$. An example of such a definition is depicted in Fig. 7.31.

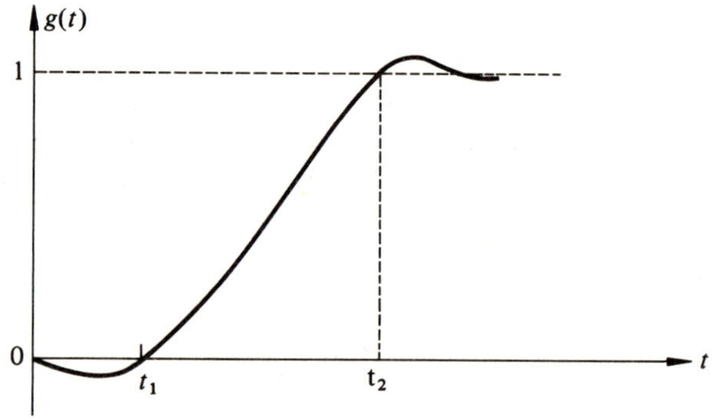

Fig. 7.31 Rise time for a response when $g_1 = 0$ and $g_2 = 1$.

(*b*) Equivalent time based on the slope of $g(t)$ at $t = T_D$. Analytically

$$T_R = \frac{1}{g'(T_D)} \tag{7.166}$$

Here T_D may be any one of the definitions given in the previous subsection. Thus this T_R is the time it takes the response to go from $g_1 = 0$ to $g_2 = 1$ had it continued at a uniform rate equal to that at $t = T_D$. This definition is illustrated in Fig. 7.32.

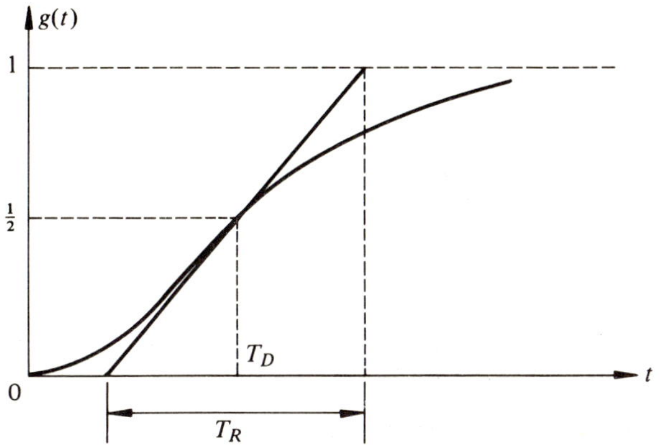

Fig. 7.32 Rise time based on the slope of the response at $t = T_D$.

(*c*) The radius of gyration of the impulse response about $t = T_D$ [EL1]. Here again, T_D may be any of the delay times defined in the provious subsection. Analytically T_R of this definition is given by

$$T_R^2 = 2\pi \frac{\int_0^\infty (t - T_D)^2 h(t)\, dt}{\int_0^\infty h(t)\, dt} \tag{7.167}$$

If T_D is given by (7.165), then it can be shown that

$$T_R = \left[2\pi \frac{\int_0^\infty t^2 h(t)\, dt}{\int_0^\infty h(t)\, dt} - T_D^2 \right]^{1/2} \tag{7.168}$$

This definition is again computationally easy to evaluate. If the inverse Laplace transform of the impulse response is given by (7.163), then

$$T_R = \sqrt{2\pi[b_1^2 - a_1^2 + 2(a_2 - b_2)]} \tag{7.169}$$

(*d*) Values of g_1 and g_2 are dependent on the amount of overshoot. A typical scheme is depicted in Fig. 7.33, in which

$$g_1 = \gamma, \qquad g_2 = 1 - \gamma \tag{7.170}$$

where γ is the per-unit overshoot. This scheme is computationally advantageous and is significant in connection with certain optimization

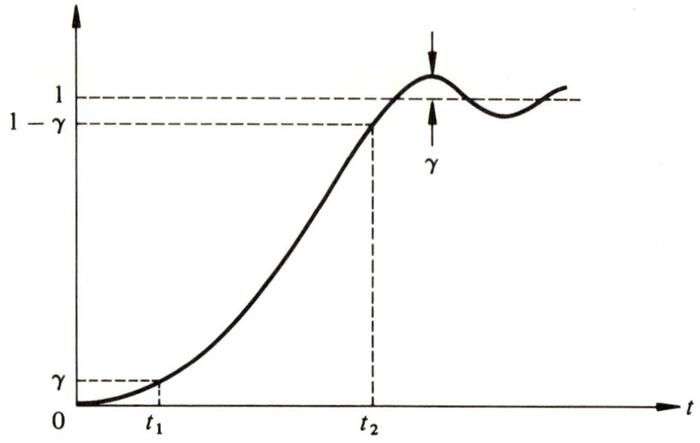

Fig. 7.33 Definition of rise time where the amount of overshoot determines g_1 and g_2.

problems. It is also very meaningful for small γ because, under that situation, $g_1 \cong 0$ and $g_2 \cong 1$. When γ is large, the rise time obtained this way tends to be inconsistent, even for a same class of networks.

(*e*) The pulse width of the impulse response. Here t_1 and t_2 are taken as those two points at which $h(t)$ is at a certain preselected level (h_1) and that lie closest to and on both sides of the main lobe of the impulse response (Fig. 7.34). When it is known that the impulse response crosses the horizontal axis, h_1 may be taken as zero. In that case t_2 corresponds to the first maximum after overshoot commences in $g(t)$.

(*f*) The reciprocal of the area under the squared impulse response. Analytically,

$$T_R = \frac{1}{\int_0^\infty h^2(t)\,dt} \tag{7.171}$$

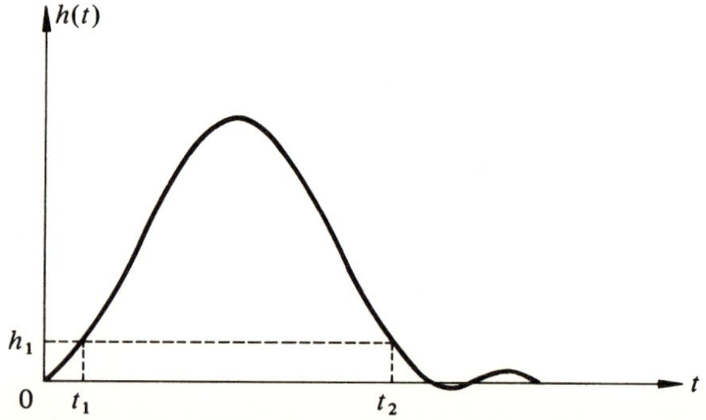

Fig. 7.34 Rise time based on two points on the impulse response.

This definition is probably a heuristic one. It is easy to see that the larger $h(t)$ is, the faster $g(t)$ rises, and hence the shorter the rise time is. Since the integral extends over the entire time range, all points on the $g(t)$ curve are taken into account.

3. Discussion on the Traditional Definitions of Delay and Rise Times

Each of the definitions enumerated in the two previous subsections suffers from one or more of the following shortcomings:

(*a*) The delay or rise time is based on the value and/or the derivative of the step or impulse response at one or two points. Thus two responses that have very similar behavior over a certain time range but with very different behavior elsewhere cannot be differentiated.

(*b*) The definition is suitable for only a limited class of responses. For example, some definitions are only meaningful for monotonic responses. It becomes either meaningless or does not give an accurate description of the response when applied to other responses.

(*c*) For very different responses the definition gives the same rise or delay time, or else it gives very different rise or delay times for similar responses.

(*d*) The definition is inherently meant to give only an approximate indication of the delay and rise time and, therefore, is not always quantitatively sound.

(*e*) The definition gives a rise or delay time that depends on the wave shape of the response itself, and is therefore inconsistent.

From these observations it is clear that none of the existing definitions of delay and rise times is satisfactory in making quantitative comparisons of different classes of constant-delay networks. Some radically different approaches must be devised so that most of the difficulties mentioned above will be alleviated.

4. New Method of Defining Delay and Rise Times

The historical and fundamental reasons for the choice of existing definitions of delay and rise times are mostly practical or rational. The difficulties do not stem from the reasons for these choices. They arose because of the necessity of comparing different types of responses. If we wish to devise a scheme to compute the delay and rise times of many different wave shapes, we must not adhere to a particular wave shape and must go back to the philosophical reasonings behind some of these definitions [SU1, SU3].

In an ideal delay network with a delay time T_D, its step response should take the form shown in Fig. 7.25(a). This response has, of course, a zero rise time. If the network has a finite rise time, T_R, then its response will have the form of $g_s(t)$ of Fig. 7.35. Verbally, the wave shape of Fig. 7.35 would be the mental image one might get when T_D and T_R of a network are given. Technically, one may say that a step response that has a given set of values of T_D and T_R should have a step response that coincides very nearly with that of Fig. 7.35.

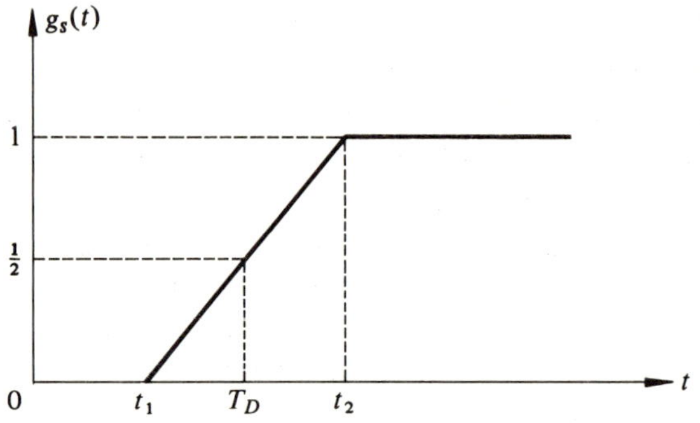

Fig. 7.35 Standard step response with only a delay and a rise time.

Thus it seems appropriate to consider that $g_s(t)$ of Fig. 7.35 as the *standard* step response of a delay network. Of course, no step response of a finite lumped network can have exactly this shape. Nevertheless, such a standard response is an ideal one in the sense that we are striving to approach it. Our achievement will fall short of attaining such a goal. But this failure is due to nature's restrictions rather than our actions.

If for an actual network a standard step response is found that best represents the network's step response, we can consider the standard response as the one the network is trying to approach. Conversely, we may consider the actual response of the network as an approximation of the standard response. Then the delay and the rise times of the standard response can be called the *nominal delay time* and the *nominal rise time* of the step response of the network.

The matter of what constitutes a "best representation" is again a matter of judicious choice. It appears that a fit in the least-mean-square-error sense is a natural one, as it takes into account the entire time axis. With this choice, for any given $g(t)$ a set of values for t_1 and t_2 is found such that the integral of the squared error between the given $g(t)$ and the standard $g_s(t)$ of Fig. 7.35,

$$E_2 = \int_0^\infty [g(t) - g_s(t)]^2 \, dt \tag{7.172}$$

is a minimum. Then the nominal delay time, T_D, and the nominal rise time, T_R, will be given by

$$T_D = \tfrac{1}{2}(t_1 + t_2) \tag{7.173}$$

and

$$T_R = t_2 - t_1 \tag{7.174}$$

And the value of the minimum E_2 is a measure of how far we have fallen short of approaching the standard response of Fig. 7.35 with these values of t_1 and t_2.

It can be expected that two step responses may differ considerably even if their T_D and T_R have the same values. However, in that case, each is trying its best to approximate the standard response with those T_D and T_R. If the error integral E_2 of both responses is the same, these two responses may be considered to be of comparable quality in their performance. On the other hand, if one network has a larger E_2 than the other, the former should be classified as being inferior to the latter as a delay network. Hence, in addition to T_R and T_D, the error integral E_2 is another important quantity for measuring the performance of the networks, as it is a quantity that gives an indication of how closely the step responses assume the ideal form of Fig. 7.35.

Since all three quantities, T_D, T_R and E_2, are proportional to the scale of t, it is sometimes more meaningful to compare the error integrals with the delay time T_D normalized to the same value. Or, we may compare their ratios of the error integral to delay time, E_2/T_D.

5. Numerical Examples

Three fifth-order low-pass functions are selected to illustrate the new definitions—the nominal rise time, T_R; the nominal delay time, T_D; and the error integral-to-delay time ratio, E_2/T_D. The data for these functions are tabulated in Table 7.17. Network function 1 is the maximally flat magnitude function (Table 7.5). Network function 2 is the maximally flat delay function (Table 7.7). Network function 3 is the function that has equiripple step response (Table 7.16).

From each function the step response has the form

$$g(t) = 1 + \sum_{i=1}^{N} C_i \epsilon^{\alpha_i t} \sin(\beta_i t + \phi_i) = 1 + f(t) \tag{7.175}$$

where each i corresponds to either a real pole ($\beta_i = 0$ and $\phi_i = \pi/2$) or a pair of complex poles. With t_1 and t_2 defined as shown in Fig. 7.35, it can be shown that

$$E_2(t_1, t_2) = \int_0^\infty f^2(t)\,dt + 2\int_0^{t_1} f(t)\,dt + \frac{2t_2}{t_2 - t_1}\int_{t_1}^{t_2} f(t)\,dt$$
$$- \frac{2}{t_2 - t_1}\int_{t_1}^{t_2} t f(t)\,dt + \frac{t_2}{3} - \frac{2t_1}{3} \tag{7.176}$$

Each integral in (7.176) can be expressed in closed form in terms of t_1 and t_2. A minimization routine was then used to determine the t_1 and t_2 which correspond to the minimum E_2. Values of t_1, t_2, and E_2 and as other pertinent parameters are summarized in Table 7.17. The actual step response of these network functions and their representative standard responses are shown in Figs. 7.36, 7.37, and 7.38.

A study of the data in Table 7.17 will lead to the conclusion that the nom-

TABLE 7.17

Data on Step Responses of Three Constant-Delay Network Functions

	Network	1	2	3
	Description	Maximally fiat magnitude	Maximally flat delay	Equiripple step response, $q_t = q_f = 0.01$
	Number of poles	5	5	5
	Locations of poles	-1.000000 -0.309017 $\pm j0.951057$ -0.809017 $\pm j0.587785$	-0.9264 -0.8516 $\pm j0.4427$ -0.5906 $\pm j0.9072$	-0.62750 -0.53927 $\pm j0.67878$ -0.29536 $\pm j1.42591$
	3-dB bandwidth	1.0000	0.6167	0.8310
Parameters based on 10–50–90% points	10% time	2.1030	2.2123	2.0070
	90% time	4.6651	5.7834	4.5935
	50% delay time	3.4960	3.8916	3.3145
	10% to 90% rise time	2.5621	3.5711	2.5864
	Delay-to-rise ratio	1.3645	1.0897	1.2815
Parameters based on the best fit of the function of Fig. 7.35	t_1	1.8905	1.7597	1.7312
	t_2	5.0144	6.1046	4.8885
	Nominal delay time (T_D)	3.4524	3.9321	3.3098
	Nominal rise time (T_R)	3.1239	4.3449	3.1573
	$0.8T_R$	2.4991	3.4759	2.5258
	T_D/T_R	1.1052	0.9050	1.0483
	$T_D/0.8T_R$	1.3815	1.1313	1.3104
	Error integral (E_2)	3.1143×10^{-2}	3.9067×10^{-3}	2.7502×10^{-3}
	E_2/T_D	9.0207×10^{-3}	9.9352×10^{-4}	8.3093×10^{-4}

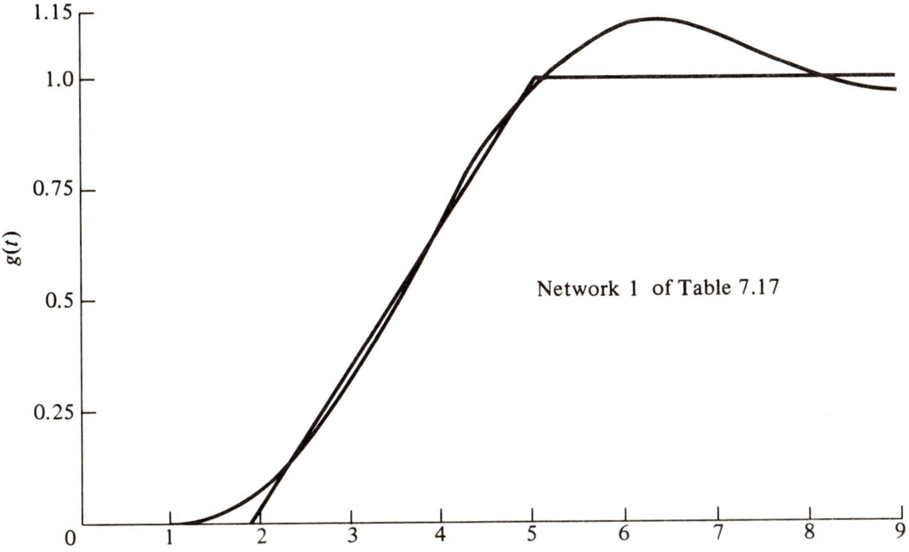

Fig. 7.36 $g(t)$ and $g_s(t)$ of the fifth-order maximally flat magnitude function.

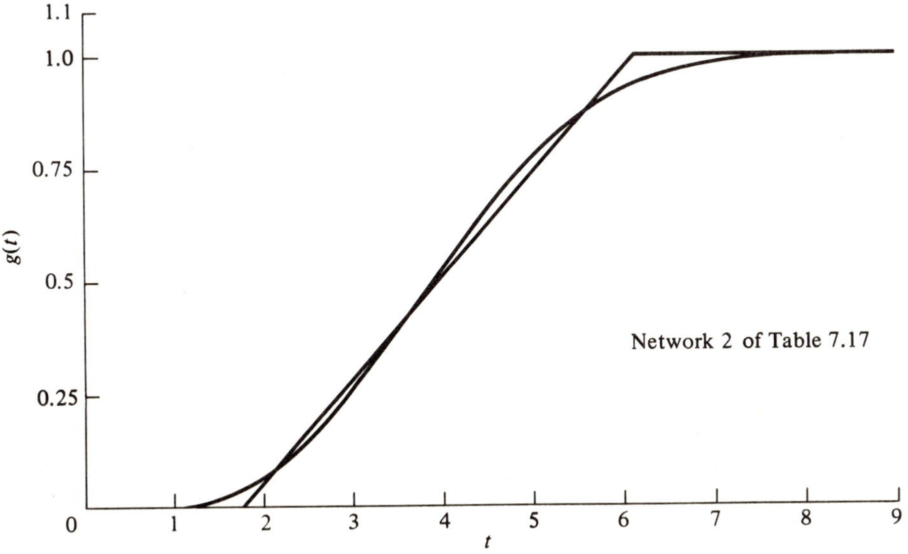

Fig. 7.37 $g(t)$ and $g_s(t)$ of the fifth-order maximally flat delay function.

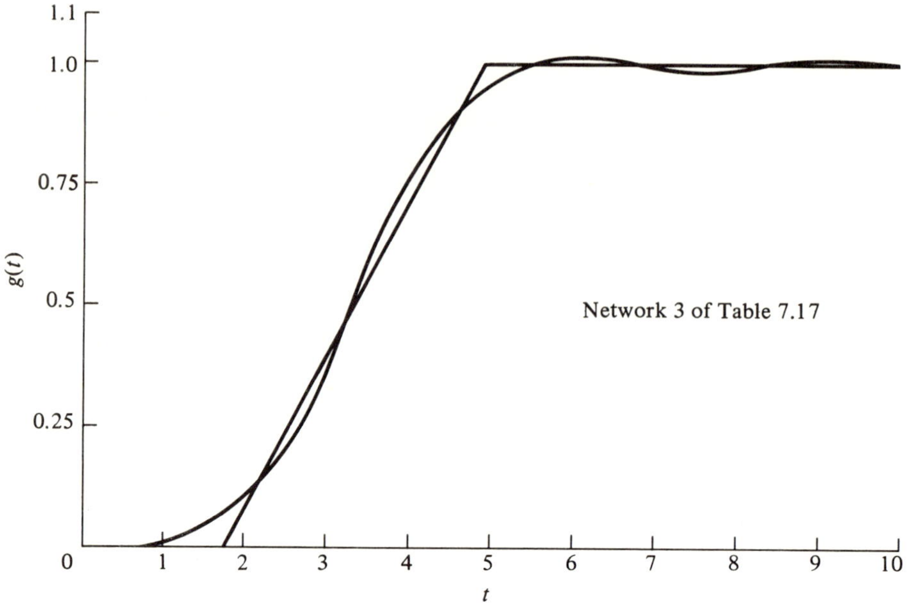

Fig. 7.38 $g(t)$ and $g_s(t)$ of the fifth-order network function with equiripple step response, $q_t = q_f = 0.01$.

inal rise time and the nominal delay time check reasonably closely with the traditional 10 to 90 percent rise time (except for a factor of 0.8, because the nominal rise time is meant to approximate the 0 to 100 percent rise time, whereas the latter is a 10 to 90 percent rise time) and the 50 percent delay time, respectively. In addition, the step response that appears to fit the standard response best has the smallest E_2/T_D ratio.

7.12 PARAMETERS OF THE STEP RESPONSES OF LUMPED CONSTANT-DELAY NETWORKS

Several important parameters of the step responses for all constant-delay network functions whose poles and zeros are tabulated in this chapter are listed in Tables 7.18 through 7.24. These parameters include the 10-50-90 percent times from which the conventional 10 to 90 percent rise time and the 50 percent delay time are obtained, the nominal delay and rise times defined in (7.173) and (7.174), the error integral defined in (7.172) and normalized with respect to the nominal delay time, the maximum percent overshoot, and the normalized frequency at which the low-pass magnitude characteristic drops to 3-dB below its dc value. These parameters give easy comparison of various types of network functions designed on different bases.

TABLE 7.18(A)

Parameters of the Responses of Maximally Flat Compromise Delay Networks
$(n = 2)$

m	10% time	50% time	90% time	Nominal delay time (T_D)	Nominal rise time (T_R)	T_R/T_D	Normalized error integral (E_2/T_D) $\times 10^{-3}$	Percent overshoot	3-dB frequency
−0.2	0.5022	1.4058	2.5376	1.4533	2.4925	1.7151	5.2523	5.91	1.0536
−0.1	0.5038	1.4196	2.5950	1.4731	2.5537	1.7335	4.3507	5.07	1.0265
0	0.5054	1.4333	2.6535	1.4929	2.6151	1.7517	3.7341	4.32	1.0000
0.1	0.5070	1.4468	2.7128	1.5127	2.6766	1.7694	3.3461	3.65	0.9742
0.2	0.5086	1.4603	2.7729	1.5324	2.7380	1.7867	3.1401	3.05	0.9492
0.3	0.5101	1.4735	2.8335	1.5520	2.7991	1.8035	3.0775	2.53	0.9251
0.4	0.5115	1.4865	2.8944	1.5715	2.8598	1.8198	3.1264	2.07	0.9019
0.5	0.5130	1.4993	2.9554	1.5906	2.9198	1.8357	3.2607	1.67	0.8798
0.6	0.5143	1.5118	3.0163	1.6095	2.9791	1.8509	3.4588	1.32	0.8589
0.7	0.5157	1.5241	3.0768	1.6281	3.0373	1.8656	3.7032	1.03	0.8390
0.8	0.5170	1.5360	3.1366	1.6462	3.0944	1.8796	3.9797	0.79	0.8203
0.9	0.5182	1.5476	3.1956	1.6640	3.1501	1.8931	4.2768	0.59	0.8026
1.0	0.5194	1.5589	3.2534	1.6812	3.2042	1.9059	4.5852	0.43	0.7862
1.1	0.5206	1.5697	3.3097	1.6979	3.2568	1.9181	4.8977	0.31	0.7708
1.2	0.5217	1.5802	3.3645	1.7140	3.3074	1.9297	5.2083	0.21	0.7564

TABLE 7.18(B)

Parameters of the Responses of Maximally Flat Compromise Delay Networks
($n = 3$)

m	10% time	50% time	90% time	Nominal delay time (T_D)	Nominal rise time (T_R)	T_R/T_D	Normalized error integral (E_2/T_D) $\times 10^{-3}$	Percent overshoot	3-dB frequency
−0.2	0.9970	2.0892	3.1522	2.0845	2.6567	1.2745	10.3602	11.16	1.0645
−0.1	1.0014	2.1121	3.2229	2.1134	2.7384	1.2957	7.5808	9.57	1.0327
0	1.0057	2.1351	3.2959	2.1428	2.8219	1.3169	5.5420	8.15	1.0000
0.1	1.0100	2.1582	3.3710	2.1726	2.9068	1.3380	4.0738	6.87	0.9669
0.2	1.0143	2.1812	3.4482	2.2026	2.9929	1.3588	3.0443	5.74	0.9338
0.3	1.0186	2.2043	3.5272	2.2328	3.0800	1.3794	2.3510	4.75	0.9013
0.4	1.0228	2.2272	3.6079	2.2632	3.1678	1.3997	1.9134	3.87	0.8696
0.5	1.0270	2.2501	3.6899	2.2936	3.2559	1.4196	1.6690	3.11	0.8391
0.6	1.0311	2.2727	3.7731	2.3239	3.3440	1.4389	1.5688	2.46	0.8102
0.7	1.0351	2.2952	3.8571	2.3542	3.4318	1.4578	1.5748	1.91	0.7829
0.8	1.0391	2.3174	3.9415	2.3842	3.5190	1.4760	1.6575	1.44	0.7575
0.9	1.0430	2.3394	4.0259	2.4139	3.6054	1.4936	1.7942	1.06	0.7338
1.0	1.0468	2.3610	4.1100	2.4432	3.6905	1.5105	1.9672	0.75	0.7119
1.1	1.0506	2.3822	4.1934	2.4721	3.7741	1.5267	2.1635	0.51	0.6917
1.2	1.0543	2.4030	4.2757	2.5004	3.8559	1.5421	2.3730	0.33	0.6731

TABLE 7.18(C)

Parameters of the Responses of Maximally Flat Compromise Delay Networks
($n = 4$)

m	10% time	50% time	90% time	Nominal delay time (T_D)	Nominal rise time (T_R)	T_R/T_D	Normalized error integral (E_2/T_D) $\times 10^{-3}$	Percent overshoot	3-dB frequency
−0.2	1.5294	2.7591	3.8134	2.7245	2.7940	1.0255	15.0797	14.92	1.0657
−0.1	1.5369	2.7895	3.8935	2.7607	2.8875	1.0460	10.7323	12.76	1.0343
0	1.5444	2.8203	3.9768	2.7976	2.9841	1.0667	7.5892	10.83	1.0000
0.1	1.5519	2.8513	4.0632	2.8352	3.0834	1.0875	5.3393	9.11	0.9634
0.2	1.5594	2.8825	4.1526	2.8735	3.1852	1.1085	3.7538	7.57	0.9251
0.3	1.5669	2.9139	4.2448	2.9123	3.2891	1.1294	2.6626	6.22	0.8862
0.4	1.5744	2.9454	4.3398	2.9516	3.3949	1.1502	1.9378	5.04	0.8476
0.5	1.5818	2.9770	4.4373	2.9912	3.5021	1.1708	1.4832	4.01	0.8104
0.6	1.5892	3.0085	4.5370	3.0311	3.6103	1.1911	1.2261	3.12	0.7752
0.7	1.5966	3.0401	4.6385	3.0711	3.7191	1.2110	1.1117	2.38	0.7425
0.8	1.6038	3.0715	4.7414	3.1111	3.8280	1.2304	1.0986	1.75	0.7125
0.9	1.6111	3.1027	4.8453	3.1511	3.9366	1.2493	1.1559	1.24	0.6852
1.0	1.6182	3.1337	4.9496	3.1908	4.0444	1.2675	1.2607	0.84	0.6604
1.1	1.6253	3.1644	5.0538	3.2302	4.1511	1.2851	1.3959	0.52	0.6379
1.2	1.6323	3.1947	5.1574	3.2691	4.2563	1.3020	1.5492	0.29	0.6175

TABLE 7.18(D)

Parameters of the Responses of Maximally Flat Compromise Delay Networks
(n = 5)

m	10% time	50% time	90% time	Nominal delay time (T_D)	Nominal rise time (T_R)	T_R/T_D	Normalized error integral (E_2/T_D) $\times 10^{-3}$	Percent overshoot	3-dB frequency
−0.2	2.0813	3.4209	4.4835	3.3658	2.9143	0.8659	18.7242	17.73	1.0637
−0.1	2.0921	3.4582	4.5723	3.4085	3.0171	0.8851	13.0386	15.12	1.0343
0	2.1030	3.4960	4.6651	3.4524	3.1239	0.9048	9.0207	12.78	1.0000
0.1	2.1139	3.5344	4.7618	3.4973	3.2346	0.9249	6.1965	10.69	0.9610
0.2	2.1248	3.5732	4.8623	3.5432	3.3490	0.9452	4.2336	8.83	0.9180
0.3	2.1358	3.6123	4.9668	3.5899	3.4668	0.9657	2.8943	7.20	0.8725
0.4	2.1468	3.6518	5.0749	3.6375	3.5875	0.9863	2.0062	5.77	0.8266
0.5	2.1578	3.6916	5.1865	3.6857	3.7106	1.0068	1.4433	4.53	0.7826
0.6	2.1687	3.7315	5.3013	3.7345	3.8358	1.0271	1.1128	3.47	0.7418
0.7	2.1797	3.7715	5.4189	3.7836	3.9624	1.0472	0.9463	2.58	0.7049
0.8	2.1906	3.8116	5.5389	3.8331	4.0898	1.0670	0.8937	1.84	0.6720
0.9	2.2015	3.8516	5.6606	3.8826	4.2175	1.0862	0.9183	1.24	0.6427
1.0	2.2123	3.8916	5.7834	3.9321	4.3449	1.1050	0.9935	0.77	0.6167
1.1	2.2231	3.9313	5.9068	3.9815	4.4714	1.1230	1.1004	0.43	0.5935
1.2	2.2338	3.9708	6.0299	4.0305	4.5965	1.1404	1.2255	0.18	0.5727

TABLE 7.18(E)

Parameters of the Responses of Maximally Flat Compromise Delay Networks
($n = 6$)

m	10% time	50% time	90% time	Nominal delay time (T_D)	Nominal rise time (T_R)	T_R/T_D	Normalized error integral (E_2/T_D) $\times 10^{-3}$	Percent overshoot	3-dB frequency
−0.2	2.6451	4.0772	5.1541	4.0064	3.0219	0.7543	21.5927	19.94	1.0606
−0.1	2.6594	4.1211	5.2515	4.0556	3.1328	0.7725	14.7136	16.94	1.0336
0	2.6737	4.1658	5.3536	4.1062	3.2488	0.7912	9.9723	14.25	1.0000
0.1	2.6881	4.2112	5.4604	4.1582	3.3699	0.8104	6.7094	11.85	0.9590
0.2	2.7027	4.2573	5.5720	4.2114	3.4957	0.8300	4.4827	9.73	0.9109
0.3	2.7173	4.3040	5.6883	4.2659	3.6260	0.8500	2.9875	7.86	0.8578
0.4	2.7320	4.3512	5.8093	4.3215	3.7603	0.8701	2.0095	6.23	0.8039
0.5	2.7467	4.3989	5.9348	4.3781	3.8982	0.8904	1.3959	4.82	0.7533
0.6	2.7614	4.4469	6.0645	4.4356	4.0391	0.9106	1.0370	3.62	0.7082
0.7	2.7762	4.4953	6.1979	4.4938	4.1822	0.9307	0.8538	2.61	0.6688
0.8	2.7910	4.5439	6.3345	4.5524	4.3268	0.9504	0.7894	1.79	0.6346
0.9	2.8058	4.5926	6.4737	4.6114	4.4722	0.9698	0.8033	1.14	0.6048
1.0	2.8206	4.6413	6.6146	4.6706	4.6176	0.9887	0.8669	0.64	0.5787
1.1	2.8353	4.6899	6.7565	4.7296	4.7623	1.0069	0.9602	0.29	0.5557
1.2	2.8500	4.7385	6.8983	4.7885	4.9056	1.0245	1.0695	0.06	0.5352

TABLE 7.18(F)

Parameters of the Responses of Maximally Flat compromise Delay Networks
($n = 7$)

m	10% time	50% time	90% time	Nominal delay time (T_D)	Nominal rise time (T_R)	T_R/T_D	Normalized error integral (E_2/T_D) $\times 10^{-3}$	Percent overshoot	3-dB frequency
−0.2	3.2170	4.7295	11.3851	4.6461	3.1195	0.6714	23.9447	21.76	1.0573
−0.1	3.2348	4.7799	5.9290	4.7016	3.2381	0.6887	15.9610	18.41	1.0327
0	3.2527	4.8313	6.0405	4.7589	3.3628	0.7066	10.5993	15.41	1.0000
0.1	3.2708	4.8837	6.1574	4.8178	3.4935	0.7251	6.9903	12.74	0.9571
0.2	3.2891	4.9370	6.2799	4.8784	3.6302	0.7441	4.5757	10.38	0.9030
0.3	3.3074	4.9911	6.4082	4.9406	3.7725	0.7636	2.9839	8.31	0.8409
0.4	3.3260	5.0460	6.5421	5.0042	3.9199	0.7833	1.9610	6.50	0.7788
0.5	3.3446	5.1015	6.6814	5.0692	4.0720	0.8033	1.3304	4.96	0.7230
0.6	3.3633	5.1577	6.8260	5.1354	4.2279	0.8233	0.9680	3.64	0.6754
0.7	3.3821	5.2143	6.9753	5.2025	4.3869	0.8432	0.7863	2.55	0.6351
0.8	3.4009	5.2713	7.1286	5.2703	4.5481	0.8630	0.7236	1.68	0.6009
0.9	3.4198	5.3286	7.2852	5.3388	4.7104	0.8823	0.7368	0.99	0.5715
1.0	3.4387	5.3861	7.4440	5.4075	4.8730	0.9012	0.7964	0.49	0.5459
1.1	3.4577	5.4437	7.6040	5.4762	5.0349	0.9194	0.8823	0.15	0.5235
1.2	3.4766	5.5012	7.7642	5.5448	5.1953	0.9370	0.9811	0.02	0.5037

TABLE 7.18(G)

Parameters of the Responses of Maximally Flat Compromise Delay Networks
($n = 8$)

m	10% time	50% time	90% time	Nominal delay time (T_D)	Nominal rise time (T_R)	T_R/T_D	Normalized error integral (E_2/T_D) $\times 10^{-3}$	Percent overshoot	3-dB frequency
−0.2	3.7948	5.3788	12.3798	5.2847	3.2090	0.6072	25.9536	23.29	1.0539
−0.1	3.8162	5.4356	6.6042	5.3465	3.3349	0.6238	16.9178	19.62	1.0317
0	3.8378	5.4937	6.7250	5.4104	3.4680	0.6410	11.0084	16.34	1.0000
0.1	3.8596	5.5530	6.8522	5.4764	3.6082	0.6589	7.1195	13.43	0.9551
0.2	3.8817	5.6135	6.9858	5.5444	3.7555	0.6773	4.5700	10.86	0.8939
0.3	3.9040	5.6750	7.1261	5.6142	3.9095	0.6963	2.9215	8.61	0.8216
0.4	3.9265	5.7376	7.2730	5.6860	4.0697	0.7158	1.8825	6.66	0.7522
0.5	3.9491	5.8010	7.4263	5.7593	4.2356	0.7354	1.2550	5.00	0.6933
0.6	3.9720	5.8653	7.5859	5.8342	4.4064	0.7553	0.9028	3.59	0.6449
0.7	3.9949	5.9302	7.7511	5.9103	4.5809	0.7751	0.7317	2.44	0.6048
0.8	4.0180	5.9957	7.9212	5.9874	4.7581	0.7947	0.6767	1.53	0.5712
0.9	4.0412	6.0617	8.0952	6.0653	4.9369	0.8140	0.6933	0.83	0.5424
1.0	4.0644	6.1280	8.2718	6.1436	5.1160	0.8327	0.7518	0.34	0.5176
1.1	4.0877	6.1944	8.4498	6.2220	5.2943	0.8509	0.8326	0.04	0.4960
1.2	4.1110	6.2610	8.6278	6.3004	5.4709	0.8683	0.9234	0.02	0.4768

TABLE 7.18(H)

Parameters of the Responses of Maximally Flat Compromise Delay Networks
($n = 9$)

m	10% time	50% time	90% time	Nominal delay time (T_D)	Nominal rise time (T_R)	T_R/T_D	Normalized error integral (E_2/T_D) $\times 10^{-3}$	Percent overshoot	3-dB frequency
−0.2	4.3770	6.0256	13.2699	5.9222	3.2917	0.5558	27.7332	24.63	1.0507
−0.1	4.4021	6.0889	12.8690	5.9904	3.4249	0.5717	17.6732	20.66	1.0306
0	4.4275	6.1537	7.4073	6.0610	3.5662	0.5884	11.2692	17.12	1.0000
0.1	4.4532	6.2199	7.5448	6.1341	3.7157	0.6057	7.1497	13.98	0.9530
0.2	4.4792	6.2876	7.6896	6.2094	3.8733	0.6238	4.5037	11.22	0.8832
0.3	4.5055	6.3566	7.8420	6.2871	4.0388	0.6424	2.8259	8.82	0.8001
0.4	4.5321	6.4269	8.0020	6.3669	4.2118	0.6615	1.7894	6.74	0.7255
0.5	4.5590	6.4983	8.1695	6.4488	4.3913	0.6810	1.1771	4.97	0.6655
0.6	4.5860	6.5707	8.3443	6.5324	4.5766	0.7006	0.8424	3.50	0.6173
0.7	4.6133	6.6440	8.5255	6.6177	4.7665	0.7203	0.6861	2.30	0.5779
0.8	4.6407	6.7180	8.7125	6.7041	4.9594	0.7398	0.6410	1.37	0.5450
0.9	4.6683	6.7927	8.9038	6.7914	5.1542	0.7589	0.6622	0.68	0.5171
1.0	4.6961	6.8679	9.0980	6.8794	5.3492	0.7776	0.7204	0.22	0.4931
1.1	4.7239	6.9433	9.2937	6.9675	5.5433	0.7956	0.7971	0.03	0.4722
1.2	4.7519	7.0190	9.4891	7.0556	5.7352	0.8129	0.8807	0.01	0.4537

TABLE 7.18(I)

Parameters of the Responses of Maximally Flat Compromise Delay Networks
($n = 10$)

m	10% time	50% time	90% time	Nominal delay time (T_D)	Nominal rise time (T_R)	T_R/T_D	Normalized error integral (E_2/T_D) $\times 10^{-3}$	Percent overshoot	3-dB frequency
−0.2	4.9640	6.6728	14.1180	6.5612	3.3718	0.5139	29.2544	25.73	1.0469
−0.1	4.9922	6.7413	13.8027	6.6346	3.5106	0.5291	18.2468	21.52	1.0291
0	5.0209	6.8117	8.0874	6.7108	3.6584	0.5451	11.4271	17.77	1.0000
0.1	5.0500	6.8838	8.2333	6.7897	3.8153	0.5619	7.1359	14.46	0.9509
0.2	5.0795	6.9575	8.3874	6.8713	3.9813	0.5794	4.4315	11.55	0.8711
0.3	5.1094	7.0329	8.5498	6.9556	4.1562	0.5975	2.7459	9.02	0.7786
0.4	5.1396	7.1098	8.7208	7.0424	4.3394	0.6162	1.7212	6.85	0.7019
0.5	5.1702	7.1881	8.9001	7.1316	4.5301	0.6352	1.1253	5.01	0.6424
0.6	5.2012	7.2677	9.0874	7.2228	4.7272	0.6545	0.8045	3.49	0.5952
0.7	5.2324	7.3484	9.2819	7.3159	4.9293	0.6738	0.6567	2.28	0.5569
0.8	5.2640	7.4301	9.4825	7.4105	5.1348	0.6929	0.6142	1.34	0.5251
0.9	5.2957	7.5125	9.6878	7.5061	5.3421	0.7117	0.6329	0.67	0.4981
1.0	5.3277	7.5957	9.8961	7.6024	5.5495	0.7300	0.6846	0.23	0.4748
1.1	5.3599	7.6792	10.1056	7.6991	5.7556	0.7476	0.7518	0.02	0.4546
1.2	5.3923	7.7631	10.3144	7.7957	5.9591	0.7644	0.8245	0.02	0.4368

TABLE 7.19(A)

Parameters of Responses of Equiripple Delay Networks
($\delta = 0.01$)

n	10% time	50% time	90% time	Nominal delay time (T_D)	Nominal rise time (T_R)	T_R/T_D	Normalized error integral (E_2/T_D) $\times 10^{-3}$	Percent overshoot	3-dB frequency
2	0.5164	1.5302	3.1073	1.6374	3.0664	1.8728	3.8403	0.91	0.8293
3	1.0283	2.2700	3.8301	2.3311	3.3976	1.4575	1.5508	1.01	0.7901
4	1.5666	2.9529	4.5147	2.9868	3.5933	1.2030	1.0200	1.03	0.7524
5	2.1117	3.6073	5.2042	3.6270	3.7685	1.0390	0.8632	0.65	0.7110
6	2.6572	4.2423	5.8820	4.2531	3.9261	0.9231	0.7922	0.43	0.6767
7	3.2020	4.8673	6.5644	4.8736	4.0881	0.8388	0.7630	0.13	0.6451
8	3.7455	5.4834	7.2369	5.4866	4.2406	0.7729	0.7343	0.08†	0.6192
9	4.2883	6.0951	7.9116	6.0971	4.3962	0.7210	0.7165	0.01	0.5949
10	4.8302	6.7017	8.5769	6.7026	4.5426	0.6777	0.6922	0.04	0.5746

†Second overshoot. (First overshoot is 0.05%.)

TABLE 7.19(B)

Parameters of Responses of Equiripple Delay Networks
($\delta = 0.02$)

n	10% time	50% time	90% time	Nominal delay time (T_D)	Nominal rise time (T_R)	T_R/T_D	Normalized error integral (E_2/T_D) $\times 10^{-3}$	Percent overshoot	3-dB frequency
2	0.5153	1.5203	3.0582	1.6224	3.0195	1.8611	3.6240	1.12	0.8450
3	1.0242	2.2509	3.7772	2.3085	3.3413	1.4474	1.4915	0.98	0.8076
4	1.5580	2.9243	4.4495	2.9551	3.5259	1.1932	0.9942	1.09	0.7665
5	2.0986	3.5726	5.1395	3.5906	3.7055	1.0320	0.8584	0.58	0.7227
6	2.6395	4.2016	5.8110	4.2110	3.8611	0.9169	0.7878	0.43	0.6875
7	3.1804	4.8229	6.4944	4.8287	4.0286	0.8343	0.7656	0.07	0.6541
8	3.7202	5.4350	7.1623	5.4376	4.1803	0.7688	0.7330	0.10†	0.6279
9	4.2599	6.0443	7.8380	6.0462	4.3406	0.7179	0.7188	0.00	0.6021
10	4.7988	6.6479	8.4993	6.6485	4.4862	0.6748	0.6908	0.06	0.5818

†Second overshoot. (First overshoot is 0.06%.)

TABLE 7.19(C)

Parameters of Responses of Equiripple Delay Networks
($\delta = 0.03$)

n	10% time	50% time	90% time	Nominal delay time (T_D)	Nominal rise time (T_R)	T_R/T_D	Normalized error integral (E_2/T_D) $\times 10^{-3}$	Percent overshoot	3-dB frequency
2	0.5145	1.5130	3.0221	1.6113	2.9846	1.8523	3.4802	1.29	0.8569
3	1.0214	2.2382	3.7431	2.2936	3.3048	1.4409	1.4591	0.93	0.8194
4	1.5524	2.9055	4.4070	2.9343	3.4820	1.1867	0.9796	1.14	0.7756
5	2.0902	3.5507	5.0995	3.5678	3.6668	1.0278	0.8581	0.51	0.7302
6	2.6283	4.1760	5.7664	4.1845	3.8203	0.9130	0.7851	0.44	0.6943
7	3.1670	4.7955	6.4519	4.8010	3.9928	0.8316	0.7687	0.03	0.6597
8	3.7045	5.4051	7.1160	5.4074	4.1430	0.7662	0.7319	0.13†	0.6335
9	4.2425	6.0133	7.7935	6.0152	4.3074	0.7161	0.7211	0.00	0.6065
10	4.7796	6.6151	8.4516	6.6155	4.4516	0.6729	0.6896	0.08	0.5863

†Second overshoot. (First overshoot is 0.07%.)

TABLE 7.19(D)

Parameters of Responses of Equiripple Delay Networks

($\delta = 0.04$)

n	10% time	50% time	90% time	Nominal delay time (T_D)	Nominal rise time (T_R)	T_R/T_D	Normalized error integral (E_2/T_D) $\times 10^{-3}$	Percent overshoot	3-dB frequency
2	0.5138	1.5069	2.9923	1.6021	2.9558	1.8449	3.3742	1.45	0.8670
3	1.0192	2.2284	3.7176	2.2822	3.2774	1.4361	1.4394	0.88	0.8286
4	1.5480	2.8912	4.3746	2.9185	3.4486	1.1816	0.9701	1.18	0.7824
5	2.0839	3.5343	5.0702	3.5509	3.6386	1.0247	0.8599	0.46	0.7359
6	2.6200	4.1570	5.7330	4.1648	3.7899	0.9100	0.7832	0.46	0.6995
7	3.1571	4.7754	6.4211	4.7808	3.9670	0.8298	0.7721	0.01	0.6638
8	3.6930	5.3832	7.0820	5.3851	4.1155	0.7642	0.7309	0.15†	0.6377
9	4.2297	5.9907	7.7615	5.9927	4.2836	0.7148	0.7236	0.00	0.6096
10	4.7656	6.5912	8.4167	6.5914	4.4262	0.6715	0.6885	0.10	0.5897

†Second overshoot. (First overshoot is 0.09%.)

TABLE 7.19(E)

Parameters of Responses of Equiripple Delay Networks
($\delta = 0.05$)

n	10% time	50% time	90% time	Nominal delay time (T_D)	Nominal rise time (T_R)	T_R/T_D	Normalized error integral (E_2/T_D) $\times 10^{-3}$	Percent overshoot	3-dB frequency
2	0.5132	1.5016	2.9666	1.5941	2.9307	1.8385	3.2927	1.60	0.8759
3	1.0174	2.2203	3.6971	2.2729	3.2553	1.4322	1.4275	0.82	0.8362
4	1.5444	2.8794	4.3480	2.9055	3.4212	1.1775	0.9640	1.23	0.7880
5	2.0788	3.5212	5.0471	3.5373	3.6164	1.0223	0.8631	0.40	0.7406
6	2.6133	4.1417	5.7062	4.1490	3.7655	0.9076	0.7819	0.48	0.7037
7	3.1492	4.7594	6.3969	4.7648	3.9468	0.8283	0.7758	0.01	0.6670
8	3.6838	5.3657	7.0547	5.3674	4.0936	0.7627	0.7301	0.17†	0.6410
9	4.2196	5.9729	7.7364	5.9749	4.2651	0.7138	0.7261	0.00	0.6120
10	4.7545	6.5722	8.3890	6.5722	4.4060	0.6704	0.6875	0.11	0.5925

†Second overshoot. (First overshoot is 0.11%.)

TABLE 7.19(F)

Parameters of Responses of Equiripple Delay Networks
($\delta = 0.06$)

n	10% time	50% time	90% time	Nominal delay time (T_D)	Nominal rise time (T_R)	T_R/T_D	Normalized error integral (E_2/T_D) $\times 10^{-3}$	Percent overshoot	3-dB frequency
2	0.5127	1.4968	2.9437	1.5870	2.9083	1.8326	3.2292	1.74	0.8840
3	1.0158	2.2133	3.6800	2.2651	3.2368	1.4290	1.4213	0.76	0.8428
4	1.5413	2.8694	4.3254	2.8944	3.3979	1.1739	0.9602	1.28	0.7926
5	2.0745	3.5102	5.0280	3.5260	3.5981	1.0204	0.8674	0.35	0.7445
6	2.6076	4.1288	5.6836	4.1357	3.7448	0.9055	0.7812	0.50	0.7072
7	3.1426	4.7461	6.3770	4.7515	3.9303	0.8272	0.7798	0.01	0.6697
8	3.6761	5.3511	7.0319	5.3526	4.0751	0.7613	0.7294	0.20†	0.6439
9	4.2112	5.9581	7.7159	5.9602	4.2500	0.7131	0.7288	0.00	0.6140
10	4.7452	6.5564	8.3658	6.5563	4.3891	0.6694	0.6867	0.13	0.5949

†Second overshoot. (First overshoot is 0.13%.)

TABLE 7.19(G)

Parameters of Responses of Equiripple Delay Networks
($\delta = 0.07$)

n	10% time	50% time	90% time	Nominal delay time (T_D)	Nominal rise time (T_R)	T_R/T_D	Normalized error integral (T_2/T_D) $\times 10^{-3}$	Percent overshoot	3-dB frequency
2	0.5122	1.4925	2.9229	1.5804	2.8879	1.8273	3.1799	1.87	0.8915
3	1.0145	2.2073	3.6654	2.2582	3.2209	1.4263	1.4195	0.70	0.8485
4	1.5386	2.8607	4.3054	2.8847	3.3773	1.1708	0.9583	1.33	0.7967
5	2.0708	3.5007	5.0118	3.5163	3.5826	1.0188	0.8726	0.30	0.7479
6	2.6027	4.1177	5.6639	4.1241	3.7269	0.9037	0.7810	0.52	0.7102
7	3.1369	4.7347	6.3601	4.7401	3.9164	0.8262	0.7840	0.01	0.6720
8	3.6695	5.3386	7.0121	5.3398	4.0592	0.7602	0.7290	0.22†	0.6464
9	4.2040	5.9455	7.6984	5.9477	4.2373	0.7124	0.7316	0.00	0.6157
10	4.7373	6.5429	8.3458	6.5426	4.3744	0.6686	0.6860	0.15	0.5970

†Second overshoot. (First overshoot is 0.15%.)

TABLE 7.19(H)

Parameters of Responses of Equiripple Delay Networks
($\delta = 0.08$)

n	10% time	50% time	90% time	Nominal delay time (T_D)	Nominal rise time (T_R)	T_R/T_D	Normalized error integral (E_2/T_D) $\times 10^{-3}$	Percent overshoot	3-dB frequency
2	0.5117	1.4885	2.9038	1.5744	2.8690	1.8223	3.1420	2.00	0.8985
3	1.0132	2.2019	3.6526	2.2521	3.2070	1.4240	1.4212	0.63	0.8536
4	1.5362	2.8528	4.2876	2.8760	3.3589	1.1679	0.9580	1.37	0.8003
5	2.0675	3.4923	4.9976	3.5078	3.5691	1.0175	0.8786	0.25	0.7510
6	2.5984	4.1079	5.6464	4.1139	3.7110	0.9021	0.7811	0.55	0.7130
7	3.1319	4.7247	6.3455	4.7301	3.9044	0.8254	0.7885	0.01	0.6740
8	3.6637	5.3275	6.9946	5.3285	4.0451	0.7591	0.7287	0.24†	0.6487
9	4.1977	5.9345	7.6832	5.9368	4.2263	0.7119	0.7346	0.00	0.6171
10	4.7303	6.5310	8.3281	6.5306	4.3616	0.6679	0.6855	0.17	0.5988

†Second overshoot. (First overshoot is 0.17%.)

TABLE 7.19(I)

Parameters of Responses of Equiripple Delay Networks

($\delta = 0.09$)

n	10% time	50% time	90% time	Nominal delay time (T_D)	Nominal rise time (T_R)	T_R/T_D	Normalized error integral (E_2/T_D) $\times 10^{-3}$	Percent overshoot	3-dB frequency
2	0.5113	1.4847	2.8860	1.5688	2.8514	1.8176	3.1140	2.13	0.9051
3	1.0121	2.1970	3.6412	2.2467	3.1946	1.4219	1.4260	0.57	0.8583
4	1.5340	2.8457	4.2713	2.8681	3.3422	1.1653	0.9592	1.42	0.8036
5	2.0645	3.4848	4.9851	3.5002	3.5572	1.0163	0.8852	0.20	0.7538
6	2.5945	4.0990	5.6306	4.1047	3.6966	0.9006	0.7817	0.57	0.7154
7	3.1274	4.7158	6.3325	4.7213	3.8938	0.8247	0.7932	0.01	0.6758
8	3.6585	5.3177	6.9789	5.3185	4.0324	0.7582	0.7286	0.27†	0.6507
9	4.1921	5.9247	7.6699	5.9271	4.2167	0.7114	0.7377	0.00	0.6184
10	4.7241	6.5205	8.3124	6.5199	4.3500	0.6672	0.6851	0.19	0.6005

†Second overshoot. (First overshoot is 0.20%.)

TABLE 7.19(J)

Parameters of Responses of Equiripple Delay Networks
($\delta = 0.1$)

n	10% time	50% time	90% time	Nominal delay time (T_D)	Nominal rise time (T_R)	T_R/T_D	Normalized error integral (E_2/T_D) $\times 10^{-3}$	Percent overshoot	3-dB frequency
2	0.5109	1.4812	2.8693	1.5635	2.8349	1.8132	3.0946	2.25	0.9113
3	1.0110	2.1925	3.6310	2.2417	3.1834	1.4201	1.4334	0.50	0.8625
4	1.5320	2.8392	4.2564	2.8609	3.3269	1.1629	0.9616	1.47	0.8066
5	2.0618	3.4780	4.9740	3.4933	3.5467	1.0153	0.8926	0.15	0.7563
6	2.5910	4.0910	5.6163	4.0963	3.6835	0.8992	0.7826	0.60	0.7176
7	3.1233	4.7078	6.3210	4.7133	3.8844	0.8241	0.7982	0.01	0.6775
8	3.6507	5.2996	6.9295	5.2968	3.9847	0.7523	0.7014	0.57	0.6648
9	4.1870	5.9158	7.6580	5.9183	4.2082	0.7110	0.7409	0.00	0.6195
10	4.7185	6.5109	8.2980	6.5102	4.3395	0.6666	0.6848	0.21	0.6021

TABLE 7.19(K)

Parameters of Responses of Equiripple Delay Networks
($\delta = 0.2$)

n	10% time	50% time	90% time	Nominal delay time (T_D)	Nominal rise time (T_R)	T_R/T_D	Normalized error integral (E_2/T_D) $\times 10^{-3}$	Percent overshoot	3-dB frequency
2	0.5077	1.4530	2.7404	1.5218	2.7048	1.7774	3.2313	3.37	0.9626
3	1.0033	2.1606	3.5653	2.2075	3.1105	1.4091	1.6158	0.04	0.8919
4	1.5174	2.7921	4.1476	2.8084	3.2148	1.1447	1.0408	1.93	0.8277
5	2.0428	3.4315	4.9018	3.4468	3.4791	1.0094	0.9962	0.03	0.7745
6	2.5662	4.0348	5.5136	4.0374	3.5896	0.8891	0.8114	0.88	0.7332
7	3.0955	4.6534	6.2463	4.6599	3.8250	0.8208	0.8607	0.01	0.6886
8	3.6213	5.2471	6.8641	5.2460	3.9387	0.7508	0.7398	0.52†	0.6662
9	4.1526	5.8565	7.5809	5.8601	4.1543	0.7089	0.7807	0.00	0.6266
10	4.6803	6.4455	8.1979	6.4435	4.2651	0.6619	0.6883	0.39	0.6137

†Second overshoot. (First overshoot is 0.44%.)

TABLE 7.19(L)

Parameters of Responses of Equiripple Delay Networks
($\delta = 0.3$)

n	10% time	50% time	90% time	Nominal delay time (T_D)	Nominal rise time (T_R)	T_R/T_D	Normalized error integral (E_2/T_D) $\times 10^{-3}$	Percent overshoot	3-dB frequency
2	0.5053	1.4319	2.6476	1.4909	2.6090	1.7499	3.7844	4.39	1.0026
3	0.9982	2.1403	3.5314	2.1869	3.0713	1.4044	1.9423	0.04	0.9090
4	1.5077	2.7608	4.0737	2.7732	3.1384	1.1317	1.2020	2.40	0.8415
5	2.0308	3.4029	4.8630	3.4192	3.4435	1.0071	1.1472	0.03	0.7866
6	2.5503	3.9989	5.4456	3.9993	3.5270	0.8819	0.8727	1.17	0.7430
7	3.0783	4.6207	6.2056	4.6283	3.7941	0.8198	0.9455	0.01	0.6954
8	3.6011	5.2084	6.7984	5.2058	3.8843	0.7462	0.7684	0.75†	0.6755
9	4.1315	5.8210	7.5385	5.8258	4.1264	0.7083	0.8339	0.00	0.6302
10	4.6567	6.4050	8.1333	6.4018	4.2162	0.6586	0.7032	0.57	0.6219

†Second overshoot. (First overshoot is 0.67%.)

TABLE 7.19(M)

Parameters of the Responses of Equiripple Delay Networks
($\delta = 0.4$)

n	10% time	50% time	90% time	Nominal delay time (T_D)	Nominal rise time (T_R)	T_R/T_D	Normalized error integral (E_2/T_D) $\times 10^{-3}$	Percent overshoot	3-dB frequency
2	0.5032	1.4144	2.5735	1.4657	2.5308	1.7267	4.6500	5.37	1.0365
3	0.9942	2.1252	3.5124	2.1725	3.0474	1.4027	2.3819	0.02	0.9194
4	1.5001	2.7366	4.0155	2.7457	3.0781	1.1211	1.4346	2.86	0.8522
5	2.0219	3.3823	4.8388	3.3997	3.4218	1.0065	1.3405	0.02	0.7960
6	2.5384	3.9717	5.3928	3.9703	3.4779	0.8760	0.9645	1.46	0.7504
7	3.0658	4.5971	6.1795	4.6061	3.7753	0.8196	1.0516	0.01	0.7003
8	3.5861	5.1796	6.7479	5.1756	3.8420	0.7423	0.8140	0.98†	0.6831
9	4.1163	5.7956	7.5108	5.8016	4.1093	0.7083	0.8999	0.00	0.6325
10	4.6394	6.3750	8.0839	6.3707	4.1781	0.6558	0.7291	0.75	0.6288

†Second overshoot. (First overshoot is 0.90%.)

TABLE 7.19(N)

Parameters of Responses of Equiripple Delay Networks
($\delta = 0.5$)

n	10% time	50% time	90% time	Nominal delay time (T_D)	Nominal rise time (T_R)	T_R/T_D	Normalized error integral (E_2/T_D) $\times 10^{-3}$	Percent overshoot	3-dB ferquency
2	0.5014	1.3993	2.5112	1.4441	2.4642	1.7064	5.7876	6.33	1.0663
3	0.9910	2.1133	3.5026	2.1618	3.0328	1.4029	2.9198	0.00	0.9248
4	1.4939	2.7165	3.9668	2.7229	3.0273	1.1118	1.7343	3.32	0.8610
5	2.0148	3.3661	4.8231	3.3850	3.4081	1.0068	1.5734	0.02	0.8037
6	2.5288	3.9496	5.3487	3.9465	3.4367	0.8708	1.0859	1.75	0.7563
7	3.0558	4.5788	6.1617	4.5892	3.7633	0.8200	1.1781	0.00	0.7043
8	3.5740	5.1564	6.7060	5.1511	3.8063	0.7389	0.8762	1.21†	0.6896
9	4.1043	5.7759	7.4915	5.7832	4.0984	0.7087	0.9785	0.00	0.6340
10	4.6256	6.3509	8.0431	6.3455	4.1461	0.6534	0.7657	0.93	0.6348

†Second overshoot (First overshoot is 1.13%.)

TABLE 7.20

Parameters of Responses of Approximate Gaussian Delay Networks

n	10% time	50% time	90% time	Nominal delay time (T_D)	Nominal rise time (T_R)	T_R/T_D	Normalized error integral (E_2/T_D) $\times 10^{-3}$	Percent overshoot	3-dB frequency
2	0.5247	1.6089	3.5167	1.7583	3.4472	1.9605	6.0888	0.05	0.7195
3	1.0669	2.4733	4.5445	2.5944	4.1221	1.5889	3.0569	0.06	0.6202
4	1.6590	3.3066	5.5140	3.4084	4.6260	1.3543	2.0539	0.03	0.5599
5	2.2771	4.1220	6.4572	4.2108	5.0262	1.1936	1.6116	0.00	0.5158
6	2.9111	4.9266	7.3853	5.0064	5.3897	1.0766	1.3750	0.00	0.4812
7	3.5561	5.7240	8.3031	5.7977	5.7230	0.9871	1.2285	0.00	0.4528
8	4.2092	6.5167	9.2128	6.5861	6.0344	0.9162	1.1273	0.00	0.4291
9	4.8688	7.3060	10.1158	7.3723	6.3287	0.8584	1.0512	0.00	0.4088
10	5.5337	8.0929	11.0129	8.1569	6.6089	0.8102	0.9905	0.00	0.3912

TABLE 7.21(A)

Parameters of the Responses of Delay Networks with Poles on a Parabolic Contour
($b = 0$)

n	10% time	50% time	90% time	Nominal delay time (T_D)	Nominal rise time (T_R)	T_R/T_D	Normalized error integral (E_2/T_D) $\times 10^{-3}$	Percent overshoot	3-dB frequency
2	0.5054	1.4333	2.6535	1.4929	2.6151	1.7517	3.7341	4.32	1.0000
3	1.0023	2.1368	3.3789	2.1592	2.9178	1.3513	1.7942	4.06	0.9771
4	1.5371	2.8346	4.1686	2.8427	3.2235	1.1339	1.0456	2.56	0.9022
5	2.0932	3.5329	4.9844	3.5360	3.5315	0.9987	0.8354	1.08	0.7718
6	2.6638	4.2327	5.8128	4.2354	3.8363	0.9058	0.8148	0.61	0.6824
7	3.2451	4.9343	6.6460	4.9386	4.1333	0.8369	0.8348	0.52	0.6256
8	3.8350	5.6373	7.4790	5.6442	4.4195	0.7830	0.8476	0.40	0.5828
9	4.4318	6.3415	8.3084	6.3511	4.6935	0.7390	0.8448	0.29	0.5480
10	5.0345	7.0466	9.1326	7.0588	4.9549	0.7020	0.8298	0.20	0.5189

TABLE 7.21(B)

Parameters of the Responses of Delay Networks with Poles on a Parabolic Contour
($b = 1.25$)

n	10% time	50% time	90% time	Nominal delay time (T_D)	Nominal rise time (T_R)	T_R/T_D	Normalized error integral (E_2/T_D) $\times 10^{-3}$	Percent overshoot	3-dB frequency
2	0.5054	1.4333	2.6535	1.4929	2.6151	1.7517	3.7341	4.32	1.0000
3	1.0022	2.1486	3.4838	2.1856	3.0287	1.3858	1.2827	1.24	0.9314
4	1.5387	2.8743	4.4689	2.9200	3.5349	1.2106	1.8647	0.34	0.7203
5	2.0998	3.6138	5.5460	3.6810	4.0876	1.1105	2.3033	0.34	0.6021
6	2.6789	4.3642	6.6040	4.4518	4.6170	1.0371	2.1650	0.12	0.5425
7	3.2722	5.1216	7.5786	5.2213	5.0822	0.9734	1.8540	0.02	0.4989
8	3.8771	5.8825	8.5000	5.9870	5.4876	0.9166	1.5946	0.02	0.4641
9	4.4916	6.6450	9.4009	6.7502	5.8523	0.8670	1.4186	0.00	0.4358
10	5.1143	7.4077	10.2953	7.5123	6.1912	0.8241	1.2999	0.00	0.4122

TABLE 7.21(C)

Parameters of the Responses of Delay Networks with Poles on a Parabolic Contour

$(b = 3)$

n	10% time	50% time	90% time	Nominal delay time (T_D)	Nominal rise time (T_R)	T_R/T_D	Normalized error integral (E_2/T_D) $\times 10^{-3}$	Percent overshoot	3-dB frequency
2	0.5054	1.4333	2.6535	1.4929	2.6151	1.7517	3.7341	4.32	1.0000
3	1.0026	2.1576	3.5582	2.2042	3.1028	1.4077	1.6329	0.04	0.8954
4	1.5417	2.9093	4.7667	2.9875	3.7950	1.2703	3.2843	0.22	0.6379
5	2.1090	3.6914	6.0769	3.8143	4.5615	1.1959	3.2783	0.05	0.5404
6	2.6983	4.4941	7.1291	4.6372	5.1887	1.1189	2.4607	0.00	0.4885
7	3.3056	5.3044	8.0846	5.4452	5.6757	1.0423	1.9266	0.01	0.4450
8	3.9274	6.1135	9.0411	6.2472	6.1013	0.9767	1.6780	0.00	0.4174
9	4.5613	6.9188	10.0077	7.0482	6.5019	0.9225	1.5416	0.00	0.3917
10	5.2050	7.7209	10.9714	7.8492	6.8839	0.8770	1.4356	0.00	0.3703

TABLE 7.22(A)

Parameters of the Responses of Delay Networks with Poles on an Elliptic Contour
($a = 0.4$)

n	10% time	50% time	90% time	Nominar delay time (T_D)	Nominal rise time (T_R)	T_R/T_D	Normalized error integral (E_2/T_D) $\times 10^{-3}$	Percent overshoot	3-dB frequency
2	0.4939	1.3379	2.2767	1.3588	2.2039	1.6219	16.3699	11.34	1.1880
3	0.9840	2.0689	3.2767	2.0932	2.8051	1.3401	1.7725	1.11	1.0672
4	1.5072	2.7494	4.0228	2.7563	3.0790	1.1171	0.9426	2.18	0.8972
5	2.0533	3.4400	4.8431	3.4439	3.4051	0.9888	0.7865	0.94	0.7889
6	2.6123	4.1244	5.6341	4.1246	3.6822	0.8927	0.7417	0.58	0.7238
7	3.1816	4.8091	6.4273	4.8080	3.9490	0.8213	0.7254	0.22	0.6689
8	3.7587	5.4931	7.2140	5.4911	4.1991	0.7647	0.7107	0.13†	0.6258
9	4.3422	6.1771	7.9971	6.1748	4.4373	0.7186	0.6967	0.09	0.5901
10	4.9309	6.8610	8.7760	6.8586	4.6644	0.6801	0.6816	0.06	0.5599

†Second overshoot. (First overshoot is 0.03%.)

TABLE 7.22(B)

Parameters of the Responses of Delay Networks with Poles on an Elliptic Contour
($a = 0.5$)

n	10% time	50% time	90% time	Nominal delay time (T_D)	Nominal rise time (T_R)	T_R/T_D	Normalized error integral (E_2/T_D) $\times 10^{-3}$	Percent overshoot	3-dB frequency
2	0.5010	1.3955	2.4959	1.4387	2.4476	1.7013	6.1392	6.58	1.0738
3	0.9997	2.1332	3.4157	2.1630	2.9569	1.3670	1.2872	2.20	0.9655
4	1.5368	2.8436	4.2240	2.8587	3.2873	1.1500	0.9230	1.94	0.8486
5	2.0965	3.5526	5.0465	3.5607	3.6021	1.0116	0.7965	1.26	0.7540
6	2.6706	4.2586	5.8608	4.2622	3.8892	0.9125	0.7453	0.80	0.6890
7	3.2553	4.9637	6.6721	4.9645	4.1597	0.8379	0.7194	0.45	0.6384
8	3.8483	5.6681	7.4796	5.6673	4.4158	0.7792	0.7023	0.21	0.5976
9	4.4479	6.3724	8.2837	6.3705	4.6600	0.7315	0.6879	0.10†	0.5637
10	5.0531	7.0765	9.0845	7.0742	4.8936	0.6918	0.6741	0.07	0.5351

†Second overshoot. (First overshoot is 0.05%.)

TABLE 7.22(C)

Parameters of the Responses of Delay Networks with Poles on an Elliptic Contour
($a = 0.6$)

n	10% time	50% time	90% time	Nominal delay time (T_D)	Nominal rise time (T_R)	T_R/T_D	Normalized error integral (E_2/T_D) $\times 10^{-3}$	Percent overshoot	3-dB frequency
2	0.5066	1.4433	2.6970	1.5075	2.6603	1.7647	3.4295	3.82	0.9810
3	1.0135	2.1945	3.5750	2.2341	3.1295	1.4008	1.3275	2.23	0.8858
4	1.5622	2.9279	4.4111	2.9514	3.4824	1.1799	0.9653	1.83	0.7960
5	2.1345	3.6562	5.2467	3.6703	3.8001	1.0354	0.8242	1.36	0.7183
6	2.7220	4.3817	6.0784	4.3897	4.0929	0.9324	0.7582	0.97	0.6583
7	3.3207	5.1058	6.9071	5.1097	4.3679	0.8548	0.7224	0.64	0.6110
8	3.9281	5.8291	7.7327	5.8304	4.6289	0.7939	0.6997	0.39	0.5725
9	4.5423	6.5520	8.5555	6.5516	4.8780	0.7445	0.6827	0.20	0.5405
10	5.1624	7.2748	9.3755	7.2733	5.1169	0.7035	0.6682	0.07†	0.5132

†Second overshoot. (First overshoot is 0.06%.)

TABLE 7.22(D)

Parameters of the Responses of Delay Networks with Poles on an Elliptic Contour
($a = 0.7$)

n	10% time	50% time	90% time	Nominal delay time (T_D)	Nominal rise time (T_R)	T_R/T_D	Normalized error integral (E_2/T_D) $\times 10^{-3}$	Percent overshoot	3-dB frequency
2	0.5110	1.4821	2.8735	1.5648	2.8390	1.8143	3.0988	2.22	0.9098
3	1.0254	2.2506	3.7369	2.3018	3.3028	1.4349	1.4393	1.90	0.8224
4	1.5843	3.0041	4.5901	3.0368	3.6697	1.2084	1.0255	1.67	0.7476
5	2.1678	3.7511	5.4406	3.7722	3.9944	1.0589	0.8609	1.36	0.6827
6	2.7675	4.4948	6.2881	4.5082	4.2929	0.9522	0.7793	1.05	0.6293
7	3.3789	5.2366	7.1327	5.2448	4.5729	0.8719	0.7329	0.77	0.5858
8	3.9993	5.9775	7.9748	5.9820	4.8384	0.8088	0.7033	0.53	0.5498
9	4.6269	6.7178	8.8145	6.7197	5.0920	0.7578	0.6822	0.34	0.5195
10	5.2605	7.4577	9.6517	7.4580	5.3355	0.7154	0.6655	0.19	0.4937

TABLE 7.22(E)

Parameters of the Responses of Delay Networks with Poles on an Elliptic Contour
($a = 0.8$)

n	10% time	50% time	90% time	Nominal delay time (T_D)	Nominal rise time (T_R)	T_R/T_D	Normalized error integral (E_2/T_D) $\times 10^{-3}$	Percent overshoot	3-dB frequency
2	0.5145	1.5133	3.0237	1.6118	2.9862	1.8527	3.4864	1.29	0.8564
3	1.0357	2.3009	3.8927	2.3641	3.4674	1.4667	1.5981	1.49	0.7717
4	1.6034	3.0730	4.7613	3.1155	3.8489	1.2354	1.1033	1.45	0.7054
5	2.1972	3.8378	5.6270	3.8669	4.1830	1.0817	0.9065	1.28	0.6492
6	2.8078	4.5986	6.4896	4.6185	4.4883	0.9718	0.8074	1.07	0.6019
7	3.4307	5.3572	7.3495	5.3706	4.7737	0.8889	0.7497	0.84	0.5622
8	4.0630	6.1144	8.2068	6.1231	5.0440	0.8238	0.7125	0.63	0.5288
9	4.7028	6.8709	9.0619	6.8762	5.3020	0.7711	0.6864	0.45	0.5004
10	5.3489	7.6269	9.9148	7.6297	5.5496	0.7274	0.6664	0.30	0.4760

TABLE 7.23(A)

Parameters of the Responses of Delay Networks with Poles on a Catenary
($b = 2$)

n	10% time	50% time	90% time	Nominal delay time (T_D)	Nominal rise time (T_R)	T_R/T_D	Normalized error integral (E_2/T_D) $\times 10^{-3}$	Percent overshoot	3-dB frequency
2	0.5054	1.4333	2.6535	1.4929	2.6151	1.7517	3.7341	4.32	1.0000
3	1.0045	2.1335	3.3077	2.1439	2.8365	1.3230	4.3752	7.24	0.9988
4	1.5408	2.8186	4.0108	2.8029	3.0318	1.0816	4.4264	8.35	0.9937
5	2.0970	3.4972	4.7250	3.4645	3.2117	0.9270	4.1305	8.78	0.9829
6	2.6658	4.1721	5.4417	4.1270	3.3800	0.8190	3.7297	8.89	0.9635
7	3.2435	4.8446	6.1582	4.7898	3.5385	0.7388	3.3330	8.85	0.9306
8	3.8278	5.5155	6.8734	5.4527	3.6888	0.6765	2.9764	8.74	0.8829
9	4.4174	6.1851	7.5870	6.1156	3.8317	0.6265	2.6674	8.60	0.8337
10	5.0114	6.8538	8.2988	6.7784	3.9682	0.5854	2.4037	8.44	0.7927

TABLE 7.23(B)

Parameters of the Responses of Delay Networks with Poles on a Catenary
($b = 2.5$)

n	10% time	50% time	90% time	Nominal delay time (T_D)	Nominal rise time (T_R)	T_R/T_D	Normalized error integral (E_2/T_D) $\times 10^{-3}$	Percent overshoot	3-dB frequency
2	0.5054	1.4333	2.6535	1.4929	2.6151	1.7517	3.7341	4.32	1.0000
3	1.0031	2.1337	3.3387	2.1498	2.8727	1.3363	2.7672	5.61	0.9911
4	1.5383	2.8228	4.0704	2.8169	3.1076	1.1032	2.1629	5.58	0.9675
5	2.0939	3.5078	4.8159	3.4886	3.3308	0.9548	1.6777	5.11	0.9175
6	2.6627	4.1907	5.5662	4.1627	3.5445	0.8515	1.3243	4.52	0.8370
7	3.2410	4.8727	6.3179	4.8385	3.7494	0.7749	1.0774	3.95	0.7632
8	3.8267	5.5541	7.0696	5.5153	3.9465	0.7155	0.9067	3.42	0.7093
9	4.4182	6.2352	7.8205	6.1929	4.1362	0.6679	0.7881	2.95	0.6674
10	5.0146	6.9162	8.5702	6.8711	4.3192	0.6286	0.7046	2.55	0.6328

TABLE 7.23(C)

Parameters of the Responses of Delay Networks with Poles on a Catenary
($b = 5$)

n	10% time	50% time	90% time	Nominal delay time (T_D)	Nominal rise time (T_R)	T_R/T_D	Normalized error integral (E_2/T_D) $\times 10^{-3}$	Percent overshoot	3-dB frequency
2	0.5054	1.4333	2.6535	1.4929	2.6151	1.7517	3.7341	4.32	1.0000
3	1.0021	2.1444	3.4491	2.1768	2.9929	1.3749	1.2663	2.05	0.9474
4	1.5376	2.8561	4.3288	2.8850	3.3967	1.1774	1.2135	0.39	0.7891
5	2.0956	3.5711	5.2411	3.6051	3.8030	1.0549	1.4401	0.46	0.6630
6	2.6693	4.2897	6.1649	4.3319	4.1993	0.9694	1.5372	0.37	0.5967
7	3.2548	5.0114	7.0851	5.0625	4.5785	0.9044	1.5202	0.21	0.5491
8	3.8499	5.7358	7.9914	5.7949	4.9366	0.8519	1.4453	0.08†	0.5111
9	4.4528	6.4622	8.8801	6.5279	5.2725	0.8077	1.3513	0.07	0.4800
10	5.0624	7.1902	9.7527	7.2610	5.5876	0.7695	1.2583	0.05	0.4541

†Second overshoot. (First overshoot is 0.04%.)

TABLE 7.24(A)

Parameters of the Responses of Delay Networks with Equiripple Step Response

$q_t = q_f = 0.01$

n	m	10% time	50% time	90% time	Nominal delay time (T_D)	Nominal rise time (T_R)	T_R/T_D	Normalized error integral (E_2/T_D) $\times 10^{-3}$	3-dB frequency
2	0	0.5159	1.5257	3.0848	1.6305	3.0449	1.8675	3.7383	0.8364
3	0	1.0087	2.1792	3.5751	2.2229	3.1259	1.4062	1.3178	0.8870
3	2	0.8900	2.1800	3.6179	2.2091	3.3235	1.5044	1.4042	0.8334
4	0	1.5086	2.7579	4.0699	2.7705	3.1300	1.1298	0.9226	0.8648
4	2	1.2402	2.7211	4.1415	2.7108	3.5330	1.3033	1.1362	0.7526
5	0	2.0070	3.3145	4.5935	3.3098	3.1573	0.9539	0.8309	0.8310
5	2	1.6386	3.2194	4.6780	3.1963	3.7066	1.1596	1.0019	0.7204
5	4	1.3304	3.0889	4.7295	3.0668	4.1535	1.3543	1.1745	0.6453
6	0	2.5073	3.8690	5.1442	3.8535	3.2133	0.8339	0.7979	0.8382
6	2	2.0902	3.7254	5.2341	3.7015	3.8374	1.0367	0.9334	0.7033
6	4	1.6079	3.5069	5.2629	3.4819	4.4497	1.2780	1.3093	0.6038
7	0	3.0131	4.4301	5.7192	4.4070	3.2924	0.7471	0.7832	0.8094
7	4	1.9982	4.0180	5.8614	3.9856	4.7028	1.1799	1.7029	0.5707
8	0	3.5273	5.0018	6.3164	4.9729	3.3888	0.6815	0.7819	0.7804
9	0	4.0518	5.5864	6.9342	5.5526	3.4984	0.6301	0.7986	0.7586

TABLE 7.24(B)

Parameters of the Responses of Delay Networks with Equiripple Step Response
($n = 5, m = 2$)

q_t	q_f	10% time	50% time	90% time	Nominal delay time (T_D)	Nominal rise time (T_R)	T_R/T_D	Normalized error integral (E_2/T_D) $\times 10^{-3}$	3-dB frequency
0.001	0.001	1.9695	3.5855	5.2921	3.6026	4.0365	1.1204	1.0508	0.6519
0.001	0.005	1.8407	3.5706	5.3399	3.5781	4.2503	1.1879	1.0852	0.6213
0.001	0.010	1.7548	3.5615	5.3777	3.5630	4.4028	1.2357	1.0793	0.6021
0.001	0.020	1.6474	3.5521	5.4369	3.5465	4.6141	1.3010	1.0298	0.5779
0.005	0.001	1.8908	3.3768	4.8347	3.3724	3.5848	1.0630	0.9191	0.7362
0.005	0.005	1.7585	3.3483	4.8713	3.3363	3.7895	1.1358	0.9642	0.7003
0.005	0.010	1.6751	3.3276	4.8995	3.3124	3.9291	1.1862	0.9557	0.6788
0.005	0.020	1.5763	3.2989	4.9428	3.2827	4.1132	1.2530	0.9003	0.6527
0.010	0.001	1.8543	3.2809	4.6248	3.2663	3.3751	1.0333	0.9604	0.7803
0.010	0.005	1.7209	3.2459	4.6556	3.2248	3.5746	1.1085	1.0113	0.7425
0.010	0.010	1.6386	3.2194	4.6780	3.1963	3.7066	1.1596	1.0019	0.7204
0.010	0.020	1.5429	3.1805	4.7099	3.1590	3.8735	1.2262	0.9462	0.6946
0.020	0.001	1.8184	3.1853	4.4120	3.1598	3.1584	0.9996	1.3517	0.8292
0.020	0.005	1.6842	3.1442	4.4368	3.1131	3.3520	1.0767	1.4365	0.7901
0.020	0.010	1.6033	3.1126	4.4529	3.0805	3.4755	1.1282	1.4498	0.7680
0.020	0.020	1.5102	3.0653	4.4725	3.0362	3.6244	1.1937	1.4343	0.7433

PROBLEMS

7.1. Using a polynomial of the fourth degree, derive an all-pass network function that has a maximally flat delay characteristic.

7.2. A transfer function has a maximally flat delay characteristic obtained by using a fourth-order Bessel polynomial. Change the constant numerator to a polynomial so that the new transfer function has a maximally flat magnitude.

7.3. Find a low-pass delay network function with poles on a catenary contour for $b = 3$, $n = 4$.

7.4. Prove the result of (7.164).

7.5. Derive explicit expressions for $E_2(t_1, t_2)$ as defined by (7.172) for a $g(t)$ with one real pole and one pair of complex-conjugate poles.

8

Application of Numerical Optimization Techniques

The availability of the modern high-speed electronic computer and its associated software offers engineers much more powerful tools for the solution of numerous engineering problems by techniques that are quite different from the traditional methods. Although the traditional solution of an engineering problem is usually theoretically sound and invariable leads to an answer determined by certain prescribed criteria, it is usually rather inflexible and becomes extremely tedious if the number of variables is large. In the area of network synthesis, many design problems lend themselves to the ready application of those computational methods that are only feasible if they are processed on high-speed computers.

In each of the approximation or design methods included in Chapters 3 through 7, the step-by-step procedure starts with a set of chosen parameters, follows a straightforward algorithm, and leads to a unique answer. During the course of carrying out the procedure, no decision need be made by the designer. If the final outcome does not meet the engineering specification of the situation, another application of the procedure may be tried. Of course, each procedure may be programmed into a computer routine. However, a routine programmed this way does no more than speed up the algorithm, which would otherwise be performed "by hand." The use of a computer in this fashion neither alters the mode of solution nor affects the outcome of the design.

The art of utilizing high-speed computational facilities to aid network design problems goes far beyond simple delegation of the computational chores to a programmer and a computer. The program can actually be designed so that certain design parameters may be modified as the design procedure is being carried out. Constraints, such as ratios or absolute values of element values, may be built into the procedure at the outset. The number of design criteria may be greatly increased over those to which classical methods are suited. Above all, these techniques can sometimes be made to render mathematically suboptimum but practically useful and acceptable designs.

On the other hand, computer-aided network designs have a number of relative disadvantages as compared to classical techniques. Such designs usually call for relatively large and complex optimization programs. Either the execution time of these programs is usually long or the amount of storage required is very large. Typically, whether a solution exists or not cannot be ascertained in advance. When a solution has been obtained, it cannot always be determined how close the solution is to the optimum.

Two features of a typical computer-aided network design procedure are worth noting. One is the fact that the success of such a procedure depends largely on the initial trial point given to the computer. The other is that sometimes it is difficult to distinguish a local optimum from an absolute optimum. In many instances the application of a simple classical procedure will provide the designer with a good starting point for the computer-aided design. When this is done, the computer-aided design becomes an extension of classical methods as a design tool. On the other hand, computer-aided design procedures do not always give desired insight into the behavior and limitations of various classes of network configurations. Thus it would seem fair to state that the two approaches tend to complement each other, both for scholastic and for practical purposes.

Mathematical optimization is a vast and rapidly developing field, as is the application of some of these techniques to network design (as well as other engineering designs). Adequate coverage of the optimization techniques and their application to network problems would entail a separate volume and is obviously beyond the intent of this book. What will be included in this chapter is a brief survey of some of the optimization methods that are especially useful for network designs, particularly in the time domain; and we shall indicate briefly by way of examples how some network design problems can be formulated to take advantage of modern computer-oriented optimization techniques.

8.1 NUMERICAL OPTIMIZATION TECHNIQUES

In this section a brief description is given for each of the more useful methods of mathematical optimization. In forming an engineering design problem to make use of an optimization technique, it is generally necessary to formulate an *objective function* $f(x_1, x_2, \ldots, x_n)$, which is, in turn, minimized with respect to a *design vector* $\bar{x} = (x_1, x_2, \ldots, x_n)$, where the x's are real variables. The objective function must somehow embody a set of design criteria and usually represents a measure of the deviation between the desired and the actual performance of the network. Typically, $f(\bar{x})$ may be the integrated squared error of the impulse responses (actual and desired) over the entire time axis,

$$f = \int_0^\infty [h^*(t) - h(t)]^2 w(t)\, dt \tag{8.1}$$

or it may be the total squared error at a set of points along the time axis,

$$f = \sum_i [h^*(t_i) - h(t_i)]^2 w(t_i) \tag{8.2}$$

where $w(t)$ is an appropriate weighting function.

The design parameters, $x_1, x_2, \ldots, x_n$, may be, for example, a set of network element values, a set of numbers which give the pole and zero positions, or a set of polynomial coefficients. If these parameters are restricted to certain allowable values, the problem is constrained; if the parameters are allowed to have any value (both positive and negative), the problem is unconstrained.

If the objective function is a linear function and the constraints are also linear, the programming technique is known as *linear programming*. Linear programming is not very useful in network problems because a network function is seldom a linear function. Almost all network problems require the use of *nonlinear programming* in their solution.

After the objective function has been judiciously formed in terms of a set of design parameters for a design problem, the next step is to find the minimum of all possible $f(\bar{x})$ for all $\bar{x}$. This minimum is known as the *global minimum*. (If the minimum is only in some neighborhood, it is a *local minimum*.)

If $f(\bar{x})$ has continuous first and second derivatives for all $\bar{x}$, the optimization problem is equivalent to the solution of the set of n equations

$$\left.\frac{\partial f}{\partial x_i}\right| = 0, \qquad i = 1, 2, \ldots, n \tag{8.3}$$

A minimum is found if the matrix of the second derivatives is positive definite.

The solution of (8.3) is usually a formidable task, because there are n simultaneous nonlinear equations. The numerous numerical methods devised for minimizing $f(\bar{x})$ are aimed specifically at circumventing this step. By the nature of the various techniques, they may be classified into the following categories:

1. Methods for Optimizing an Unconstrained Problem

(*a*) *Systematic search* Schemes have been developed to systematically scan the behavior of the objective function and then estimate the location of the global minimum [WI2]. This group of methods usually is not very efficient even when high-speed computers are employed, especially when the number of design parameters is large.

(*b*) *Methods using the gradient* If $f(\bar{x})$ is continuous and differentiable, a number of techniques using the gradient of $f(\bar{x})$ are available. Starting with an initial point, these methods estimate the successive points that are closer to the true minimum. Some of the more powerful methods in this category are the following:

1. *Method of steepest descent* [CU2]. Starting with an initial point, the gradient of $f(\bar{x})$ at this point is computed. The next estimate of the minimum point is taken along the direction of the negative gradient. The methods of locating this second estimated point along the negative gradient direction vary according to the interpolation or estimation method employed. When a method is used to locate the second point at the point along the gradient at which the objective function is a minimum, the method is said to be *optimum steepest descent*.
2. *Second-order gradient method.* This method is based on the assumption that at the minimum the first partial derivatives vanish. Thus the function behaves like a quadratic near the minimum [SH3]. This scheme usually requires fewer steps to reach the minimum than the steepest descent method.
3. *Steepest descent with metric.* In this method the direction of the second estimated minimum point is located along the direction

$$-\frac{[H(\bar{x})]^{-1}\nabla f(\bar{x})}{|[H(\bar{x})]^{-1}\nabla f(\bar{x})|} \tag{8.4}$$

   where $\nabla f(\bar{x})$ is the gradient of $f(x)$ and

$$[H(\bar{x})] = \left[\frac{\partial^2 f}{\partial x_i \partial x_j}\right]$$

   is the Hessian matrix of the second derivatives [CR1]. Another version of this method employs a variable metric to avoid the necessity of calculating the second derivatives [DA2].

   For a quadratic function, this method finds the minimum in a single step. For other functions, the direction given by (8.4) is more nearly in the direction of the true minimum, in the vicinity of the minimum.
4. *Fletcher-Powell Method.* Instead of using the second derivatives, the Fletcher-Powell method uses the knowledge of the function and its first derivatives from the previous step to determine the direction of the next point [FL1]. This method is one of the most powerful iteration techniques known and has proved to be very efficient in terms of execution time.

2. Methods for Optimizing a Constrained Problem

A number of methods are available to locate the minimum of $f(\bar{x})$ subject to a set of constraints placed on the independent parameters. These constraints may generally be described by

$$
\begin{aligned}
g_1(\bar{x}) &= 0 \\
g_2(\bar{x}) &= 0 \\
&\vdots \\
g_r(\bar{x}) &= 0 \\
g_{r+1}(\bar{x}) &\leq 0 \\
&\vdots \\
g_m(\bar{x}) &\leq 0
\end{aligned}
\tag{8.5}
$$

Some of the better known methods are summarized as follows:

(*a*) *Methods of feasible directions* These methods start with a point that satisfies the constraints and then finds a direction along which a small move does not violate any of the constraints and, at the same time, reduces the objective function. A new point along such a direction is then chosen by specific criteria and the process is repeated.

A direction along which a small move can be made without violating any of the constraints is called a *feasible* direction. A feasible direction along which a small move decreases the objective function is called a *usable feasible direction.* The different methods in this group differ in their techniques of finding usable feasible directions [ZO1, RO1, RO2].

(*b*) *Penalty-function techniques* The idea behind this technique is to deal with, instead of the objective function itself, a modified objective function

$$\psi(x) = f(\bar{x}) + \sum_{i=1}^{m} \phi_i[g_i(\bar{x})] \tag{8.6}$$

where each ϕ_i is constructed such that it becomes very large when $\bar{x}$ approaches or falls outside the boundary of that constraint and remains zero or very nearly zero when $\bar{x}$ is within the boundary. The original constrained optimization of $f(\bar{x})$ is now replaced by the unconstrained optimization of $\psi(x)$.

(*c*) *Fiacco–McCormick Method* [F12] The Fiacco–McCormick method greatly resembles the penalty-function technique. The constrained optimization problem is replaced by the unconstrained optimization of a new function,

$$\psi(\bar{x}, r) = f(\bar{x}) + r\sum_{i=1}^{m} \frac{1}{|g_i(\bar{x})|} \tag{8.7}$$

where r is a scalar. For any $r > 0$, if $\psi(\bar{x}, r)$ is minimized subject to no constraint with an initial point within the set of constraints, a minimum can be found inside the constraint set. After this minimum has been found, r may be reduced and the process repeated to bring the point closer to the true minimum. As r approaches zero, the true minimum will be approached from the inside of the constraint boundary.

3. Dynamic Programming

Dynamic programming is a general approach to the solution of sequential decision processes. Instead of optimizing a single function as in the previously discussed methods, optimization is performed on

$$\psi_i(\bar{x}) = \min_{\bar{x}_j} |f_i(\bar{x}) + \psi_{i-1}(\bar{x})|, \qquad i = 1, 2, \ldots, N$$

where $f_i(\bar{x})$ is the objective function for the ith stage defined over some range of interest and $\bar{x}_j$ is the combination of $(x_1, x_2, \ldots, x_n)$ that optimizes $\psi_i(\bar{x})$. The objective in the procedure is to determine the value of $\bar{x}_j$ in the ith stage which optimizes $\psi_i(\bar{x})$ with regard to the stage resulting from the $(i - 1)$th decision. This decision process is performed N times to obtain the desired optimization. The approach can be extended to the constrained case.

A number of techniques of optimization can be formed on the basis of this approach [BE4, BE6, BE7].

4. Remarks

The summary given above is only meant to indicate, in skeleton form, some of the numerical optimization techniques available to engineers. The reader should refer to the literature if he wishes more detailed information on the subject [HA3, LA2, WI2, WI3].

The purpose of this summary is to provide the necessary background in the discussion of how network problems can be formulated to take advantage of these techniques. Obviously the possibility of utilizing these optimization techniques in network synthesis is virtually unlimited, both in the types of problems and in the manner of application. In the following sections, several typical problems, the nature of which make them readily amenable to the utilization of these optimization techniques, are described. These problems should be regarded merely as examples of how network problems in the time domain may be formulated to utilize the numerical optimization techniques. Only certain key points, such as the desirable design parameters, the appropriate design criterion, suggested ways of forming the objective function and possible methods of computing its approximate derivatives, and the particular suitable numerical techniques, are presented in each problem. It is hoped that these examples will serve to illustrate the powerful potential of applying these optimization techniques to problems considered in this volume.

8.2 OPTIMUM DELAY NETWORKS FOR SPECIFIC WAVE FORMS

In Chapter 7 a number of schemes for designing a lumped constant-delay network are discussed. Network functions achieved there are not necessarily optimum in any specific way. Although a systematic comparison of the step responses of several constant-delay networks is made in Section 7.12, this comparison is made on the step responses of these functions and is only very general in nature.

In many practical situations it is desired to design a constant-delay network for signals with a specific wave form other than a step function. In such situations, not only does comparison made on the basis of step responses become less precise quantitatively, but criteria for optimizing the network performance can be better defined. A procedure for optimizing delay network functions under these situations is described here.

We shall assume that the excitation to be delayed has a wave form such that its Laplace transform may be expressed as

$$F_1(s) = \sum_{\lambda=1}^{l} R_\lambda(s)\epsilon^{-sT_\lambda} \tag{8.8}$$

in which the $R_\lambda(s)$ are rational functions of s none of which has right-halfplane poles. It is desired to find a network function $H^*(s)$ such that the actual response rendered by this function,

$$f_2^*(t) = \mathscr{L}^{-1}[F_1(s)H^*(s)] \tag{8.9}$$

approximates the ideal delayed response,

$$f_2(t) = f_1(t - T_D) \tag{8.10}$$

in the least-squares sense. Equivalently, the problem is to find a set of parameters of the network function $H^*(s)$ such that the integral of the squared error

$$E_2 = \int_0^\infty [f_2^*(t) - f_1(t - T_D)]^2\, dt \tag{8.11}$$

is minimum. Applying Parseval's theorem, we have

$$E_2 = \frac{1}{2\pi j}\int_{-j\infty}^{j\infty} F_1(s)F_1(-s)[H^*(s) - \epsilon^{-sT_D}][H^*(-s) - \epsilon^{sT_D}]\, ds \tag{8.12}$$

where the contour of integration is the j axis of the s plane. The integrand in (8.12) involves only rational functions and exponential factors and can generally be written in the form

$$E_2 = \sum_{i=1}^{N} \left[\frac{1}{2\pi j} \int_{-j\infty}^{j\infty} G_i(s) \epsilon^{-sT_i}\, ds \right] \tag{8.13}$$

where each $G_i(s)$ may include factors from $R_\lambda(s)$, $F_1(\pm s)$, and $H^*(\pm s)$ and each T_i is the sum of quantities from among T_λ and $\pm T_D$. Each integral in (8.13) can be evaluated in terms of the residues and expressed in closed form for a given $f_1(t)$ and an assumed $H^*(s)$.

In particular, if the network is assumed to be an all-pass one,†

$$H^*(s) = (-1)^n \prod_{\nu=1}^{n} \frac{s + p_\nu}{s - p_\nu} \tag{8.14}$$

Since

$$H^*(s)H^*(-s) = 1$$

and

$$\frac{1}{2\pi j} \int_{-j\infty}^{j\infty} F_1(s)F_1(-s)H(s)\epsilon^{sT_D}\, ds = \frac{1}{2\pi j} \int_{-j\infty}^{j\infty} F_1(s)F_1(-s)H(-s)\epsilon^{-sT_D}\, ds$$

(8.12) reduces to

$$E_2(x_1, x_2, \ldots, x_n) = \frac{2}{2\pi j} \int_{-j\infty}^{j\infty} F_1(s)F_1(-s)[1 - H^*(s)\epsilon^{sT_D}]\, ds \tag{8.15}$$

The expression for the error integral is greatly simplified. Here the x_i represent the real and imaginary (when applicable) parts of the poles.

To find an optimum all-pass network function of a given degree n, we simply apply any of the optimization methods to minimize the objective function $E_2(x_1, x_2, \ldots, x_n)$ with $\bar{x} = (x_1, x_2, \ldots, x_n)$ as the design vector. If any of the optimization techniques that require the derivatives of $E_2(x_1, x_2, \ldots, x_n)$ is used, these derivatives may be obtained by differentiating after the integral sign,

$$\frac{\partial E_2}{\partial x_\nu} = -\frac{2}{2\pi j} \int_{-j\infty}^{j\infty} F_1(s)F_1(-s) \left(\frac{\partial H^*}{\partial x_\nu}\right) \epsilon^{sT_D}\, ds \tag{8.16}$$

As an example, suppose it is desired to find the optimum all-pass network functions to delay a triangular signal (Fig. 8.1) by $T_D = 1$ second. For this signal

$$F_1(s) = \frac{1}{s^2}(1 - 2\epsilon^{-s} + \epsilon^{-2s}) \tag{8.17}$$

†The factor $(-1)^n$ makes the step response of this all-pass function have a final value of $+1$ for either even or odd n.

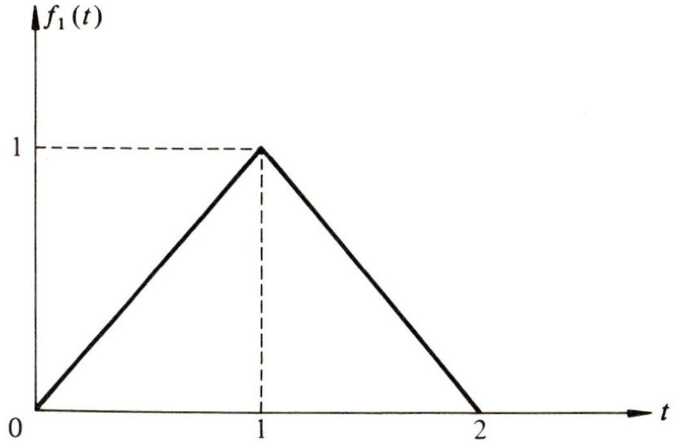

Fig. 8.1 A signal with the triangular waveform.

The first part of E_2, using the form of (8.15), may be readily evaluated using Parseval's theorem:

$$\frac{2}{2\pi j}\int_{-j\infty}^{j\infty} F_1(s)F_1(-s)\,ds = 2\int_0^{\infty} f_1^2(t)\,dt = \tfrac{4}{3} \tag{8.18}$$

The second part is

$$-\frac{2}{2\pi j}\int_{-j\infty}^{j\infty} \frac{\epsilon^{-s} - 4 + 6\epsilon^{s} - 4\epsilon^{2s} + \epsilon^{3s}}{s^4} H^*(s)\,ds \tag{8.19}$$

If we now let

$$Q(s) = \frac{H^*(s)}{s^4} \tag{8.20}$$

then

$$\begin{aligned} E_2(x_1, x_2, \ldots, x_n) &= \frac{4}{3} - \frac{2}{2\pi j}\int_{-j\infty}^{j\infty} Q(s)(\epsilon^{-s} - 4 + 6\epsilon^{s} - 4\epsilon^{2s} + \epsilon^{3s})\,ds \\ &= \tfrac{4}{3} - 12q(1) + 8q(2) - 2q(3) \end{aligned} \tag{8.21}$$

where

$$q(t) = \frac{1}{2\pi j}\int_{-j\infty}^{j\infty} Q(s)\epsilon^{st}\,ds = \mathscr{L}^{-1}[Q(s)]$$

The derivatives of the objective function are

$$\frac{\partial E_2}{\partial x_\nu} = -\frac{2}{2\pi j}\int_{-j\infty}^{j\infty} \frac{\partial Q}{\partial x_\nu}\frac{\epsilon^{-s} - 4 + 6\epsilon^{s} - 4\epsilon^{2s} + \epsilon^{3s}}{s^4}\,ds \tag{8.22}$$

and may also be expressed in terms of the inverse Laplace transforms of another set of rational functions.

Meyer [ME2] used these expressions to obtain the optimum all-pass network functions. The first few steps of the optimization were obtained by the method of steepest descent with quadratic interpolation.† Near the region of the minimum, the Newton–Raphson interpolation method was used to reach the minimum at a faster rate. The results of his computation are tabulated in Table 8.1.

TABLE 8.1

Optimum Pole Locations for All-Pass Delay Network Functions for a Triangular Pulse

n	Poles	E_2 (optimum)	E_2 (maximally flat delay)
1	-0.21597	138.72	215.29
2	$-0.34696 \pm j0.34453$	5.042	18.031
3	-0.53779 $-0.43737 \pm j0.68418$	0.453	1.281
4	$-0.62765 \pm j0.34269$ $-0.43671 \pm j1.05174$	0.082	0.223
5	-0.72405 $-0.64599 \pm j0.69876$ $-0.39670 \pm j1.47520$	0.022	0.069
6	$-0.77772 \pm j0.34917$ $-0.62930 \pm j1.07710$ $-0.36296 \pm j1.92298$	0.008	0.028

The error is seen to be greatly reduced from a network function with maximally flat delay characteristics. Similar results were also obtained for a step function $[f_1(t) = u_{-1}(t)]$, a rectangular pulse $[f_1(t) = u_{-1}(t) - u_{-1}(t-1)]$, and a squared sine pulse $[f_1(t) = \sin^2(\pi t/2)\, u_{-1}(2-t)]$.

8.3 OPTIMIZATION OF A GENERAL NETWORK FUNCTION

The procedure outlined in Section 8.2 deals with a somewhat specialized time-domain problem. What makes the procedure specialized are the assumptions made on the excitation waveform (8.8) and the impulse response (8.10) of the

†Along the direction of steepest descent, the objective function is a one-dimensional function of the distance from an $\bar{x}_i$. A parabola is fitted to the function in the region in which it goes through a minimum. The point at which the parabola is lowest will be taken as the next point, $\bar{x}_{i+1}$.

desired network. In a more general time-domain synthesis problem, this same approach can also be useful [MC1].

In a more general problem, we may assume a general excitation function $f_1(t)$. If the closed-form expression of $f_1(t)$ is not readily Laplace transformable, an approximate $\hat{f}_1(t)$ may be found and substituted for $f_1(t)$. We shall assume that

$$F_1(s) = \mathscr{L}[f_1(t)] \tag{8.23}$$

can be obtained. Similarly, the desired response $f_2(t)$, or an approximating $\hat{f}_2(t)$, is known and we have

$$F_2(s) = \mathscr{L}[f_2(t)] \tag{8.24}$$

We then may assume an appropriate rational expression for $H^*(s)$. Let

$$H^*(s) = \frac{a_0 + a_1 s + a_2 s^2 + \dots + a_m s^m}{1 + b_1 s + b_2 s^2 + \dots + b_n s^n} = \frac{P(s)}{Q(s)} \tag{8.25}$$

Then the error function is

$$\epsilon(t) = \mathscr{L}^{-1}[F_1(s)H^*(s) - F_2(s)] \tag{8.26}$$

Again, assuming a least-squares criterion for optimization, we wish to minimize the objective function,

$$E_2(a_0, a_1, a_2, \dots, a_m, b_1, b_2, \dots, b_n) = E_2(\bar{x}) = \int_0^\infty \epsilon^2(t)\, dt \tag{8.27}$$

If any of the minimization methods that make use of the gradient of E_2 is used, the components of the gradient can be obtained through the following development.

Since

$$\frac{\partial E_2}{\partial a_i} = 2\int_0^\infty \epsilon(t)\, \frac{\partial \epsilon(t)}{\partial a_i}\, dt \tag{8.28}$$

$$\frac{\partial E_2}{\partial b_i} = 2\int_0^\infty \epsilon(t)\, \frac{\partial \epsilon(t)}{\partial b_i}\, dt \tag{8.29}$$

and

$$\mathscr{L}\left[\frac{\partial \epsilon(t)}{\partial a_i}\right] = \frac{\partial}{\partial a_i}[F_1(s)H^*(s) - F_2(s)] = F_1(s)\frac{s^i}{Q(s)} \tag{8.30}$$

$$\mathscr{L}\left[\frac{\partial \epsilon(t)}{\partial b_i}\right] = \frac{\partial}{\partial b_i}[F_1(s)H^*(s) - F_2(s)] = -F_1(s)\frac{P(s)s^i}{Q^2(s)} \tag{8.31}$$

we have

$$\frac{\partial E_2}{\partial a_i} = 2\int_0^\infty \epsilon(t)\mathscr{L}^{-1}\left[F_1(s)\frac{s^i}{Q(s)}\right] dt \tag{8.32}$$

$$\frac{\partial E_2}{\partial b_i} = -2\int_0^\infty \epsilon(t)\mathscr{L}^{-1}\left[F_1(s)\frac{P(s)s^i}{Q^2(s)}\right] dt \tag{8.33}$$

For a given $f_1(t)$ and an initial $H^*(s)$, each of the inverse Laplace transforms in the integrand can be computed by various numerical methods [BE5]. Once the inverse Laplace-transform term has been computed, $\partial E_2/\partial x_i$ can then be computed by numerical integration of (8.32) and (8.33) using a sufficiently high upper limit for the integral.

The foregoing procedure is straightforward in principle. However, the rate of convergence can be quite slow. The following linearized version of these equations sometimes proves to be more efficient [MC1].

We assume a new error function,

$$\epsilon_1(t) = \mathscr{L}^{-1}\left[\frac{P_j(s)}{Q_{j-1}(s)}F_1(s) - \frac{Q_j(s)}{Q_{j-1}(s)}F_2(s)\right] \tag{8.34}$$

in which the subscripts of $P(s)$ and $Q(s)$ refer to the iteration numbers. That this error function is approximately equal to $\epsilon(t)$ provided $Q_{j-1}(s) \cong Q_j(s)$ is seen by comparing (8.34) with (8.26). In each step of this iteration procedure, the $Q(s)$ obtained from the previous step is regarded as fixed. To obtain the derivatives of an objective function to be defined shortly, we set

$$\frac{\partial \epsilon_1(t)}{\partial a_i} = \mathscr{L}^{-1}\left[\frac{s^i}{Q_{j-1}(s)}F_1(s)\right] \triangleq p_{ai}(t), \qquad i = 0, 1, \ldots, m \tag{8.35}$$

$$\frac{\partial \epsilon_1(t)}{\partial b_i} = \mathscr{L}^{-1}\left[-\frac{s^i}{Q_{j-1}(s)}F_2(s)\right] \triangleq p_{bi}(t), \qquad i = 1, 2, \ldots, n \tag{8.36}$$

In addition, we also define

$$p_{b0}(t) \triangleq \mathscr{L}^{-1}\left[-\frac{F_2(s)}{Q_{j-1}(s)}\right] \tag{8.37}$$

The Laplace transform of $\epsilon_1(t)$ as defined by (8.34) is then expressible in terms of these functions:

$$\mathscr{L}[\epsilon_1(t)] = \frac{b_n s^n + \dots + b_1 s + 1}{Q_{j-1}(s)} F_1(s) - \frac{a_m s^m + \dots + a_1 s + a_0}{Q_{j-1}(s)} F_1(s)$$
$$= b_n P_{bn}(s) + \dots + b_1 P_{b1}(s) + P_{b0}(s) + a_m P_{am}(s) + \dots$$
$$+ a_1 P_{a1}(s) + a_0 P_{a0}(s) \tag{8.38}$$

where

$$P_{ai}(s) = \mathscr{L}[p_{ai}(t)] \tag{8.39}$$
$$P_{bi}(s) = \mathscr{L}[p_{bi}(t)] \tag{8.40}$$

The inverse transform of (8.38) renders an expression for $\epsilon_1(t)$ in the following convenient form:

$$\epsilon_1(t) = a_m p_{am}(t) + \dots + a_1 p_{a1}(t) + a_0 p_{a0}(t)$$
$$+ b_n p_{bn}(t) + \dots + b_1 p_{b1}(t) + p_{b0}(t) \tag{8.41}$$

If an objective function based on this new error function is defined as

$$E_{12} = \int_0^\infty \epsilon_1^2(t)\, dt \tag{8.42}$$

the derivative of E_{12} with respect to any of the coefficients is given by

$$\frac{\partial E_{12}}{\partial a_i} = 2\int_0^\infty \epsilon_1(t) \frac{\partial \epsilon_1}{\partial a_i}\, dt$$
$$= 2\int_0^\infty [a_m p_{am}(t) + \dots + a_1 p_{a1}(t) + a_0 p_{a0}(t)$$
$$+ b_n p_{bn}(t) + \dots + b_1 p_{a1}(t) + p_{b0}(t)] p_{ai}(t)\, dt, \qquad i = 0, 1, \dots, m \tag{8.43}$$

or

$$\frac{\partial E_{12}}{\partial b_i} = 2\int_0^\infty \epsilon_1(t) \frac{\partial \epsilon_1}{\partial b_i}\, dt$$
$$= 2\int_0^\infty [a_m p_{am}(t) + \dots + a_1 p_{a1}(t) + a_0 p_{a0}(t)$$
$$+ b_n p_{bn}(t) + \dots + b_1 p_{b1}(t) + p_{b0}(t)] p_{bi}(t)\, dt \qquad i = 1, 2, \dots, n \tag{8.44}$$

Setting $\partial E_{12}/\partial a_i = 0$ and $\partial E_{12}/\partial b_i = 0$ for all applicable i yields a set of $m + n + 1$ linear equations in a_i and b_i. These equations may be written in matrix form (leaving out the factor 2) as

$$\begin{bmatrix} \int_0^\infty p_{a0}(t)p_{am}(t)\,dt & \cdots & \int_0^\infty p_{a0}^2(t) & \int_0^\infty p_{a0}(t)p_{bn}(t)\,dt & \cdots & \int_0^\infty p_{a0}(t)p_{b1}(t)\,dt \\ & & \cdots\cdots\cdots\cdots & & & \\ \int_0^\infty p_{am}^2(t)\,dt & \cdots & \int_0^\infty p_{a0}(t)p_{am}(t)\,dt & \int_0^\infty p_{am}(t)p_{bn}(t)\,dt & \cdots & \int_0^\infty p_{am}(t)p_{b1}(t)\,dt \\ \int_0^\infty p_{b1}(t)p_{am}(t)\,dt & \cdots & \int_0^\infty p_{b1}(t)p_{a0}(t)\,dt & \int_0^\infty p_{b1}(t)p_{bn}(t)\,dt & \cdots & \int_0^\infty p_{b1}^2(t)\,dt \\ & & \cdots\cdots\cdots\cdots & & & \\ \int_0^\infty p_{bn}(t)p_{am}(t)\,dt & \cdots & \int_0^\infty p_{bn}(t)p_{a0}(t)\,dt & \int_0^\infty p_{bn}^2(t)\,dt & \cdots & \int_0^\infty p_{b1}(t)p_{bn}(t)\,dt \end{bmatrix} \times$$

$$\begin{bmatrix} a_0 \\ \vdots \\ a_m \\ b_1 \\ \vdots \\ b_n \end{bmatrix} = \begin{bmatrix} \int_0^\infty p_{a0}(t)p_{b0}(t)\,dt \\ \vdots \\ \int_0^\infty p_{am}(t)p_{b0}(t)\,dt \\ \int_0^\infty p_{b1}(t)p_{b0}(t)\,dt \\ \vdots \\ \int_0^\infty p_{bn}(t)p_{b0}(t)\,dt \end{bmatrix} \tag{8.45}$$

All integrals in the matrix equation are computable by selecting an upper limit large enough for the functions defined by (8.35) and (8.36). The a's and b's are then solved from (8.45). Thus a new $H^*(s)$ is found. Whether a minimum in E_{12} has been reached or not may be determined by comparing the new E_{12} to its previous value. If two successive values of E_{12} differ only slightly, chances are that E_{12} has reached its minimum value.

This iteration technique usually gives a fairly fast convergence, although such convergence is not guaranteed. In addition, although E_{12} does converge to its minimum, its gradient does not converge to zero. Thus in the vicinity of the minimum of E_{12} the convergence is somewhat erratic, and whether a true minimum has been reached or not is somewhat difficult to ascertain.

A simple remedy to improve the convergence near the minimum is to switch the definition of $p_{bi}(t)$ over to $\hat{p}_{bi}(t)$ at some stage of the iteration process, where

$$\hat{p}_{bi}(t) = \mathscr{L}^{-1}\left[\frac{s^i P_{j-1}(s)}{Q_{j-1}^2(s)}\right] F_1(s) \tag{8.46}$$

However, this switchover should be done only when the point is fairly close to the minimum, because the use of this new $\hat{p}_{bi}(t)$ will not always converge to the minimum if the initial guess is poor. With this new $\hat{p}_{bi}(t)$, the gradient of E_{12}

does approach that of E_2, which is zero at the minimum, and the iteration precedure is identical to that described previously.

8.4 FORMULATION OF TIME-DOMAIN PROBLEMS IN TERMS OF NETWORK ELEMENT VALUES

In the precedures outlined in Sections 8.2 and 8.3, the time-domain error is expressed in terms of the pole-zero locations or the coefficients of the frequency-domain function $H^*(s)$. Once a set of acceptable coefficients has been obtained for $H^*(s)$, the synthesis of a network to realize this $H^*(s)$ remains to be carried out. There is no reason why the time-domain error cannot be related directly to the element values for a specific network configuration. This approach would seem especially desirable when a network of fairly good design has already been obtained by the classical methods and further improvement is sought through adjusting its element values.

In expressing a time-domain approximation problem directly in terms of a set of network element values, the state-variable method is the most convenient formulation to use. The significance of the notation to be used here is illustrated in Fig. 8.2. The excitation is represented by $f_1(t)$ and the response by $f_2^*(t)$. The internal state of the network can be represented by a properly chosen set of capacitance voltages $v_1(t), v_2(t), \ldots, v_m(t)$ and a set of inductance currents $i_{m+1}(t), i_{m+2}(t), \ldots, i_{m+n}(t)$.

Let

$$[v] = \begin{bmatrix} v_1 \\ v_2 \\ \vdots \\ v_m \end{bmatrix} \tag{8.47}$$

$$[i] = \begin{bmatrix} i_{m+1} \\ i_{m+2} \\ \vdots \\ i_{m+n} \end{bmatrix} \tag{8.48}$$

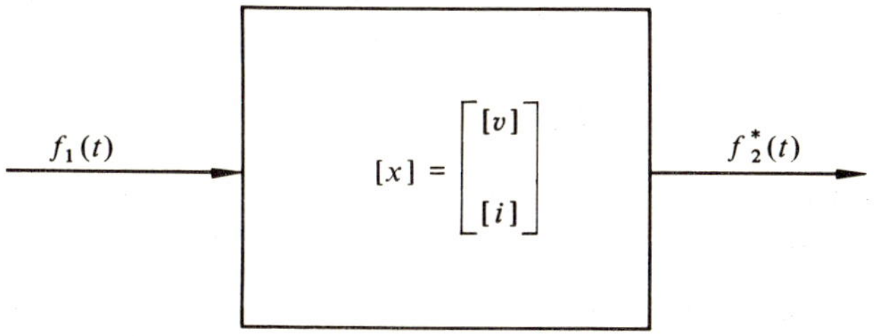

Fig. 8.2 Notation for state-variable representation of a network.

Then

$$[x] = \begin{bmatrix} [v] \\ [i] \end{bmatrix} \tag{8.49}$$

is the state vector. Following Bashkow's procedure [BA3] of using only capacitance currents ($= C_i\, dv_i/dt$) as tree-branch currents and only inductance voltages ($= L_j\, di_j/dt$) as link voltages†, the state equation can always be written in the form

$$\begin{bmatrix} [\tilde{C}] & 0 \\ 0 & [\tilde{L}] \end{bmatrix} [\dot{x}] = \begin{bmatrix} [A_{11}] & [A_{12}] \\ [A_{21}] & [A_{22}] \end{bmatrix} [x] + \begin{bmatrix} [B_1] \\ [B_2] \end{bmatrix} [u] \tag{8.50}$$

In (8.50) $[\tilde{C}]$ is an $m \times m$ diagonal matrix; $[\tilde{L}]$ is an $n \times n$ one. Submatrix $[A_{11}]$ is $m \times m$, $[A_{22}]$ is $n \times n$, and $[A_{21}] = -[A_{12}]_t$. The matrix

$$[A] = \begin{bmatrix} [A_{11}] & [A_{12}] \\ [A_{21}] & [A_{22}] \end{bmatrix}$$

is commonly known as Bashkow's $[A]$ matrix. Submatrices $[B_1]$ and $[B_2]$ are column matrices. The matrix $[u]$ includes all forcing functions. In our problems all elements of $[u]$ are zero except one, which is $f_1(t)$, and the last term may be reduced to a scalar $f_1(t)$ times a column matrix $[B]$.

For any given set of initial conditions, (8.50) may be solved for the state vector $[x]$. Once $[x]$, that is, $[v]$ and $[i]$, is known, $f_2^*(t)$ can generally be obtained through

$$f_2^*(t) = [C][x] + Df_1(t) \tag{8.51}$$

where $[C]$ is a row matrix of real numbers and operators and D is a scalar, which can be either a real number or an operator.

The foregoing outline is a manner by which a network behavior in the time domain may be analyzed effectively and without going via frequency domain and inverse Laplace transform.

In a time-domain synthesis problem, we should like to be able to determine how we may adjust a set of design parameters to reduce the error

$$\epsilon(t) = f_2(t) - f_2^*(t) \tag{8.52}$$

by some criteria. In using the state-variable approach, it is desired that the set of design parameters be the network element values themselves. Let these ele-

†When "excess" capacitances of inductances are present, a capacitance may be added across each excess inductance and an inductance added in series with each excess capacitance.

ment values be denoted by $c_1, c_2, \ldots, c_e$. The key to many computational techniques lies in the availability of the derivatives

$$\frac{\partial \epsilon(t)}{\partial c_i} = \frac{\partial f_2^*(t)}{\partial c_i} \tag{8.53}$$

To find $\partial f_2^*(t)/\partial c_i$ we begin by differentiating (8.50) with respect to c_i to get

$$\begin{bmatrix} \left[\frac{\partial \tilde{C}}{\partial c_i}\right] & 0 \\ 0 & \left[\frac{\partial \tilde{L}}{\partial c_i}\right] \end{bmatrix} [\dot{x}] + \begin{bmatrix} [\tilde{C}] & 0 \\ 0 & [\tilde{L}] \end{bmatrix} \left[\frac{\partial \dot{x}}{\partial c_i}\right]$$
$$= \begin{bmatrix} \left[\frac{\partial A_{11}}{\partial c_i}\right] & \left[\frac{\partial A_{12}}{\partial c_i}\right] \\ \left[\frac{\partial A_{21}}{\partial c_i}\right] & \left[\frac{\partial A_{22}}{\partial c_i}\right] \end{bmatrix} [x] + \begin{bmatrix} [A_{11}] & [A_{12}] \\ [A_{21}] & [A_{22}] \end{bmatrix} \left[\frac{\partial x}{\partial c_i}\right] + \begin{bmatrix} \left[\frac{\partial B_1}{\partial c_i}\right] \\ \left[\frac{\partial B_2}{\partial c_i}\right] \end{bmatrix} [u] \tag{8.54}$$

Equation (8.54) may be rearranged to read

$$\begin{bmatrix} [\tilde{C}] & 0 \\ 0 & [\tilde{L}] \end{bmatrix} [\dot{S}] = \begin{bmatrix} [A_{11}] & [A_{12}] \\ [A_{21}] & [A_{22}] \end{bmatrix} [S] + \begin{bmatrix} [\hat{B}_1] \\ [\hat{B}_2] \end{bmatrix} [\hat{u}] \tag{8.55}$$

where

$$[S] = \frac{\partial}{\partial c_i}[x] \tag{8.56}$$

$$[\hat{B}] = \begin{bmatrix} [\hat{B}_1] \\ [\hat{B}_2] \end{bmatrix} = \begin{bmatrix} -\left[\frac{\partial \tilde{C}}{\partial c_i}\right] & 0 & \left[\frac{\partial A_{11}}{\partial c_i}\right] & \left[\frac{\partial A_{12}}{\partial c_i}\right] & \left[\frac{\partial B_1}{\partial c_i}\right] \\ 0 & -\left[\frac{\partial \tilde{L}}{\partial c_i}\right] & \left[\frac{\partial A_{21}}{\partial c_i}\right] & \left[\frac{\partial A_{22}}{\partial c_i}\right] & \left[\frac{\partial B_2}{\partial c_i}\right] \end{bmatrix} \tag{8.57}$$

$$[\hat{u}] = \begin{bmatrix} [\dot{x}] \\ [x] \\ [u] \end{bmatrix} \tag{8.58}$$

Since $[\dot{x}]$, $[x]$, and $[u]$ are all known, $[\hat{u}]$ may be regarded as a new set of excitations. Matrix $[\hat{B}]$ is known and will generally be composed of ± 1 and 0 only. Equation (8.55) may be solved for $[S]$ in the same way (8.50) would be.

Differentiating (8.51) with respect to c_i yields

$$\frac{\partial f_2^*(t)}{\partial c_i} = [C][S] + \left[\frac{\partial C}{\partial c_i}\right][x] + \frac{\partial D}{\partial c_i} f_1(t) \tag{8.59}$$

Generally, $[\partial C/\partial c_i]$ and $\partial D/\partial c_i$ will be made up of ± 1 and 0. Hence $\partial f_2^*(t)/\partial c_i$ can be found from (8.59) by substituting the matrix $[S]$ found from (8.55).

8.5 DIRECT TIME-DOMAIN DISCRETE CHEBYSHEV APPROXIMATION BY CONSTRAINED OPTIMIZATION TECHNIQUES

As another example of the application of optimization techniques to time-domain approximation problems, let us apply the formulation of the previous section to the approximation at discrete equally spaced points along the time axis [AT1, AT2]. Suppose it is desired to produce an $f_2(t)$ for a given $f_1(t)$. Let $f_2(t)$ be known at $t = 0, t_1, t_2, \ldots, t_{M-1}$. If the actual response function is $f_2^*(t)$, then at these instants the errors are

$$\epsilon_k = f_2(t_k) - f_2^*(t_k), \qquad k = 0, 1, 2, \ldots, M-1 \tag{8.60}$$

In the discrete Chebyshev approximation, the maximum error

$$\epsilon_\infty = \max_k |\epsilon_k| \tag{8.61}$$

is minimized. We shall use the network element values themselves as the design parameters. The response function $f_2^*(t)$ may be expressed as

$$f_2^*(t) = f_2^*(c_1, c_2, \ldots, c_e, t) = f_2^*(\bar{c}, t) \tag{8.62}$$

Further, let it be assumed that the network elements are constrained to be not only nonnegative but also within certain bounds for practical reasons, or

$$0 \leq c_{i,\min} \leq c_i \leq c_{i,\max}, \qquad i = 1, 2, \ldots, e \tag{8.63}$$

To formulate the minimization of the maximum error of (8.61) into a constrained optimization problem, we first introduce two envelope functions

$$f_2(t) \pm w(t)\eta \tag{8.64}$$

where $w(t)$ is an arbitrary weighting function and η is a positive real number. When $w(t) = 1$, these envelopes are equally distant from $f_2(t)$. This is depicted in Fig. 8.3. If the approximating function $f_2^*(t)$ is to lie within these two envelopes, it is required that

$$\begin{aligned} \Phi_1(\bar{c}, \eta, t) &= f_2(t) + w(t)\eta - f_2^*(\bar{c}, t) \geq 0 \\ \Phi_2(\bar{c}, \eta, t) &= f_2^*(\bar{c}, t) - f_2(t) + w(t)\eta \geq 0 \end{aligned} \tag{8.65}$$

Equation (8.65) can alternatively be expressed in discrete form by letting $t = 0, t_1, t_2, \ldots, t_{M-1}$.

Now the problem of minimizing ϵ_∞ may be viewed as another problem of minimizing η with respect to $\bar{c}$ and η under the constraints of (8.63) and

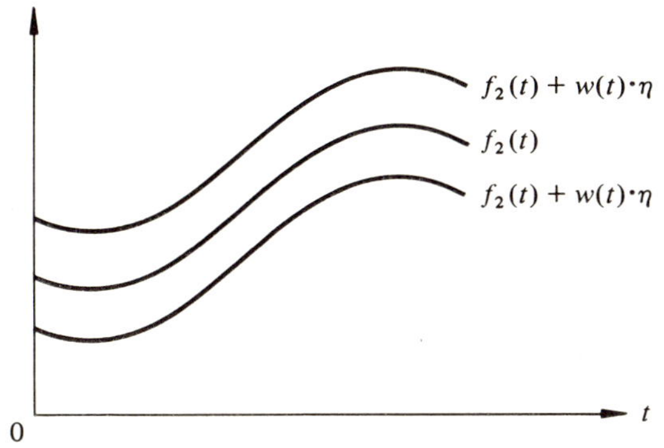

Fig. 8.3 A given $f_2(t)$ and two envelope functions.

$$\left.\begin{aligned}\Phi_1(\bar{c}, \eta, t_k) \geq 0\\ \Phi_2(\bar{c}, \eta, t_k) \geq 0\end{aligned}\right\} \qquad k = 0, 1, 2, \ldots, M - 1 \tag{8.66}$$

Thus the approximation problem is transformed into a constrained optimization problem. As outlined in Section 8.1, several methods are available for this purpose and can all be used with varying effectiveness.

For instance, the Fiacco-McCormick method can be used to transform this constrained optimization problem into an unconstrained one. The objective function

$$\begin{aligned} P(\bar{c}, \eta, r) = \eta &+ r \sum_{k=0}^{M-1} \left[\frac{1}{\Phi_1(\bar{c}, \eta, t_k)} + \frac{1}{\Phi_2(\bar{c}, \eta, t_k)}\right] \\ &+ r \sum_{i=1}^{e} \left[\frac{1}{c_i - c_{i,\min}} + \frac{1}{c_{i,\max} - c_i}\right] \end{aligned} \tag{8.67}$$

may be formed and treated as an unconstrained optimization problem.

From the physical consideration of the problem, it is easy to expect that if an initial value of η is chosen large enough, the constraints of (8.63) and (8.66) can all be satisfied. From that point on, r can be monotonically decreased until the change in P fails to exceed a specified value over a predetermined number of iterations

In minimizing $P(\bar{c}, \eta, r)$ as an unconstrained optimization problem, many different techniques are available and may be used. Several of these techniques call for the first derivatives of P with respect to the element values. When such derivatives are desired, they may be found by the following equations:

$$\begin{aligned} \frac{\partial P}{\partial c_i} = r &\sum_{k=0}^{M-1} \left\{\left[\frac{1}{\Phi_1^2(\bar{c}, \eta, t_k)} - \frac{1}{\Phi_2^2(\bar{c}, \eta, t_k)}\right] \frac{\partial f_2^*(\bar{c}, t_k)}{\partial c_i}\right\} \\ &+ r\left[\frac{1}{(c_{i,\max} - c_i)^2} - \frac{1}{(c_i - c_{i,\min})^2}\right] \end{aligned} \tag{8.68}$$

and

$$\frac{\partial P}{\partial \eta} = 1 + r \sum_{k=0}^{M-1} \left\{ \left[\frac{1}{\Phi_1^2(\bar{c}, \eta, t_k)} - \frac{1}{\Phi_2^2(\bar{c}, \eta, t_k)} \right] w(t_k) \right\} \tag{8.69}$$

All quantities in (8.68) and (8.69) are computable. In particular, $\partial f_2^*/\partial c_i$ can be found by the procedure described in Section 8.4.

8.6 CONCLUSIONS

This chapter gives a bird's-eye view of the application of computer-oriented optimization techniques to some time-domain synthesis problems. In a very skeleton fashion, several more powerful methods of optimization techniques are mentioned. Then a few typical problems are outlined to illustrate how some time-domain network problems can readily be formulated to take advantage of these techniques. It is hoped that this brief presentation will serve to illustrate how classical network theory and computer-aided design techniques complement each other in producing a better design or arriving at an acceptable design more easily than if either technique were used alone.

Generally speaking, in a time-domain synthesis problem the methods developed under classical network theory are most helpful when nothing or very little is known about the network configuration, complexity, or element values. They are also suitable for cases in which the network function can reasonably be expected to be fairly simple. On the other hand, optimization techniques are most advantageous when the network elements are numerous or when an initial design of the network is already available. For further reading on this topic, the reader is referred to [TE2, WA2].

PROBLEMS

8.1. Let $f_1(t) = u_{-1}(t) - u_{-1}(t - 1)$. Find a three-pole all-pass function such that $f_2^*(t)$ approximates $f_2(t) = f_1(t - T_D)$ in such a way that the error integral E_2 as defind in (8.11), with $T_D = 1$, is a minimum.

8.2. Repeat Problen 8.1 for a four-pole all-pass network function with $T_D = 0.5$.

8.3. Derive Equation (8.15).

8.4. Show that $\epsilon(t)$ of (8.26) and $\epsilon_1(t)$ of (8.34) are equal if $Q_{j-1}(s) = Q_j(s)$.

8.5. For the two-port network of Fig. P8.1, Let $f_1(t) = e_1(t), f_2^*(t) = v_3(t)$ and $\bar{c} = C_1, C_2, C_3, L_4, L_5$. The dissipation factor δ is a constant. Letting $\bar{x} = v_1, v_2, v_3, i_4, i_5$, write the state equation for the circuit in the form of (8.50) and the output equation in the form of (8.51). Then find the state and output equations in the form of (8.55) from whose solution $\partial f_2^*(t)/\partial c_i$ can be obtained through (8.59).

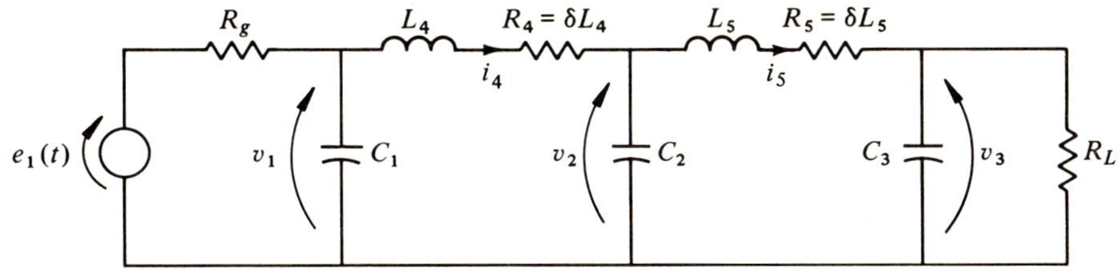

Fig. P8.1

APPENDIX A

Coefficients of

$$(1+x)^m(1-x)^n = c_0 + c_1x + c_2x + c_3x^3 + \ldots + c_{m+n}x^{m+n}$$

The following tables are useful in conjunction with the method for testing the absolute integrability of an impulse response described in Section 3.5.

$m + n = 2$

m	n	c_0	c_1	c_2
2	0	1	2	1
1	1	1	0	−1
0	2	1	−2	1

$m + n = 3$

m	n	c_0	c_1	c_2	c_3
3	0	1	3	3	1
2	1	1	1	−1	−1
1	2	1	−1	−1	1
0	3	1	−3	3	−1

$$m+n=4$$

m	n	c_0	c_1	c_2	c_3	c_4
4	0	1	4	6	4	1
3	1	1	2	0	−2	−1
2	2	1	0	−2	0	1
1	3	1	−2	0	2	−1
0	4	1	−4	6	−4	1

$$m+n=5$$

m	n	c_0	c_1	c_2	c_3	c_4	c_5
5	0	1	5	10	10	5	1
4	1	1	3	2	−2	−3	−1
3	2	1	1	−2	−2	1	1
2	3	1	−1	−2	2	1	−1
1	4	1	−3	2	2	−3	1
0	5	1	−5	10	−10	5	1

$$m+n=6$$

m	n	c_0	c_1	c_2	c_3	c_4	c_5	c_6
6	0	1	6	15	20	15	6	1
5	1	1	4	5	0	−5	−4	−1
4	2	1	2	−1	−4	−1	2	1
3	3	1	0	−3	0	3	0	−1
2	4	1	−2	−1	4	−1	−2	1
1	5	1	−4	5	0	−5	4	−1
0	6	1	−6	15	−20	15	−6	1

$$m+n=7$$

m	n	c_0	c_1	c_2	c_3	c_4	c_5	c_6	c_7
7	0	1	7	21	35	35	21	7	1
6	1	1	5	9	5	−5	−9	−5	−1
5	2	1	3	1	−5	−5	1	3	1
4	3	1	1	−3	−3	3	3	−1	−1
3	4	1	−1	−3	3	3	−3	−1	1
2	5	1	−3	1	5	−5	−1	3	−1
1	6	1	−5	9	−5	−5	9	−5	1
0	7	1	−7	21	−35	35	−21	7	−1

$m + n = 8$

m	n	c_0	c_1	c_2	c_3	c_4	c_5	c_6	c_7	c_8
8	0	1	8	28	56	70	56	28	8	1
7	1	1	6	14	14	0	−14	−14	−6	−1
6	2	1	4	4	−4	−10	−4	4	4	1
5	3	1	2	−2	−6	0	6	2	−2	−1
4	4	1	0	−4	0	6	0	−4	0	1
3	5	1	−2	−2	6	0	−6	2	2	−1
2	6	1	−4	4	4	−10	4	4	−4	1
1	7	1	−6	14	−14	0	14	−14	6	−1
0	8	1	−8	28	−56	70	−56	28	−8	1

$m + n = 9$

m	n	c_0	c_1	c_2	c_3	c_4	c_5	c_6	c_7	c_8	c_9
9	0	1	9	36	84	126	126	84	36	9	1
8	1	1	7	20	28	14	−14	−28	−20	−7	−1
7	2	1	5	8	0	−14	−14	0	8	5	1
6	3	1	3	0	−8	−6	6	8	0	−3	−1
5	4	1	1	−4	−4	6	6	−4	−4	1	1
4	5	1	−1	−4	4	6	−6	−4	4	1	−1
3	6	1	−3	0	8	−6	−6	8	0	−3	1
2	7	1	−5	8	0	−14	14	0	−8	5	−1
1	8	1	−7	20	−28	14	14	−28	20	−7	1
0	9	1	−9	36	−84	126	−126	84	−36	9	−1

$m + n = 10$

m	n	c_0	c_1	c_2	c_3	c_4	c_5	c_6	c_7	c_8	c_9	c_{10}
10	0	1	10	45	120	210	252	210	120	45	10	1
9	1	1	8	27	48	42	0	−42	−48	−27	−8	−1
8	2	1	6	13	8	−14	−28	−14	8	13	6	1
7	3	1	4	3	−8	−14	0	14	8	−3	−4	−1
6	4	1	2	−3	−8	2	12	2	−8	−3	2	1
5	5	1	0	−5	0	10	0	−10	0	5	0	−1
4	6	1	−2	−3	8	2	−12	2	8	−3	−2	1
3	7	1	−4	3	8	−14	0	14	−8	−3	4	−1
2	8	1	−6	13	−8	−14	28	−14	−8	13	−6	1
1	9	1	−8	27	−48	42	0	−42	48	−27	8	−1
0	10	1	−10	45	−120	210	−252	210	−120	45	−10	1

$m + n = 11$

m	n	c_0	c_1	c_2	c_3	c_4	c_5	c_6	c_7	c_8	c_9	c_{10}	c_{11}
11	0	1	11	55	165	330	462	462	330	165	55	11	1
10	1	1	9	35	75	90	42	−42	−90	−75	−35	−9	−1
9	2	1	7	19	21	−6	−42	−42	−6	21	19	7	1
8	3	1	5	7	−5	−22	−14	14	22	5	−7	−5	−1
7	4	1	3	−1	−11	−6	14	14	−6	−11	−1	3	1
6	5	1	1	−5	−5	10	10	−10	−10	5	5	−1	−1
5	6	1	−1	−5	5	10	−10	−10	10	5	−5	−1	1
4	7	1	−3	−1	11	−6	−14	14	6	−11	1	3	−1
3	8	1	−5	7	5	−22	14	14	−22	5	7	−5	1
2	9	1	−7	19	−21	−6	42	−42	6	21	−19	7	−1
1	10	1	−9	35	−75	90	−42	−42	90	−75	35	−9	1
0	11	1	−11	55	−165	330	−462	462	−330	165	−55	11	−1

$m + n = 12$

m	n	c_0	c_1	c_2	c_3	c_4	c_5	c_6	c_7	c_8	c_9	c_{10}	c_{11}	c_{12}
12	0	1	12	66	220	495	792	924	792	495	220	66	12	1
11	1	1	10	44	110	165	132	0	−132	−165	−110	−44	−10	−1
10	2	1	8	26	40	15	−48	−84	−48	15	40	26	8	1
9	3	1	6	12	2	−27	−36	0	36	27	−2	−12	−6	−1
8	4	1	4	2	−12	−17	8	28	8	−17	−12	2	4	1
7	5	1	2	−4	−10	5	20	0	−20	−5	10	4	−2	−1
6	6	1	0	−6	0	15	0	−20	0	15	0	−6	0	1
5	7	1	−2	−4	10	5	−20	0	20	−5	−10	4	2	−1
4	8	1	−4	2	12	−17	−8	28	−8	−17	12	2	−4	1
3	9	1	−6	12	−2	−27	36	0	−36	27	2	−12	6	−1
2	10	1	−8	26	−40	15	48	−84	48	15	−40	26	−8	1
1	11	1	−10	44	−110	165	−132	0	132	−165	110	−44	10	−1
0	12	1	−12	66	220	495	−792	924	−792	495	−220	66	−12	1

$$m + n = 13$$

m	n	c_0	c_1	c_2	c_3	c_4	c_5	c_6	c_7	c_8	c_9	c_{10}	c_{11}	c_{12}	c_{13}
13	0	1	13	78	286	715	1287	1716	1716	1287	715	286	78	13	1
12	1	1	11	54	154	275	297	132	−132	−297	−275	−154	−54	−11	−1
11	2	1	9	34	66	55	−33	−132	−132	−33	55	66	34	9	1
10	3	1	7	18	14	−25	−63	−36	36	63	25	−14	−18	−7	−1
9	4	1	5	6	−10	−29	−9	36	36	−9	−29	−10	6	5	1
8	5	1	3	−2	−14	−5	25	20	−20	−25	5	14	2	−3	−1
7	6	1	1	−6	−6	15	15	−20	−20	15	15	−6	−6	1	1
6	7	1	−1	−6	6	15	−15	−20	20	15	−15	−6	6	1	−1
5	8	1	−3	−2	14	−5	−25	20	20	−25	−5	14	−2	−3	1
4	9	1	−5	6	10	−29	9	36	−36	−9	29	−10	−6	5	−1
3	10	1	−7	18	−14	−25	63	−36	−36	63	−25	−14	18	−7	1
2	11	1	−9	34	−66	55	33	−132	132	−33	−55	66	−34	9	−1
1	12	1	−11	54	−154	275	−297	132	132	−297	275	−154	54	−11	1
0	13	1	−13	78	−286	715	−1287	1716	−1716	1287	−715	286	−78	13	−1

$m + n = 14$

m	n	c_0	c_1	c_2	c_3	c_4	c_5	c_6	c_7	c_8	c_9	c_{10}	c_{11}	c_{12}	c_{13}	c_{14}
14	0	1	14	91	364	1001	2002	3003	3432	3003	2002	1001	364	91	14	1
13	1	1	12	65	208	429	572	429	0	−429	−572	−429	−208	−65	−12	−1
12	2	1	10	43	100	121	22	−165	−264	−165	22	121	100	43	10	1
11	3	1	8	25	32	−11	−88	−99	0	99	88	11	−32	−25	−8	−1
10	4	1	6	11	−4	−39	−38	27	72	27	−38	−39	−4	11	6	1
9	5	1	4	1	−16	−19	20	45	0	−45	−20	19	16	−1	−4	−1
8	6	1	2	−5	−12	9	30	−5	−40	−5	30	9	−12	−5	2	1
7	7	1	0	−7	0	21	0	−35	0	35	0	−21	0	7	0	−1
6	8	1	−2	−5	12	9	−30	−5	40	−5	−30	9	12	−5	−2	1
5	9	1	−4	1	16	−19	−20	45	0	−45	20	19	−16	−1	4	−1
4	10	1	−6	11	4	−39	38	27	−72	27	38	−39	4	11	−6	1
3	11	1	−8	25	−32	−11	88	−99	0	99	−88	11	32	−25	8	−1
2	12	1	−10	43	−100	121	−22	−165	264	−165	−22	121	−100	43	−10	1
1	13	1	−12	65	−208	429	−572	429	0	−429	572	−429	208	−65	12	−1
0	14	1	−14	91	−364	1001	−2002	3003	−3432	3003	−2002	1001	−364	91	−14	1

$$m + n = 15$$

m	n	c_0	c_1	c_2	c_3	c_4	c_5	c_6	c_7	c_8	c_9	c_{10}	c_{11}	c_{12}	c_{13}	c_{14}	c_{15}
15	0	1	15	105	455	1365	3003	5005	6435	6435	5005	3003	1365	455	105	15	1
14	1	1	13	77	273	637	1001	1001	429	−429	−1001	−1001	−637	−273	−77	−13	−1
13	2	1	11	53	143	221	143	−143	−429	−429	−143	143	221	143	53	11	1
12	3	1	9	33	57	21	−99	−187	−99	99	187	99	−21	−57	−33	−9	−1
11	4	1	7	17	7	−43	−77	−11	99	99	−11	−77	−43	7	17	7	1
10	5	1	5	5	−15	−35	1	65	45	−45	−65	−1	35	15	−5	−5	−1
9	6	1	3	−3	−17	−3	39	25	−45	−45	25	39	−3	−17	−3	3	1
8	7	1	1	−7	−7	21	21	−35	−35	35	35	−21	−21	7	7	−1	−1
7	8	1	−1	−7	7	21	−21	−35	35	35	−35	−21	21	7	−7	−1	1
6	9	1	−3	−3	17	−3	−39	25	45	−45	−25	39	3	−17	3	3	−1
5	10	1	−5	5	15	−35	−1	65	−45	−45	65	−1	−35	15	5	−5	1
4	11	1	−7	17	−7	−43	77	−11	−99	99	11	−77	43	7	−17	7	−1
3	12	1	−9	33	−57	21	99	−187	99	99	−187	99	21	−57	33	−9	1
2	13	1	−11	53	−143	221	−143	−143	429	−429	143	143	−221	143	−53	11	−1
1	14	1	−13	77	−273	637	−1001	1001	−429	−429	1001	−1001	637	−273	77	−13	1
0	15	1	−15	105	−455	1365	−3003	5005	−6435	6435	−5005	3003	−1365	455	−105	15	−1

APPENDIX B

Smoothing of a Periodic Function by Local Averaging and the Approximation of Impulse Trains by Trigonometric Series

Engineers are familiar with the power of the Fourier-series method of analysis and approximation. They are also aware of some of the dangers and potential pitfalls associated with the application of this method when a series is either truncated or used to represent a function at a point of discontinuity. The following methods devised by Lanczos are useful to reduce the potential hazard encountered in both of these situations [LA1].

For many Fourier-series analysis problems in which either discontinuities or slow convergence is involved, it is advantageous to replace a given function

$$f(x) = \frac{a_0}{2} + \sum_{k=1}^{\infty} (a_k \cos kx + b_k \sin kx), \qquad -\pi < x < \pi \tag{B-1}$$

by a smoothed function

$$\bar{f}(x) = \frac{1}{2\pi/m} \int_{-\pi/m}^{\pi/m} f(x + t)\, dt \tag{B-2}$$

The significance of (B-2) can be interpreted as follows: The value of the function at x is replaced by that of a new function whose value is equal to the average of the original function in the vicinity of x. The interval of averaging is the period of the mth harmonic and is centered at x.

It is easy to see that this local averaging process does not affect the low harmonics materially; it completely eliminates all harmonics that are also harmonics of the mth harmonic, and greatly reduces the magnitude of higher harmonics, while preserving their frequency-domain spectrum to some extent.

Analytically, (B-2) may be explicitly expressed as

$$\bar{f}(x) = \frac{1}{2\pi/m}\int_{x-\pi/m}^{x+\pi/m} f(t)\,dt$$

$$= \frac{a_0}{2} + \frac{m}{2\pi}\sum_{k=1}^{\infty}\left[a_k\frac{\sin kt}{k} - b_k\frac{\cos kt}{k}\Big|_{x-\pi/m}^{x+\pi/m}\right]$$

$$= \frac{a_0}{2} + \frac{m}{2\pi}\sum_{k=1}^{\infty}\left\{\frac{a_k}{k}\left[\sin k\left(x+\frac{\pi}{m}\right) - \sin k\left(x-\frac{\pi}{m}\right)\right]\right.$$

$$\left. - \frac{b_k}{k}\left[\cos k\left(x+\frac{\pi}{m}\right) - \cos k\left(x-\frac{\pi}{m}\right)\right]\right\}$$

$$= \frac{a_0}{2} + \sum_{k=1}^{\infty}\frac{\sin(k\pi/m)}{k\pi/m}(a_k\cos kx + b_k\sin kx) \tag{B-3}$$

The factors

$$_m\sigma_k = \frac{\sin(k\pi/m)}{k\pi/m} \tag{B-4}$$

are known as the Lanczos σ factors and are discrete ordinates of the sampling function

$$Sa\,(x) = \frac{\sin x}{x} \tag{B-5}$$

The Fourier series (B-3) is said to be the *σ-modified series* of (B-1). The σ-modified series has less truncation errors than the unmodified series because the higher harmonics in the modified series are made relatively less important. However, these two series do not approach the same function. The unmodified series approaches $f(x)$ while the modified series approaches $\bar{f}(x)$, which has been smoothed by local averaging as described by (B-2).

In a *truncated σ-modified Fourier series*, the highest harmonic retained is the mth one (which, of course, has a zero σ factor). Such a series shall be denoted by $\bar{f}_m(x)$.

To illustrate the effect of modifying a trigonometric series by these σ factors, let us look at the series of square waves of Fig. B-1. It is well known that the function $f(x)$ as defined by Fig. B-1 has a Fourier series

$$f(x) = \frac{4}{\pi}\left(\sin x + \tfrac{1}{3}\sin 3x + \tfrac{1}{5}\sin 5x + \ldots\right) \tag{B-6}$$

The truncated series for $m = 8$ for (B-6) is

$$f_8(x) = \frac{4}{\pi}\left(\sin x + \tfrac{1}{3}\sin 3x + \tfrac{1}{5}\sin 5x + \tfrac{1}{7}\sin 7x\right) \tag{B-7}$$

and its corresponding σ-modified series is

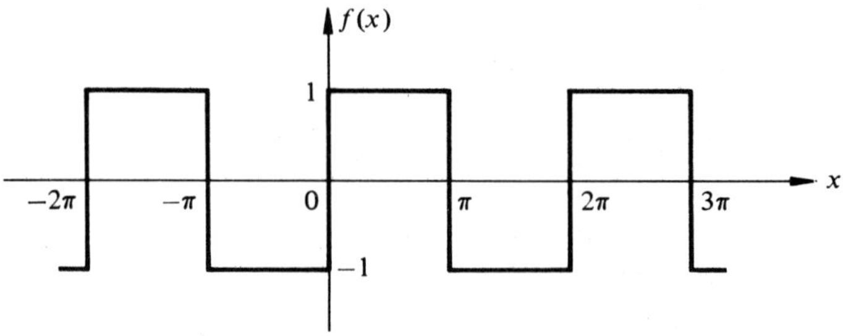

Fig. B-1 Function $f(x)$ whose Fourier series is given by (B-6).

$$\begin{aligned}\bar{f}_8(x) &= \frac{4}{\pi}\left({}_8\sigma_1 \sin x + \frac{1}{3}\,{}_8\sigma_3 \sin 3x + \tfrac{1}{5}\,{}_8\sigma_5 \sin 5x + \frac{1}{7}\,{}_8\sigma_7 \sin 7x\right) \\ &= \frac{4}{\pi}\left(0.9745 \sin x + \frac{0.7842}{3} \sin 3x + \frac{0.4705}{5} \sin 5x \right. \\ &\qquad \left. + \frac{0.1392}{7} \sin 7x\right) \qquad \text{(B-8)}\end{aligned}$$

The variations of $f_8(x)$ and $\bar{f}_8(x)$ are plotted in Fig. B-2 for the region $0 < x < \pi/2$. It is seen from these plots that the overshoot is greatly reduced by the introduction of σ factors. The overshoot of the unmodified series is

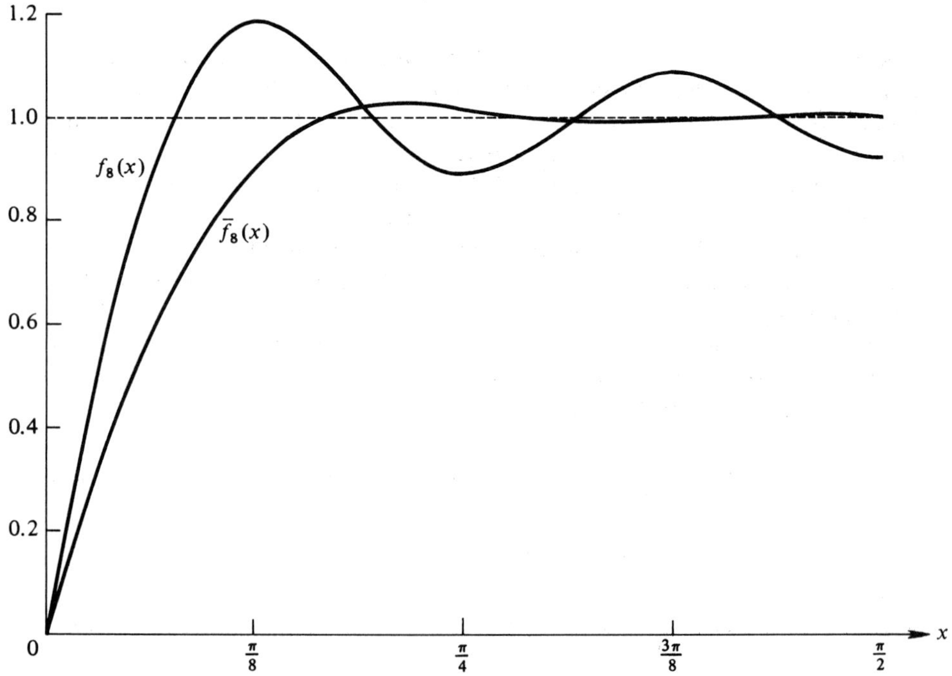

Fig. B-2 Variation of $f_8(x)$ and $\bar{f}_8(x)$ of (B-7) and (B-8).

approximately 18 percent, while that of the modified series is only slightly over 2 percent. This reduction of overshoot is achieved at the expense of a slight increase in rise time. However, it should be noted that the 18 percent overshoot (known as the Gibbs phenomenon) cannot be reduced merely by including more terms in the unmodified series.

Suppose the Fourier series of a function is given to be

$$f(x) = \frac{a_0}{2} + \sum_{k=1}^{\infty} (a_k \cos kx + b_k \sin kx) \tag{B-9}$$

and we let

$$f_m(x) = \frac{a_0}{2} + \sum_{k=1}^{m} (a_k \cos kx + b_k \sin kx) \tag{B-10}$$

The error introduced by representing $f(x)$ by $f_m(x)$ is

$$\epsilon_m(x) = f(x) - f_m(x) = \sum_{k=m+1}^{\infty} (a_k \cos kx + b_k \sin kx) \tag{B-11}$$

When $f_m(x)$ and $f(x)$ are both differentiated term by term, the error between the two derivatives is

$$\epsilon'_m(x) = f'(x) - f'_m(x) = \sum_{k=m+1}^{\infty} k(-a_k \sin kx + b_k \cos kx) \tag{B-12}$$

Although in most series a_k and b_k may be quite small, the coefficients in $\epsilon'_m(x)$ may be quite large, because the k are large for high harmonics.

Now, if the function is smoothed by local averaging, we have

$$\bar{f}(x) = \frac{a_0}{2} + \sum_{k=1}^{\infty} {}_m\sigma_k(a_k \cos kx + b_k \sin kx) \tag{B-13}$$

$$\bar{f}_m(x) = \frac{a_0}{2} + \sum_{k=1}^{m} {}_m\sigma_k(a_k \cos kx + b_k \sin kx) \tag{B-14}$$

and

$$\bar{\epsilon}_m(x) = \bar{f}(x) - \bar{f}_m(x) = \sum_{k=m+1}^{\infty} {}_m\sigma_k(a_k \cos kx + b_k \sin kx) \tag{B-15}$$

Now the error caused by truncating the modified Fourier series in the derivative is

$$\bar{\epsilon}'_m(x) = \bar{f}'(x) - \bar{f}'_m(x) = \sum_{k=m+1}^{\infty} k\, {}_m\sigma_k(-a_k \sin kx + b_k \cos kx) \tag{B-16}$$

and

$$|k\ {}_m\sigma_k| = \left|k\,\frac{\sin(k\pi/m)}{k\pi/m}\right| = \left|\frac{m}{\pi}\sin\frac{k\pi}{m}\right| \leq \frac{m}{\pi}$$

Hence the error introduced in approximating $\bar{f}'(x)$ by $\bar{f}'_m(x)$ is of the same order as the error that is present when $\bar{f}(x)$ is approximated by $\bar{f}_m(x)$. The error introduced by replacing $f'(x)$ by $f'_m(x)$ can, however, be much larger than the difference between $f_m(x)$ and $f(x)$. For this reason, the truncated σ-modified Fourier series is considerably safer to differentiate than the unmodified series without local averaging.

Suppose the square-wave $f(x)$ of Fig. B-1 is the function in question. The derivative of this function is the series of impulse functions as shown in Fig. B-3. The Fourier series of $f(x)$ is given in (B-6). If this series is differentiated term by term, we have

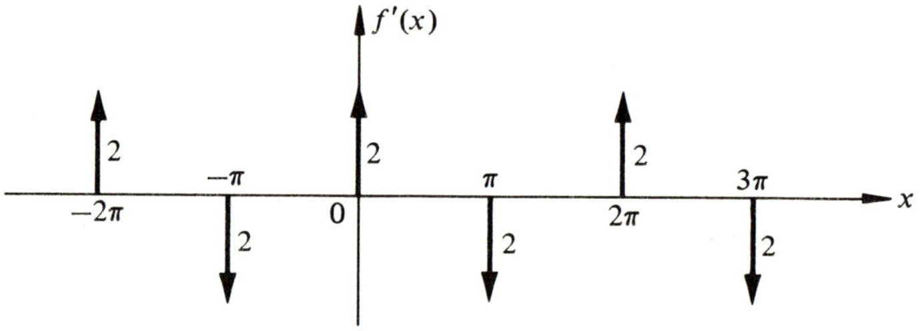

Fig. B-3 Derivative of $f(x)$ of Fig. B-1.

$$f'(x) = \frac{4}{\pi}(\cos x + \cos 3x + \cos 5x + \ldots) \tag{B-17}$$

An example of this series truncated after the eighth term, $f'_8(x)$, is shown in Fig. B-4.

If the square-wave $f(x)$ of Fig. B-1 had been replaced by its σ-modified series or

$$\bar{f}_m(x) = \frac{4}{\pi}\left[{}_m\sigma_1 \sin x + {}_m\sigma_3\frac{\sin 3x}{3} + {}_m\sigma_5\frac{\sin 5x}{5} + \ldots + {}_m\sigma_{(m-1)}\frac{\sin(m-1)x}{m-1}\right] \tag{B-18}$$

its derivative obtained by differentiating term by term is

$$\bar{f}'_m(x) = \frac{4}{\pi}[{}_m\sigma_1 \cos + {}_m\sigma_3 \cos 3x + \ldots + {}_m\sigma_{(m-1)} \cos(m-1)x] \tag{B-19}$$

Equation (B-19) should approximate the impulse train of Fig. B-3 with less ripples than the series in (B-17) does. This is illustrated by a comparison

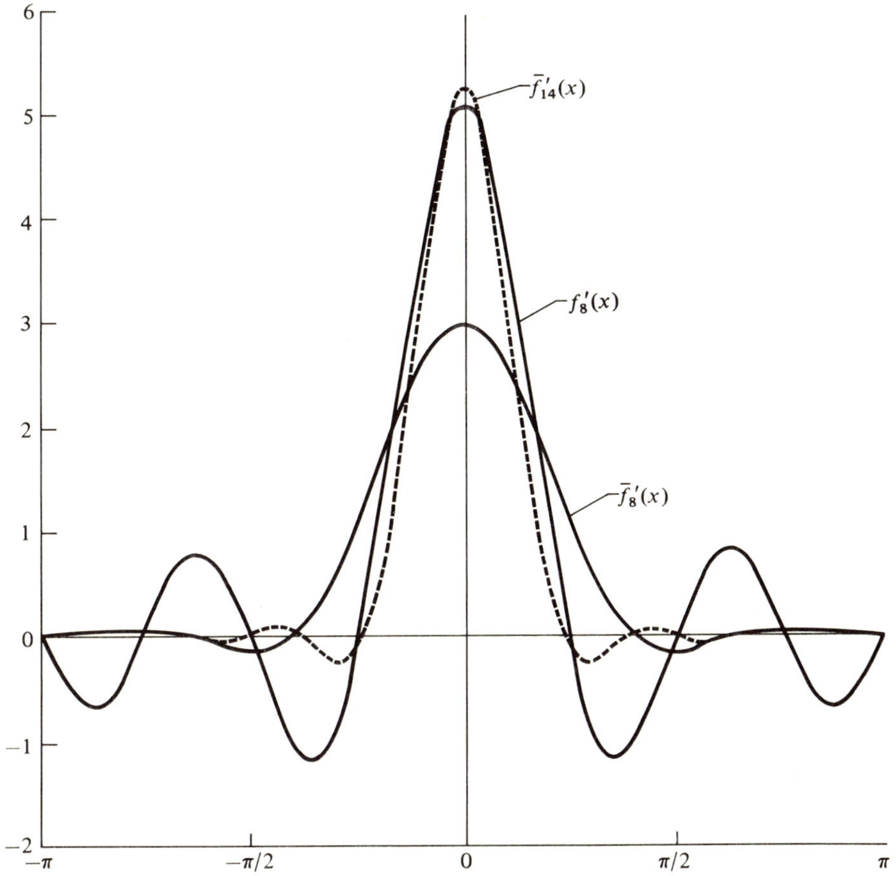

Fig. B-4 Comparison of $f'_8(x)$, $\bar{f}'_8(x)$, and $\bar{f}'_{14}(x)$.

between $\bar{f}'_8(x)$ and $f'_8(x)$ in Fig. B-4. There it is seen that the σ factors greatly reduce the ripples with an accompanying "flattening out" of the impulse. This latter degradation of the main pulse can be somewhat remedied by increasing m. For instance, $\bar{f}'_{14}(x)$ as shown in Fig. B-4 has almost the same main pulse as $f'_8(x)$. But the associated side ripples of $\bar{f}'_{14}(x)$ are far less than those of $f'_8(x)$. This reduction in ripples is due partly to the increase in m and partly to the σ factors.

In a similar manner, if a sequence of positive unit impulses as shown in Fig. B-5 is given, straightforward Fourier analysis gives

$$g(x) = \frac{1}{2\pi} + \frac{1}{\pi}(\cos x + \cos 2x + \cos 3x + \dots) \tag{B-20}$$

When the series of (B-20) is truncated, an approximation of the impulse train of Fig. B-5 is obtained. A plot of a typical cycle of $g_8(x)$, whose highest-order term is cos $8x$, is shown in Fig. B-6.

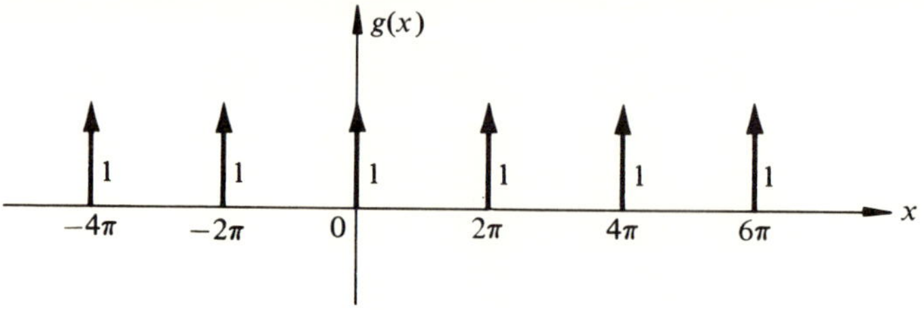

Fig. B-5 A positive impulse train, $g(x)$.

In some applications it is better to represent $g(x)$ by $\bar{g}_m(x)$—the smoothed approximation—than by $g_m(x)$. Here

$$\bar{g}_m(x) = \frac{1}{2\pi} + \frac{1}{\pi}\left[{}_m\sigma_1 \cos x + {}_m\sigma_3 \cos 3x + \dots + {}_m\sigma_{(m-1)} \cos (m-1)x\right] \tag{B-21}$$

A typical cycle of $\bar{g}_8(x)$ is also shown in Fig. B-6. The striking contrast between the two curves is the fact the oscillation in the smoothed wave is much less than that of the former. Again, the "flattening out" of the main pulse may be restored by increasing m of the modified series.

The significance of the Lanczos σ factors can also be interpreted in the following way. Suppose we define an operator D_m such that it implies the operation

$$D_m[f(t)] = \frac{f(t + \pi/m) - f(t - \pi/m)}{2\pi/m} \tag{B-22}$$

This operation appears, at first, to be an approximate method of computing the slope of $f(t)$ at t. This is indeed true for low-frequency components, as is illustrated by Fig. B-7.

For a higher-frequency harmonic, say $(m + k)$th, it is expedient to consider the harmonic in question as a modulated wave with a carrier frequency equal to that of the mth harmonic. In other words, let

$$\begin{aligned} a_{m+k} \cos (m + k)t + b_{m+k} \sin (m + k)t &= a_{m+k} \cos kt \cos mt + b_{m+k} \cos kt \sin mt \\ &\quad - a_{m+k} \sin kt \sin mt + b_{m+k} \sin kt \cos mt \end{aligned} \tag{B-23}$$

It suffices to show the effect of applying the operator D_m on one of the terms of (B-23):

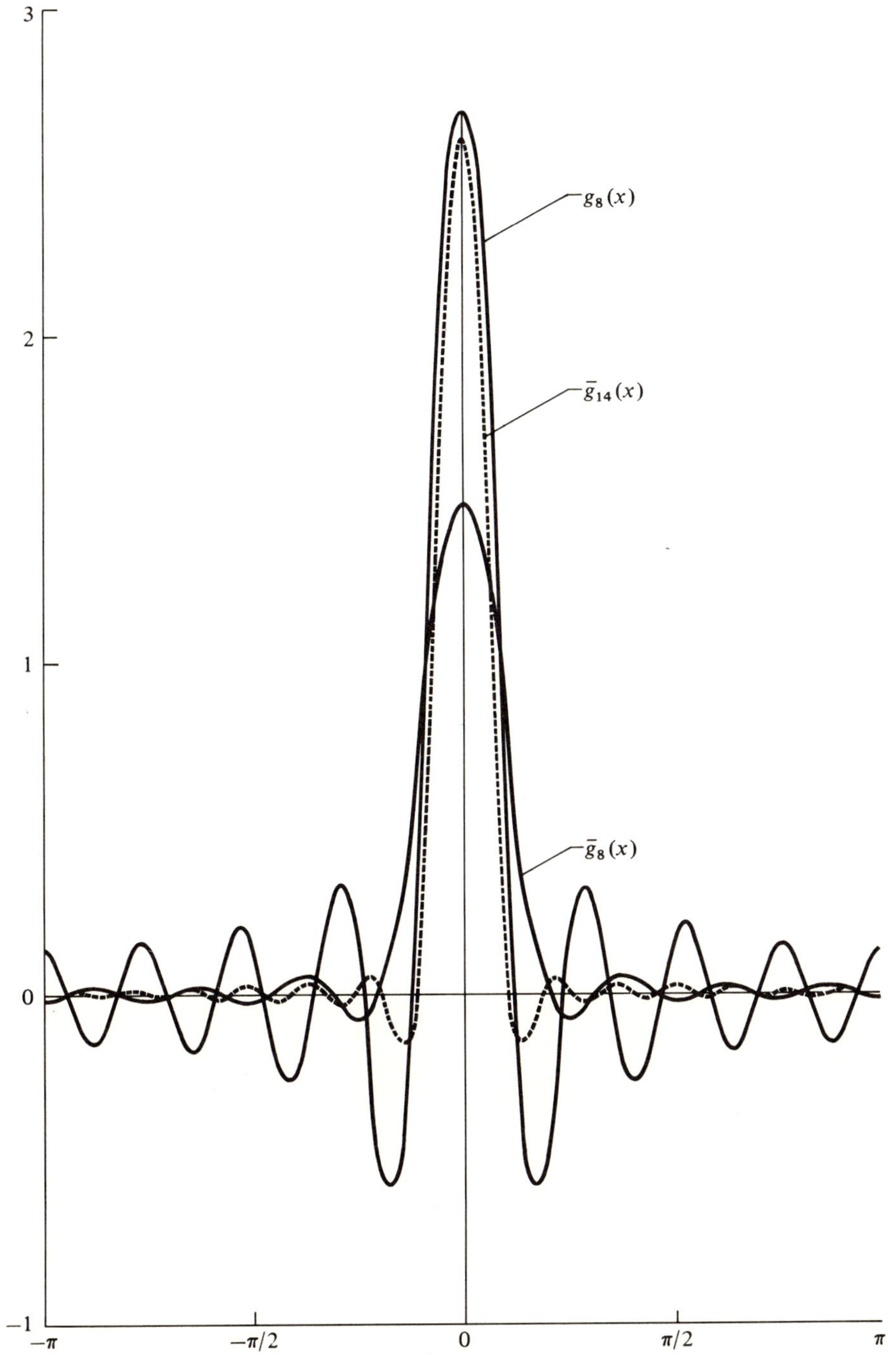

Fig. B-6 Comparison of $g_8(x)$, $\bar{g}_8'(x)$, and $\bar{g}_{14}'(x)$.

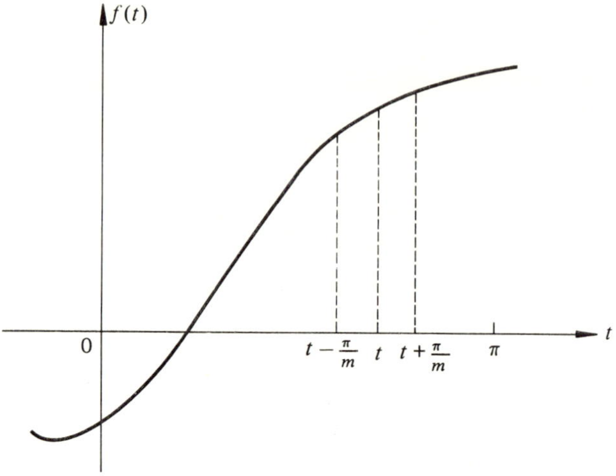

Fig. B-7 Interpretation of the operator D_m.

$$D_m[\cos kt \cos mt]$$
$$= \frac{\cos k(t + \pi/m) \cos (mt + \pi) - \cos k(t + \pi/m) \cos (mt - \pi)}{2\pi/m}$$
$$= \frac{-\cos k(t + \pi/m) + \cos k(t - \pi/m)}{2\pi/m} \cos mt \tag{B-24}$$

The effect of applying D_m to high-frequency terms can be compared to differentiation if we compare (B-24) to (B-12). In (B-12) the carrier and the modulation are not separated and the error is greatly magnified because of the drastic change which exists typically in the carrier wave form. In (B-24), however, the carrier is not being differentiated, while the modulation is treated much the same way as low-frequency components are (except for the sign reversal). Hence $D_m[\epsilon_m(t)]$ is generally much less than $d[\epsilon_m(t)]/dt$ in magnitude.

In carrying out the operation defined by (B-22) on a typical Fourier-series cosine term, we have

$$D_m[\cos kt] = \frac{\cos (kt + k\pi/m) - \cos (kt - k\pi/m)}{2\pi/m}$$
$$= -\frac{\sin (k\pi/m)}{\pi/m} \sin kt \tag{B-25}$$

But

$$\frac{d}{dt}[\cos kt] = -k \sin kt$$

Hence

$$D_m[\cos kt] = \frac{d}{dt}[{}_m\sigma_k \cos kt] \tag{B-26}$$

Similarly, on the sine term

$$\begin{aligned} D_m[\sin kt] \\ &= \frac{\sin(kt + k\pi/m) - \sin(kt - k\pi/m)}{2\pi/m} \\ &= \frac{\sin(k\pi/m)}{\pi/m}\cos kt \end{aligned} \tag{B-27}$$

But

$$\frac{d}{dt}[\sin kt] = k \cos kt$$

Hence

$$D_m[\sin kt] = \frac{d}{dt}[{}_m\sigma_k \sin kt] \tag{B-28}$$

In summary, the process of differentiating the σ-modified version of a Fourier series is equivalent to obtaining the finite differential $\Delta f/\Delta t$, where Δt is taken as the period of the mth harmonic. For high-frequency harmonics this differential is equivalent to the derivative of the envelope of the carrier, which has a frequency equal to the frequency of the mth harmonic.

Furthermore, the process of smoothing a curve by local averaging is equivalent to first finding the finite differential $\Delta f/\Delta t$ of the function $f(t)$ and then integrating this differential.

APPENDIX C

The Chebyshev Rational Function

The properties and applications of the Chebyshev polynomials, $T_n(x)$, in various engineering approximation problems are well known. The magnitude of these polynomials can be made to oscillate between $+1$ and -1 within the limit $-1 < x < +1$. The use of these polynomials usually leads to the equiripple-error type of approximation.

The fact that $T_n(x)$ are polynomials makes them undesirable for certain applications. This is because a polynomial necessarily has all its poles located at infinity. It is sometimes desirable to be able to achieve a similar magnitude variation as the Chebyshev polynomial with functions whose poles are finite and, to some degree, arbitrary. A function that has these properties is the Chebyshev rational function.

The Chebyshev rational function can best be derived by the aid of the conformal transformation between $p = x + jy$ and $s = \sigma + j\omega$ [SH1, UL1]:

$$p^2 = \frac{s^2 + 1}{s^2} \tag{C-1}$$

or

$$s^2 = \frac{1}{p^2 - 1} \tag{C-2}$$

The relationships in the two planes are shown in Fig. C-1.

Since this transformation is irrational and involves square roots, each plane shown in Fig. C-1 is one half of the Reimann surface, which has two layers. In this particular case it is convenient to define that each corresponding quadrant is the image of the others. In other words, all points in the s plane at which

$$\sigma \gtrless 0 \qquad \text{and} \qquad \omega \gtrless 0 \tag{C-3}$$

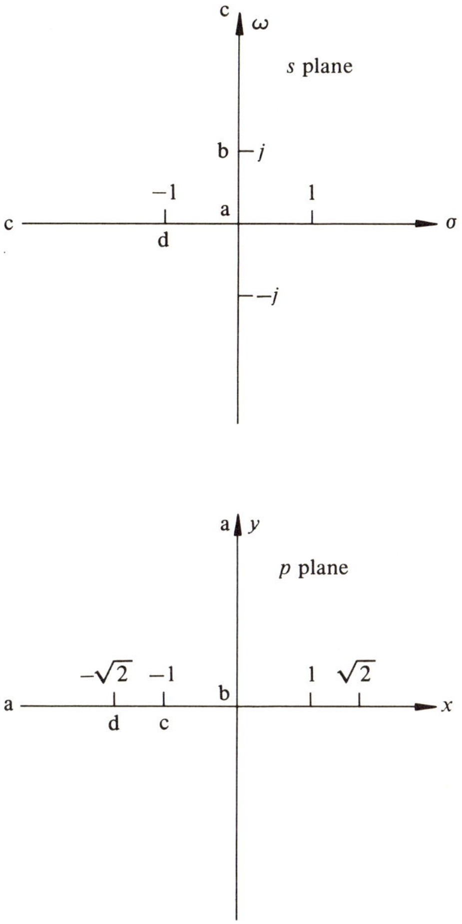

Fig. C-1 Conformal mapping properties of Equations (C-1) and (C-2).

are mapped to points given by (C-1) and

$$x \gtrless 0 \qquad \text{and} \qquad y \gtrless 0 \tag{C-4}$$

respectively.

Thus the transformation given by (C-1) maps the region $-1 < \omega < +1$ on the $j\omega$ axis onto the entire jy axis in the p plane according to the relationship

$$y(\omega) = \sqrt{\frac{1-\omega^2}{\omega}} \tag{C-5}$$

The point at infinity in the s plane is mapped to ± 1 in the p plane.

Suppose now that an arbitrary set of N zeros is given in the s plane at $s = s_1, s_2, \ldots, s_N$. Further, Re $(s_\nu) < 0$. Their corresponding locations in the p plane can be found by (C-1) and may be designated $p_1, p_2, \ldots, p_N$, respectively, and Re $(p_\nu) < 0$.

Suppose now that a Hurwitz polynomial with real coefficients is formed with these zeros, or

$$Q(p) = (p - p_1)(p - p_2)\ldots(p - p_\nu)\ldots(p - p_N) \tag{C-6}$$

and the function $w(p)$ is formed

$$w(p) = \frac{Q(-p)}{Q(p)} \tag{C-7}$$

$w(p)$ is obviously an all-pass function. In other words,

$$|w(jy)| \equiv 1 \tag{C-8}$$

Furthermore, poles and zeros are arranged with quadrantal symmetry in the p plane. A brief observation of the phase angle of $w(p)$ will show that as y is increased from 0 toward infinity along the jy axis, each pair of real singularities symmetric about the jy axis will contribute a net change of $-180°$, while each quadrantal group of four singularities will contribute a net change of $-360°$ to this phase angle. Hence, if we define

$$w(p)|_{p=p(s)} \triangleq w_s(s) \tag{C-9}$$

then arg $[w_s(j\omega)]$ will vary from 0 to $-N\pi$ when ω is varied from 1 to 0.

Now if we form a new function $R(y)$ which is equal to the real part of $w(jy)$, or

$$R(y) = \text{Re}\,[w(jy)] = \frac{1}{2}\left[\frac{Q(p)}{Q(-p)} + \frac{Q(-p)}{Q(p)}\right]_{p=jy} \tag{C-10}$$

it is seen that $R(y)$ always starts with $+1$, goes through N zero values, and approaches either $+1$ (N even) or -1 (N odd) as y varies from 0 to ∞.

If $R(y)$ is transformed back to the s plane through transform (C-1), the same variation must take place along the $j\omega$ axis from $0 \leq \omega \leq 1$. Here the value of $R_s(\omega) \triangleq R\,[y(\omega)]$ starts with either $+1$ (even N) or -1 (odd N), oscillates between $+1$ and -1 by going through N zeros, and always attains the value of $+1$ at $\omega = 1$. $R_s(\omega)$ is obviously an even function of ω. An example for $N = 4$ is shown in Fig. C-2.

Thus the function

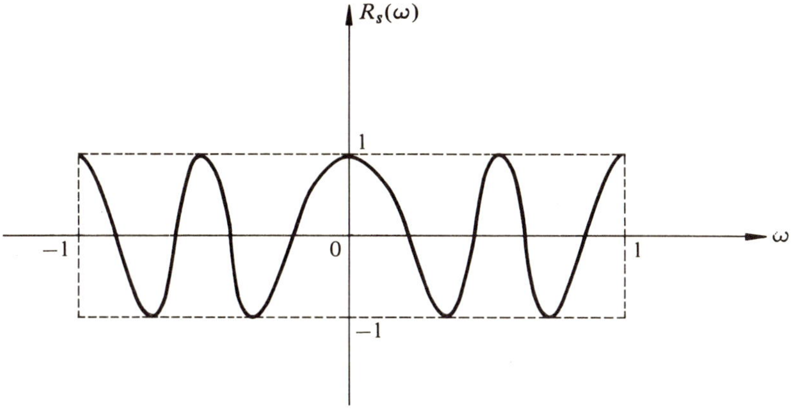

Fig. C-2 Typical variation of a Chebyshev rational function along the $j\omega$ axis in the region $-1 < \omega < 1$, $N = 4$.

$$R(s) = \frac{1}{2}\left(\prod_{\nu=1}^{N} \frac{-p - p_\nu}{p - p_\nu} + \prod_{\nu=1}^{N} \frac{p - p_\nu}{-p - p_\nu}\right)_{p^2=(s^2+1)/s^2} \tag{C-11}$$

is defined as the *Chebyshev rational function* because it has the equiripple behavior in a finite interval much like the behavior of a Chebyshev polynomial. Since $Q(p)$ is a Hurwitz polynomial with real coefficients, $R(s)$ is a real rational function.

Outside the interval of $|\omega| < 1$ it is difficult to state how $R(j\omega)$ might behave, as the latter depends on the locations of p_ν's. In some cases $R(j\omega)$ may have a number of minima and maxima before approaching a finite value. In other instances $R(j\omega)$ may approach a finite value monotonically. However, if unbounded values of $R(j\omega)$ is to be avoided for any reason, it is clear that no p_ν should be permitted in the interval on the negative real axis in which $|p| \leq 1$.

When a set of poles are specified in the s plane, the Chebyshev rational function with these poles may be obtained through (C-11). A somewhat more convenient form of that equation might be

$$R(s) = \frac{1}{2}\,\frac{\prod_{\nu=1}^{N}(p - p_\nu)^2 + \prod_{\nu=1}^{N}(p + p_\nu)^2}{\prod_{\nu=1}^{N}(p^2 - p_\nu^2)} \tag{C-12}$$

$$p_\nu = \frac{\sqrt{s_\nu^2 + 1}}{s_\nu} \tag{C-13}$$

in which it is understood that for $Re\,(p_\nu) > 0$, the corresponding $\text{Re}\,(s_\nu) > 0$. Substitution of (C-13) into (C-12) gives

$$R(s) = \frac{1}{2} \frac{\prod_{\nu=1}^{N} \left(\frac{\sqrt{s^2+1}}{s} - \frac{\sqrt{s_\nu^2+1}}{s_\nu}\right)^2 + \prod_{\nu=1}^{N} \left(\frac{\sqrt{s^2+1}}{s} + \frac{\sqrt{s_\nu^2+1}}{s_\nu}\right)^2}{\prod_{\nu=1}^{N} \left(\frac{s_\nu^2+1}{s_\nu^2} - \frac{s^2+1}{s^2}\right)}$$

$$= \frac{1}{2} \frac{\prod_{\nu=1}^{N} (s_\nu^2\sqrt{s^2+1} - s\sqrt{s_\nu^2+1})^2 + \prod_{\nu=1}^{N} (s_\nu\sqrt{s^2+1} + s\sqrt{s_\nu^2+1})^2}{\prod_{\nu=1}^{N} (s^2 - s_\nu^2)} \tag{C-14}$$

There are occasions in which the variation in the equiripple region is desired to be between +1 and zero. Such a function can, of course, be obtained easily by squaring $R(s)$. $R^2(s)$, however, will require a duplicate set of poles (double poles) and invariably increases the number of ripples.

An alternative is to construct a new function $G(s)$ such that

$$G(s) = \frac{1}{2}[R(s) + 1] \tag{C-15}$$

This new function can be expressed as

$$G(s) = \frac{\left[\prod_{\nu=1}^{N} (s\sqrt{s_\nu^2+1} - s_\nu\sqrt{s^2+1}) + \prod_{\nu=1}^{N} (s\sqrt{s_\nu^2+1} + s_\nu\sqrt{s^2+1})\right]^2}{4\prod_{\nu=1}^{N} (s^2 - s_\nu^2)} \tag{C-16}$$

References

[AB 1] Abele, Th.-A., "Übertragungsfaktoren mit Tschebyscheffscher Approximation konstanter Gruppenlaufzeit," *Archiv der Elektrischen Übertragung*, vol. 16, pp. 9-18, January 1962.

[AB 2] Abramowitz, M., and I. A. Stegun, *Handbook of Mathematical Functions*, Dover Publications, Inc., New York, 1965.

[AI 1] Aigrain, P. R., and E. M. Williams, "Synthesis of *n*-reactance networks for desired transient response," *Journal of Applied Physics*, vol. 20, p. 597, June 1949.

[AI 2] Aigrain, P. R., and E. M. Williams, "Design of optimum transient response amplifiers," *Proceedings of the IRE*, vol. 37, pp. 873-879, August 1949.

[AL 1] Allemandou, P., "Low-pass filters approximating—in modulus and phase—the exponential function," *IEEE Transactions on Circuit Theory*, vol. CT-13, pp. 298-301, September 1966.

[AN 1] Anderson, G. G., and S. C. Dutta Roy, "Improving the transient-response characteristics of pulse filters with parabolic distribution of poles," *Electronics Letters*, vol. 4, pp. 173-174, May 3, 1968.

[AR 1] Armstrong, H. L., "On finding an orthonormal basis for representing transients," *IRE Transactions on Circuit Theory*, vol. CT-4, pp. 286-287, September 1957.

[AR 2] Armstrong, H. L., "On the representation of transient by series of orthogonal functions," *IRE Transactions on Circuit Theory*, vol. CT-6, pp. 351-354, December 1959.

[AT 1] Athanassopoulos, I. A., J. D. Schoeffler, and A. D. Waren, "Time domain synthesis by nonlinear programming," *Proceedings Fourth Allerton Conference on Circuit and System Theory*, pp. 766-775, 1965.

[AT 2] Athanassopoulos, I. A., J. D. Schoeffler, and A. D. Waren, "Optimal synthesis in the time domain," *Proceedings of the IEEE*, vol. 54, pp. 1947-1948, December 1966.

[BA 1] Ba Hli, F., "A general method for time domain network synthesis," *Transactions of the IRE Professional Group on Circuit Theory*, vol. CT-1, pp. 21-28, September 1954.

[BA 2] Balabanian, N., *Network Synthesis*, Prentice-Hall, Inc., Englewood Cliffs, N.J., 1958.

[BA 3] Bashkow, T. R., "The *A* matrix, new network description," *IRE Transactions on Circuit Theory*, vol. CT-4, pp. 117-119, September 1957.

[BA 4] Bangert, J. T., "Practical applications of time domain theory," *IRE Wescon Convention Record*, vol. 3, pt. 3, pp. 29-38, 1959.

[BE 1] Beletskii, A. F., "Synthesis of filters with linear phase characteristics," *Telecommunications*, vol. 1961, pp. 408-419, April 1961.

[BE 2] Bennett, B. J., "Synthesis of electric filters with arbitrary phase characteristics," *1953 IRE Convention Record*, pt. 5, pp. 19-26.

[BE 3] Bergholm, J. O., and K. A. Fegley, "Time domain synthesis using linear and quadratic programming," *1967 IEEE International Convention Record*, pt. 5, pp. 121-126.

[BE 4] Bellman, R. E., *Adaptive Control Processes*, Princeton University Press, Princeton, N.J., 1961.

[BE 5] Bellman, R. E., R. E. Kalaba, and J. A. Lockett, *Numerical Inversion of the Laplace Transform*, American Elsevier Publishing Company, Inc., New York, 1966.

[BE 6] Bellman, R. E., and S. E. Dreyfus, *Applied Dynamic Programming*, Princeton University Press, Princeton, N.J., 1962.

[BE 7] Bellman, R. E., *Dynamic Programming*, Princeton University Press, Princeton, N.J., 1957.

[BO 1] Bode, H. W., *Network Analysis and Feedback Amplifier Design*, Van Nostrand Reinhold Company, New, York, 1953.

[BR 1] Bradley, W. E., "Design of a simple bandpass amplifier with approximate ideal frequency characteristics," *Transactions of the IRE Professional Group on Circuit Theory*, vol. PGCT-2, pp. 30-38, December 1953.

[BR 2] Brockhaus, M., and H. W. Schüssler, "New results in the design of pulse-forming networks," presented at the 1966 Conference on Networks, University of Newcastle-upon-Tyne, England, September 1966.

[BR 3] Brulé, J. D., "Comments on an approximation procedure," *IRE Transactions on Circuit Theory*, vol. CT-8, pp. 81-82, March 1961.

[BU 1] Burchnall, J. L., "The Bessel polynomials," *Canadian Journal of Mathematics*, vol. 3, pp. 62-68, 1951.

[CA 1] Calahan, D. A., "On magnitude functions with equiripple response," *IEEE Transaction on Circuit Theory*, vol, CT-11, pp. 176-177, March 1964.

[CA 2] Cauer, W., *Synthesis of Linear Communication Networks*, McGraw-Hill, Inc., New York, 1958.

[CE 1] Cerrillo, M. V., "Study of transient effects by a new method of integral approximation," *Paper 29, IRE National Convention*, New York, March 7, 1950.

[CH 1] Chen, W. H., *Linear Network Design and Synthesis*, McGraw-Hill, Inc, New York, 1964.

[CO 1] Corrington, M. S., and R. W. Sonnenfeldt, "Synthesis of constant time delay networks," *RCA Review*, vol. 15, pp. 163-186, June 1954; also *Proceedings of the National Electronics Conference*, vol. 9, pp. 50-63, 1953.

[CR 1] Crockett, J. B., and H. Chernoff, "Gradient method of maximization," *Pacific Journal of Mathematics*, vol. 5, pp. 33-50, 1955.

[CU 1] Cunningham, W. J., "Time delay networks for an analog computer," *IRE Transactions on Electronic Computers*, vol. PGEC-3, pp. 16-18, December 1954.

[CU 2] Curry, H. B., "The method of steepest descent for nonlinear minimization problems," *Quarterly Applied Mathematics*, vol. 2, pp. 258-261, October 1944.

[DA 1] Darlington, S., "Network synthesis using Tchebyscheff polynomial series," *Bell System Technical Journal*, vol. 31, pp. 613-665, July 1952.

[DA 2] Davidon, W. C., "Variable metric method for minimization," *Argonne National Laboratory, ANL-5990 Rev.*, November 1959.

[DE 1] Deutsch, S., "Maximally flat time delay ladders," *IRE Transactions on Circuit Theory*, vol. CT-6, pp. 214-218, June 1959.

[DE 2] Dewsnap, G. D., "The approximation of a time delay," *Proceedings of IREE (Australia)*, vol. 25, pp. 168-174, March 1964.

[DI 1] Dishal, M., "Gaussian-response filter design," *Electrical Communication*, vol. 36, pp. 3-26, 1959.

[DU 1] Dutta Roy, S. C., "Reducing overshoot and ringing in filters," *Proceedings of the IEEE*, vol. 54, pp. 1189-1190, September 1966.

[DU 2] Dutta Roy, S. C., "Transient response characteristics of a third-order filter," *IEEE Transactions on Circuit Theory*, vol. CT-15, pp. 69-71, March 1968.

[DU 3] Dutta Roy, S. C., "On the transient response of all-pole low-pass filters," *IEEE Transactions on Circuit Theory*, vol. CT-15, pp. 485-487, December 1968.

[EL 1] Elmore, W. C., "The transient response of damped linear network with particular regard to wideband amplifiers," *Journal of Applied Physics*, vol. 19, pp. 55-63, January 1948.

[FE 1] Fetzer, V., "The transfer function of networks with prescribed transient response," *Nachichtentechnische Zeitschrift*, vol. 9, pp. 462-468, October 1956.

[FI 1] Fischl, R., "Optimal Chebyshev approximation at discrete points in the time domain," *Proceedings Seventh Midwest Symposium on Circuit Theory*, pp. 33-43, 1964.

[FI 2] Fiacco, A. V., and G. P. McCormick, "Time sequential unconstrained minimization technique for nonlinear programming, a primal dual method," *Management Science*, vol. 10, pp. 360-366, January 1964.

[FL 1] Fletcher, R., and M. J. D. Powell, "A rapidly convergent descent method for minimization," *British Computer Journal*, vol. 6, pp. 163-168, 1963.

[FR 1] Fredendall, G.L., and R.C. Kennedy, "Linear phase shift video amplifiers," *RCA Review*, vol. 11, pp. 418-430, September 1950.

[FR 2] Frobenius, G., "Über Relationen zwischen den Näherungsbruchen von Potenzreihen," *Journal fur di renine und angewandte Mathematik*, vol. 90, pp. 1-17, 1881.

[GE 1] Geffe, P., "On the approximation for bandpass delay lines," *Proceedings of the IRE*, vol. 50, p. 1986, September 1962.

[GH 1] Ghausi, M. S., *Principles and Design of Linear Active Circuits*, McGraw-Hill, Inc., New York, 1964, chap. 16.

[GH 2] Ghausi, M. S., and M. Adamowicz, "A new class of filters for pulse applications," *Journal of the Franklin Institute*, vol. 282, pp. 20-30, July 1966.

[GI 1] Gibson, J. E., Z. V. Rekasius, E. S. McVey, R. Sridhar, and C. D. Leedham, "A set of standard specifications for linear automatic control systems," *Transactions of the AIEE*, vol. 80, pt. 2, pp. 65-74, May 1961.

[GO 1] Gouriet, G. G., "Two theorems concerning group delay with practical application to delay correction," *Proceedings of the IEE* (*London*), vol. 105, pt. C, pp. 240-244, March 1958 (Monograph 275R, December 1957).

[GR 1] Gray, R. L., "All-pole delay network—a time domain approximation," *IEEE Transactions on Circuit Theory*, vol. CT-11, pp. 410-412, September 1964.

[GR 2] Grinich, V. H., "Approximating band pass attenuation and phase functions," *IRE Convention Record*, vol. 2, pt. 2, pp. 96-104, 1954.

[GU 1] Guillemin, E. A., *Synthesis of Passive Networks*, John Wiley & Sons, Inc., New York, 1957.

[HA 1] Hamming, R. W., *Numerical Methods for Scientists and Engineers*, McGraw-Hill, Inc., New York, 1962.

[HA 2] Hazony, D., *Elements of Network Synthesis*, Van Nostrand Reinhold Company, New York, 1963.

[HA 3] Hadley, C., *Nonlinear and Dynamic Programming*, Addison-Wesley Publishing Company, Inc., Reading, Mass., 1964.

[HE 1] Hellman, D., "Comments on 'Synthesis of electric filters with arbitrary phase characteristics,'" *IRE Transactions on Circuit Theory*, vol. CT-2, pp. 217-218, June 1955.

[HE 2] Hellman, D., "Tschebyscheff approximations for amplitude and delay with rational functions," *Proceedings of the Symposium on Modern Network Synthesis*, Microwave Research Institute Symposia Series, Polytechnic Institute of Brooklyn, vol. 5, pp. 385-402, April 1955.

[HE 3] Henderson, K. W., and W. H. Kautz, "Transient responses of conventional filters," *IRE Transactions on Circuit Theory*, vol. CT-5, pp. 333-347 December 1958.

[HE 4] Herrmann, O., and H. W. Schüssler, "Zur Auswahl von Filtern mit günstigem Einschwingverhalten," *Archiv der Elektrischen Übertragung*, vol. 14, pp. 183-189, April 1960.

[HI 1] Hibino, M., Y. Ishizaki, and H. Watanabe, "Design of Chebyshev filters with flat group-delay characteristics," *IEEE Transactions on Circuit Theory*, vol. CT-15, pp. 316-325, December 1968.

[HI 2] Hindin, H. J., "Impulse response and transfer phase of transitional Butterworth-Thompson filters," *IEEE Transactions on Circuit Theory*, vol. CT-15, pp. 471-474, December 1968.

[HO 1] Holt, A. G. J., J. P. Gray, and J. K. Fidler, "Transient response of elliptic function filters," *IEEE Transactions on Circuit Theory*, vol. CT-15, pp. 71-73, March 1968.

[HU 1] Huggins, W. H., "Signal theory," *IRE Transactions on Circuit Theory*, vol. CT-3, pp. 210-216, December 1956.

[HU 2] Humpherys, D. S., "A relationship between impulse response overshoot and phase ripple," *IEEE Transaction on Circuit Theory*, vol. CT-10, pp. 450-452, September 1963.

[IE 1] Iedokoro, T., Y. Ishizaki, and H. Watanabe, "On the approximation and synthesis of Tchebycheff filters with flat group delay characteristics," *Proceedings First Allerton Conference on Circuit and System Theory*, pp. 16-28, 1963.

[JE 1] Jess, J., "Über Impulsfilter mit Tschebyscheffschem Verhalten im Ziet- und Frequenzbereich," *Archiv der Elektrischen Übertragung*, vol. 16, pp. 391-401, August 1963.

[JE 2] Jess, J., and H. W. Schüssler, "Über Filter mit günstigem Einschwingverhalten," *Archiv der Elektrischen Übertragung*, vol. 16, pp. 116-128, March 1962.

[JE 3] Jess, J., and H. W. Schüssler, "A class of pulse-forming networks," *IEEE Transactions on Circuit Theory*, vol. CT-12, pp. 296-299, June 1965.

[JE 4] Jess, J., and H. W. Schüssler, "On the design of pulse-forming networks," *IEEE Transactions on Circuit Theory*, vol. CT-12, pp. 393-400, September 1965.

[JO 1] Johnson, W. C., *Transmission Lines and Networks*, McGraw-Hill, Inc., New York, 1950.

[JO 2] Jones, S. G., "Pulse networks with parabolic distribution of poles," *IRE Transactions on Circuit Theory*, vol. CT-9, p. 297, September 1962.

[KA 1] Karni, S., and N. Ahmed, "The derivation of minimum phase network functions from time-delay specifications," *Proceedings Second Allerton Conference on Circuit Theory and Systems*, pp. 220–232, 1964.

[KA 2] Kautz, W. H., "Application of integral approximation method of transient evaluation," *Paper 30, IRE National Convention*, New York, March 7, 1950.

[KA 3] Kautz, W. H., "Network synthesis for specified transient response," *Tech. Rept. 209, Research Laboratory of Electronics, Massachusetts Institute of Technology*, April 23, 1952; also *Paper 134, IRE National Convention*, New York, March 1952.

[KA 4] Kautz, W. H., "Transient synthesis in the time domain," *Transactions of the IRE Professional Group on Circuit Theory*, vol. CT-1, pp. 29–38, September 1954.

[KE 1[Kerr, R. B., and W. H. Surber, Jr., "Precision of impulse-response identification based on short, normal operation records," *IRE Transactions on Automatic Control*, vol. AC-6, pp. 173–182, May 1961.

[KH 1] Khovanskii, A. N., *The Application of Continued Fractions and Their Generalizations to Problems in Approximation Theory*, P. Noodhoff, N.V., Groningen, 1963.

[KI 1] Kiseda, J. R., and D. J. Ford, "Ripple type time delay networks using elliptic functions," *Transactions of the AIEE*, vol. 78, pt. 1, pp. 996–1002, 1960.

[KR 1] Krall, H. L., and O. Fink, "A new class of orthogonal polynomials: the Bessel polynomials," *Transactions of the American Mathematical Society*, vol. 65, pp. 100–115, January 1949.

[KU 1] Ku, Y. H., and M. Durbin, "Network synthesis using Legendre and Hermite polynomials," *Journal of the Franklin Institute*, vol. 273, pp. 138–157, February 1962.

[KU 2] Kuh, E. S., "Synthesis of lumped parameter precision delay line," *Proceedings of the IRE*, vol. 45, pp. 1632–1642, December 1957.

[KU 3] Kuh, E. S., and D. O. Pederson, *Principles of Circuit Synthesis*, McGraw-Hill, Inc., New York, 1959.

[KU 4] Kulmann, C. A., "An iterative approximation to flat delay in an equiripple manner," *ERL Rept. 89* (Series 60, Issue 247), *University of California, Berkeley*, June 1959.

[KU 5] Kuo, F. F., "A method to reduce overshoot in filters," *IRE Transactions on Circuit Theory*, vol. CT-9, pp. 413–414, December 1962.

[KU 6] Kuo, F. F., *Network Analysis and Synthesis*, John Wiley & Sons., Inc., New York, 1962.

[KU 7] Kuo, F. F., and M. Karnaugh, "Approximation of linear phase filters with Gaussian damped impulse response," *IEEE Transactions on Circuit Theory*, vol. CT-11, pp. 255–260, June 1964.

[KW 1] Kwan, R. K., and G. G. Bach, "Equiripple approximations of group delay and attenuation in network synthesis," *Proceedings Fourth Annual Allerton Conference on Circuit and System Theory*, pp. 186–196, 1966.

[LA 1] Lanczos, C., *Applied Analysis*, Prentice-Hall, Inc., Englewood Cliffs, N.J., 1956.

[LA 2] Lavi, A., and T. P. Vogl, *Recent Advances in Optimization*, Prentice-Hall, Inc., Englewood Cliffs, N.J., 1966.

[LE 1] Lee, Y. W., "Synthesis of electrical networks by means of the Fourier transforms of Laguerre's functions," *Journal of Mathematical Physics*, vol. 11, pp. 83–113, 1932.

[LE 2] Lewis, N. W., "Waveform computations by the time-series method," *Proceedings of the IEE* (*London*), vol. 99, pt. 3, pp. 294–306, September 1952.

[LI 1] Liu, B., "A time domain approximation method and its application to lumped delay lines," *IRE Transactions on Circuit Theory*, vol. CT-9, pp. 256–261, September 1962.

[LI 2] Liu, B., "Approximation to a specified time response," *IEEE Transactions on Circuit Theory*, vol. CT-10, p. 115, March 1963.

[LI 3] Liu, B., and H. E. Meadows, "A method of time-domain approximation for arbitrary input," *1964 IEEE International Convention Record*, pt. 1, pp. 338–343, 1964.

[MA 1] Macnee, A. B., "Chebyshev approximation of a constant group delay," *IEEE Transactions on Circuit Theory*, vol. CT-10, pp. 284–285, June 1963.

[MA 2] Mathers, G. W. C., "The synthesis of lumped-element circuits for optimum transient response," *Tech. Rept. 38, Electronics Research Laboratory, Stanford University*, November 1, 1951.

[MC 1] McBride, L. E., Jr., H. W. Schaefgen, and K. Stieglits, "Time-domain approximation by iterative method," *IEEE Transactions on Circuit Theory*, vol. CT-13, pp. 381–387, December 1966.

[MC 2] McGee, W. F., "Band-pass Bessel polynomial filters and delay approximation," *IEEE Transactions on Circuit Theory*, vol. CT-12, pp. 622–623, December 1965.

[ME 1] Meyer, E. R., "A note on the step response of high-pass filters," *IEEE Transactions on Circuit Theory*, vol. CT-18, pp. 481–482, December 1968.

[ME 2] Meyer, P. A., "Über den Entwurf von Vergögerungsschaltungen im Zeitbereich," *Archiv der Elektrischen Übertragung*, vol. 20, pp. 610–616, November 1966.

[MI 1] Mitra, S. K., *Analysis and Synthesis of Linear Active Networks*, John Wiley & Sons, Inc., New York, 1969.

[MO 1] Moore, A. K., "An application of Prony's method to time domain network synthesis," *Tech. Rept. 42, Electronic Research Laboratories, Stanford University*, February 1, 1952.

[MO 2] Morrill, W. D., "A Sub-audio time delay circuit," *IRE Transactions on Electronic Computers*, vol. PGEC-3, pp. 45-49, June 1954.

[MO 3] Morrison, R., "Linear phase versus step response," *IEEE Transactions on Circuit Theory*, vol. CT-14, pp. 429-430, December 1967.

[MU 1] Muller, F. A., "High-frequency compensation of *RC* amplifiers," *Proceedings of the IRE*, vol. 42, pp. 1271-1276, August 1954.

[MU 2] Mullick, S. K., "Pulse networks with parabolic distribution of poles," *IRE Transactions on Circuit Theory*, vol. CT-8, pp. 302-305, September, 1961.

[MU 3] Mullick, S. K., "Optimum interval approximation in the time domain," *IRE Transactions on Circuit Theory*, vol. CT-8, pp. 369-371, September 1961.

[MU 4] Mulligan, J. H., "The effect of pole and zero locations on the transient response of linear dynamic systems," *Proceedings of the IRE*, vol. 37, pp. 516-529, May 1949; correction p. 1181, October 1949.

[NA 1] Nadler, M., "The synthesis of electric networks according to prescribed transient conditions," *Proceedings of the IRE*, vol. 37, pp. 627-630, June 1949.

[NA 2] Nadler, M., "A note on the synthesis of electric networks according to prescribed transient response," *Proceedings of the IRE*, vol. 38, p. 441, April 1950.

[NE 1] Neirynck, J. J., "Transient behavior of systems with equal-ripple delay," *IEEE Transactions on Circuit Theory*, vol. 11, pp. 302-303, June 1964.

[NE 2] Neirynck, J. J., and Ph. van Bastalaer, "A class of filters with optimal transient behavior," presented at the 1965 Symposium on Network Theory, Cranfield, England, September 1965.

[NE 3] Newcomb, R. W., *Linear Multiport Synthesis*, McGraw-Hill, Inc., New York, 1966.

[NE 4] Newcomb, R. W., *Active Integrated Circuit Synthesis*, Prentice-Hall, Inc., Englewood Cliffs, N.J., 1968.

[OR 1] Orchard, H. J., "The roots of the maximally flat delay polynomials," *IEEE Transactions on Circuit Theory*, vol. CT-12, pp. 452-454, September 1965.

[PA 1] Padé, H., "Sur la représentation approchée d'une function par des fractions rationelles," *Annales Scientifiques de l'École Normale Superieure*, ser. 3, vol. 9, pp. 1-93 (suppl.), 1892.

[PA 2] Padulo, L., "The time-domain inversion of convolution," Ph.D. Thesis, Georgia Institute of Technology, May 1966.

[PA 3] Papoulis, A., "Network response in terms of behavior at imaginary frequencies," *Proceeding of the Symposium on Modern Network Synthesis*,

Microwave Research Institute Symposia Series, Polytechnic Institute of Brooklyn, vol. 5, pp. 403–424, April 1955.

[PA 4] Papoulis, A., "The approximation problem in lumped delay lines," *IRE National Convention Record*, vol. 8, pt. 2, pp. 102–108, 1958.

[PA 5] Papoulis, A., "Optimum filters with monotonic response," *Proceedings of the IRE*, vol. 46, pp. 606–609, March 1958.

[PA 6] Papoulis, A.. "On monotonic response of filters," *Proceedings of the IRE*, vol. 47, pp. 332–333, February 1959.

[PE 1] Peless, Y., and T. Murakami, "Analysis and synthesis of transitional Butterworth-Thompson filters and bandpass amplifiers," *RCA Review*, vol. 18, pp. 60–94, March 1957.

[PO 1] Poschenrieder, W., and H. Sontheim, "Ein Analogiegerät für Aufgaban der Filterenwicklung," *Frequenz*, vol. 13, pp. 379–385, December 1959.

[PO 2] Postl, W., "Ein Energiekriterium zur Beurteilung von Impulsfiltern," *Archiv der Elektrischen Übertragung*, vol. 16, pp. 351–355, July 1962.

[RA 1] Rakovich, B. D., "Pulse networks with elliptic distribution of poles," *Electronics Letters*, vol. 3, pp. 531–533, December 1967.

[RA 2] Rakovich, B. D., "Transfer functions approximating a constant group delay," *Electronic Engineering*, vol. 40, pp. 242–246, May 1968; pp. 326–328, June 1968.

[RA 3] Rakovich, B. D., and D. M. Rabrenovich, "Transfer functions with poles located on parabolic and elliptic contours," *Electronics Letters*, vol. 3, pp. 201–202, May 1967.

[RA 4] Rakovich, B. D., and D. M. Rabrenovich, "On the problem of constant group delay approximation," *IEEE Transactions on Circuit Theory*, vol. CT-15, pp. 66–67, March 1968.

[RE 1] Reddick, H. W., and F. H. Miller, *Advanced Mathematics for Engineers*, John Wiley & Sons, Inc., New York, 1948.

[RO 1] Rosen, J. B., "The gradient projection method for nonlinear programming: Part I, Linear constraints," *SIAM Journal*, vol. 8, pp. 181–217, 1960.

[RO 2] Rosen, J. B., "The gradient projection method for nonlinear programming: Part II, Nonlinear constraints," *SIAM Journal*, vol. 9, pp. 515–523, 1961.

[SC 1] Scanlan, J. O., "A class of transfer functions having approximately maximally flat group-delay characteristics," *IEEE Transactions on Circuit Theory*, vol. CT-10, p. 122, March 1963.

[SC 2] Scanlan, J. O., "The linear phase approximation problem," *IEEE International Convention Record*, pt. 1, pp. 289–295, 1964.

[SC 3] Scanlan, J. O., "Transfer functions with elliptic distribution of poles at equal spacings," *IEEE Transactions on Circuit Theory*, vol. CT-12, pp. 206–266, June 1965.

[SC 4] Schultz, W. C., and V. C. Rideout, "A general criterion for servo performance," *Proceedings of the National Electronics Conference*, vol. 13, pp. 549–560, 1957.

[SC 5] Schultz, W. C., and V. C., Rideout, "The selection and use of servo performance criteria," *Transactions of the AIEE*, vol. 76, pt. 2 (Application and Industry), pp. 383–388, January 1958.

[SC 6] Schultz, W. C., and V. C. Rideout, "Performance measures: past, present and future," *IRE Transactions on Automatic Control*, vol. AC-6, pp. 22–35. February 1961.

[SH 1] Sharpe, C. B., "A general Tchebycheff rational function," *Proceedings of the IRE*, vol. 42, pp. 454–457, February 1954.

[SH 2] Shenoi, B. A., "Pulse networks with parabolic distribution of poles and Butterworth networks compared," *IEEE Transactions on Circuit Theory*, vol. CT-11, p. 278, June 1964.

[SH 3] Shah, B. V., R. J. Buehler, and O. Kempthorne, "Some algorithms for minimizing a function of several variables," *SIAM Journal*, vol. 12, pp. 74–92, March 1964.

[SK 1] Skwirzynski, J. K. and J. Zdunek, "Design of networks with prescribed delay and amplitude characteristics," *The Marconi Review*, vol. 23, pp. 115–139, 1960.

[SM 1] Smith, G. W., Jr., "Dual requirements for optimizing orthonormal approximations of impulse responses," *IRE Transactions on Circuit Theory*, vol. CT-8, pp. 486–487, December 1961.

[SP 1] Spaulding, D. A., "Synthesis of pulse-shaping networks in the time domain," *Bell System Technical Journal*, vol. 48, pp. 2425–2444, September 1969.

[ST 1] Stewart, J. L., "Quasi-distortionless filter functions," *IRE Transactions of the Professional Group on Circuit Theory*, vol. PGCT-2, pp. 39–54, December 1953.

[ST 2] Storch, L., "Synthesis of constant-time delay ladder networks using Bessel polynomials," *Proceedings of the IRE*, vol. 42, pp. 1666–1675, December 1954.

[ST 3] Storer, J. E., *Passive Network Synthesis*, McGraw-Hill, Inc., New York, 1957.

[ST 4] Strieby, N., "A Fourier method for time domain synthesis," *Proceedings of the Symposium on Modern Network Synthesis*, Microwave Research Institute Symposia Series, Polytechnic Institute of Brooklyn, vol. 5, pp. 197–209, April 1955.

[SU 1] Su, K. L., "The constant-delay networks: how to compare them," *Proceedings of the 1968 Summer School on Circuit Theory*, Prague, Czechoslovakia.

[SU 2] Su, K. L., *Active Network Synthesis*, McGraw-Hill, Inc., New York, 1965.

[SU 3] Su, K. L., "The nominal delay and rise times of lumped delay networks," *IEEE Transactions on Circuit Theory*, vol. CT-16, pp. 574–577, November 1969.

[TA 1] Tang, D. T., "The Tchebycheff approximation of a prescribed impulse response with *RC* network realization," *IRE Convention Record*, vol. 9, pt. 4, pp. 214–220, 1961.

[TE 1] Teasdale, R. D., "Time domain approximation by use of Padé approximants," *IRE Convention Record*, vol. 1, pt. 5, pp. 89–94, 1953.

[TE 2] Temes, G. C., and D. A. Calahan, "Computer-aided network optimization—the state-of-the-art," *Proceedings of the IEEE*, vol. 55, pp. 1832–1863, November 1967.

[TH 1] Thomson, W. E., "Delay networks having maximally flat frequency characteristics," *Proceedings of the IEE*, vol. 96, pt. 3, pp. 487–490, November 1949.

[TH 2] Thomson, W. E., "Networks with maximally-flat delay," *Wireless Engineering*, vol. 29, pp. 256–263, October 1952.

[TH 3] Thomson, W. E., "Maximally-flat delay networks," *IRE Transactions on Circuit Theory*, vol. CT-6, p. 235, June 1959.

[TU 1] Tustin, A., "A method of analysing the behaviour of linear systems in terms of time series," *Journal of the IEE (London)*, vol. 94, pt. IIA, pp. 130–142, 1947.

[TU 2] Tuttle, D. F., Jr., *Network Synthesis*, vol. I. John Wiley & Sons, Inc., New York, 1958.

[TU 3] Tuttle, D. F., Jr., "Network synthesis for prescribed transient response," Sc. D. Thesis, Massachusetts Institute of Technology, 1948.

[UH 1] Uhl, K., "Ein Filtersystem mit günstigen Impulseigenschaften," *Archiv der Elektrischen Übertragung*, vol. 15, pp. 109–114, March 1961.

[UL 1] Ulbrich, E., and H. Piloty, "Über der Entwurf von Allpässen, Tiefpässen und Bandpassen mit einer im Tschebyscheffschen Sinne approximierten konstanten Grupenlaufzeit," *Archiv der Elektrischen Übertragung*, vol. 14, pp. 451–467, October 1960.

[UL 2] Ulstad, M. S., "Time domain approximations and an active network realization of transfer functions derived from ideal filters," *IEEE Transactions on Circuit Theory*, vol. CT-15, pp. 205–210, September 1968.

[UN 1] Unbehauen, R., and H. Griesinger, "Zur Synthese von Reaktanzvierpolen mit vorgeschriebenem Zeitverhalter," *Archiv der Elektrischen Übertragung*, vol. 21, pp. 161–169, April 1967.

[UN 2] Unbehauen, R., "Low-pass filters with predetermined phase or delay and Chebyshev stop band attenuation," *IEEE Transactions on Circuit Theory*, vol. CT-15, pp. 337–341, December 1968.

[VA 1] Valand, J., "On the linear phase approximation," *Proceedings of the IEEE*, vol. 55, pp. 1627–1628, September 1967.

[VA 2] Valand, J., "On the linear-slope delay approximation," *Proceedings of the IEEE*, vol. 55, pp. 2059–2060, November 1967; Addendum, vol. 56, pp. 748–750, April 1968.

[VA 3] Valand, J., "A variation of the linear phase approximation technique," *Proceedings of the IEEE*, vol. 56, pp. 1141–1143, June 1968.

[VA 4] Valand, J., "A further note on the linear-slope delay approximation," *Proceedings of the IEEE*, vol. 56, pp. 1356–1957, August 1968.

[VA 5] Van Valkenburg, M. E., *Modern Network Synthesis*, John Wiley & Sons, Inc., New York, 1960.

[VA 6] Vasiliu, C. G., "A practical method for time-domain network synthesis," *IEEE Transactions on Circuit Theory*, vol. CT-12, pp. 234–241, June 1965.

[VI 1] Vich, R., "Approximation in the time domain," *Proceedings of the Third Colloquium on Microwave Communication*, Budapest pp. 359–365, April 1968.

[VI 2] Villes, J. A., and J. Bouzitat, "Sur un type de signaux practiquement bornés en temp et en frequence," *Cables et Transmissions*, vol. 9, pp. 293–303, October 1955.

[VO 1] Voss, H. H., "Realisierbare Tiefpässe and Bandpässe minimaler Phase mit geebneter Laufzeit und aperiodischem Einschwing verhalten," *Frequenz*, vol. 8, pp. 97–102, April 1954.

[WA 1] Wall, H., *Analytic Theory of Continued Fraction*, Van Nostrand Reinhold Company, New York, 1950.

[WA 2] Waren, A. D., and L. S. Lasdon, "Practical filter design using mathematical optimization," *Proceedings Third Allerton Conference on Circuit and Systems Theory*, pp. 677–687, 1965.

[WA 3] Watson, G. N., *A Treatise on the Theory of Bessel Functions*, Cambridge University Press, New York, 1952.

[WE 1] Weinberg, L., "Network design by use of modern synthesis techniques and tables," *Proceedings of the National Electronics Conference*, vol. 12, pp. 794–817, 1956.

[WE 2] Weinberg, L., *Network Analysis and Synthesis*, McGraw-Hill, Inc., New York, 1962.

[WI 1] Wiener, N., *Extrapolation, Interpolation and Smoothing of Stationary Time Series*, John Wiley & Sons, Inc., New York, 1949.

[WI 2] Wilde, D. J., *Optimum Seeking Methods*, Prentice-Hall, Englewood Cliffs, N.J., 1964.

[WI 3] Wilde, D. J., and C. S. Beightler, *Foundation of Optimization*, Prentice-Hall, Englewood Cliffs, N.J., 1967.

[WO 1] Wood, I. E., "Note on maximally-flat delay networks," *IRE Transactions on Circuit Theory*, vol. CT-5, pp. 363-364, December 1958.

[YE 1] Yengst, W. C., "Approximation to a specified time response," *IRE Transactions on Circuit Theory*, vol. CT-9, pp. 152-162, June 1962; comments by B. Liu, vol. CT-10, p. 115, March 1963.

[YE 2] Yengst, W. C., *Proceedures of Modern Network Synthesis*, The Macmillan Company, New York, 1964.

[YO 1] Young, T. Y., and W. H. Huggins, "On the representation of electrocardiograms," *IEEE Transactions on Bio-Medical Engineering*, vol. BME-10, pp. 86-95, July 1963.

[ZE 1] Zemanian, A. H., "On transfer functions and transients," *Quarterly of Applied Mathematics*, vol. 16, pp. 273-394, October 1958.

[ZE 2] Zemanian, A. H., "Network realizability in the time domain," *IRE Transactions in Circuit Theory*, vol. CT-6, pp. 288-291, September 1959.

[ZE 3] Zemanian, A. H., "The properties of pole and zero locations for nondecreasing step responses," *Transactions of the AIEE*, vol. 79, pt. 1 (Communication and Electronics), pp. 421-426, September 1960.

[ZE 4] Zemanian, A. H., "The zeros of transient response," *IRE Transactions on Circuit Theory*, vol. CT-8, no. 2, pp. 109-113, June 1961.

[ZO 1] Zoutendijk, G., *Methods of Feasible Directions*, American Elsevier Publishing Company, Inc., New York, 1960.

[ZY 1] Zysman, G. I., and H. J. Carlin, "Restrictions on linear phase, low-pass networks with prescribed parasitic loads," *IEEE Transactions on Circuit Theory*, vol. CT-12, pp. 387-392, September 1965.

Index

Index

D

E

J

K

L

M

N